Bacillus thuringiensis,
An Environmental Pesticide: Theory and Practice

Bacillus thuringiensis, An Environmental Biopesticide: Theory and Practice

Edited by
PHILIP F. ENTWISTLE, JENNY S. CORY, MARK J. BAILEY and STEPHEN HIGGS*

*Institute of Virology and Environmental Microbiology, Natural Environment Research Council, Oxford, UK, and *College of Veterinary Medicine and Biomedical Sciences, Colorado State University, USA*

JOHN WILEY & SONS
Chichester · New York · Brisbane · Toronto · Singapore

Other Wiley Editorial Offices

John Wiley & Sons, Inc., 605 Third Avenue,
New York, NY 10158-0012, USA

Jacaranda Wiley Ltd, G.P.O. Box 859, Brisbane,
Queensland 4001, Australia

John Wiley & Sons (Canada) Ltd, 22 Worcester Road,
Rexdale, Ontario M9W 1L1, Canada

John Wiley & Sons (SEA) Pte Ltd, 37 Jalan Pemimpin #05-04,
Block B, Union Industrial Building, Singapore 2057

Library of Congress Cataloging-in-Publication Data

Bacillus thuringiensis, an environmental biopesticide; theory and practice edited by
    Philip F. Entwistle, Jenny S. Cory, Mark J. Bailey, Stephen Higgs.
        p. cm.
    Includes bibliographical references and index.
    ISBN 0 471-93306-6
    1. Bacillus thuringiensis. 2. Bacterial insecticides.
    I. Entwistle, P. F. (Philip Frank)
    QR82.B3B336   1993
    632'.96—dc20   92-27213
                                                                    CIP

British Library Cataloguing in Publication Data

A catalogue record for this book is available from the British Library

ISBN 0 471 93306 6

Typeset in 10/12 pt Optima by Mathematical Composition Setters Ltd, Salisbury
Printed and bound in Great Britain by Biddles Ltd, Guildford, Surrey

This book is dedicated to Helen R. Whiteley in admiration of not only her contributions to the study of *Bacillus thuringiensis* but also her great abilities as a scientist.

Contents

Contributors

N. Becker
German Mosquito Control Association (KABS), Europaplatz 5, 6700 Ludwigshafen, Germany

K. Bernhard
Ciba-Geigy AG, Plant Protection Division, CH-4002 Basel, Switzerland

D. Bosch
Centre for Plant Breeding and Reproduction Research CPRO, PO Box 16, 6700 AA Wageningen, The Netherlands

H. D. Burges
c/o Horticultural Research International, Worthing Road, Littlehampton, West Sussex BN16 3PU, UK

A. Delécluse
Unité de Biochimie Microbienne, Institut Pasteur, 25 rue du Docteur Roux, 75724 Paris Cedex 15, France

S. Ely
Plant Science Center, Cornell University, Ithaca, NY, USA

W. Gelernter
Mycogen Corporation 5451 Oberlin Drive, San Diego, CA 92121, USA

G. Honée
Department of Phytopathology, Agricultural University, Binnenhaven 9, 6709 PD Wageningen, The Netherlands

B. Keller
Federal Biological Research Centre for Agriculture and Forestry, Institute for Biological Control, Heinrichstrasse 243, D-6100, Darmstadt, Germany

G. A. Langenbruch
Federal Biological Research Centre for Agriculture and Forestry, Institute for Biological Control, Heinrichstrasse 243, D-6100, Darmstadt, Germany

M.-M. Lecadet
Unité de Biochimie Microbienne, Institut Pasteur, 25 rue du Docteur Roux, 75724 Paris Cedex 15, France

D. Lereclus
Unité de Biochimie Microbienne, Institut Pasteur, 25 rue du Docteur Roux, 75724 Paris Cedex 15, France

S. C. MacIntosh
Entotech Inc., 1497 Drew Avenue, Davis, CA 95616-4880, USA

J. Margalit
Ben-Gurion University of the Negev, Laboratory for Biological Control, Department of Life Sciences, Beer-Sheva, Israel

P. G. Marrone
Novo Nordisk Entotech Inc., 1497 Drew Avenue, Davis, CA 95616-4880, USA

M. P. Meadows
Energy Technology Support Unit, Department of the Environment, B156 Harwell Laboratory, Oxfordshire OX11 0RA, UK

O. N. Morris
Agriculture Canada Research Station, 195 Dafoe Road, Winnipeg, Manitoba R3T 2M9, Canada

A. Navon
Department of Entomology, Agricultural Research Organization, The Volcani Center, Bet Dagan 50250, Israel

H. S. Salama
National Research Centre, Tahrir Street, Dokki, Cairo, Egypt

G. E. Schwab
Mycogen Corporation, 5451 Oberlin Drive, San Diego, CA 92121, USA

R. Utz
Ciba-Geigy AG, Plant Protection Division, CH-4002 Basel, Switzerland

K. van Frankenhuyzen
Forest Pest Management Institute, PO Box 490, 1219 Queen Street East, Sault Ste. Marie, Ontario, Canada

B. Visser
Centre for Plant Breeding and Reproduction Research CPRO, PO Box 16, 6700 AA Wageningen, The Netherlands

Preface

The bacterium *Bacillus thuringiensis* (*B. t.*) has attracted the attention of both microbiologists and entomologists for the greater part of this century. Its unique capacity to synthesize insecticidal protein crystals has spurred investigations into its use as a natural biological control agent in agriculture, forestry and human health for the elimination of disease vectors. Over the last decade research activity into the molecular and genetic make-up of this bacterial group has intensified leading to a comprehensive understanding of toxin gene control and expression, the structure and function of the toxin molecule itself, improved efficiency of biocontrol formulations, knowledge of the ecology of the bacteria and the construction of more effective toxins and delivery systems. Observing the progression in the field of *B. t.* research both at the molecular level and as ecologists concerned with biological control, it seemed fitting by early 1990 that an international symposium should be held. Such a meeting was organized in Oxford in the summer of 1991. This first international *B. t.* conference brought together research scientists from both universities and commerce, scientists whose backgrounds ranged from microbial and entomological ecology to molecular biologists and protein crystallographers. Though not intended as a proceedings of the conference, the contributions to this timely book were drawn from the key speakers who presented at the meeting. We hope the book will be both informative and interesting to experts and newcomers in the field of *B. t.* research. We have attempted to bring together in one volume examples of all disciplines essential to the future success of *B. t.* If there are omissions then they perhaps will be corrected at a future date.

We would like to take this opportunity to thank all the authors who contributed to this book.

P. F. Entwistle
J. S. Cory
M. J. Bailey
S. Higgs

Foreword

Bacillus thuringiensis (*B. t.*) is a microbe of outstanding scientific interest and also the leading organism used in industrial microbial pesticides. Research and industrial development have catalysed each other. Both have increased exponentially so that now the subject merits a whole meeting and an entire book. Such a meeting was held over 3 days at Oxford in July 1991 and the present book comprises invited papers given at that meeting.

Discovered to science in 1911, this insect pathogen appeared to be a simple, spore-forming, infectious bacterium, easy to grow on laboratory media. Later, recognition of the proteinous toxin crystal that accompanied the spore extended consideration of the species in comparative bacteriology. A whole family of unique crystalline δ-endotoxins is now known to attack the insect gut upon ingestion, causing a characteristic gut syndrome even when the gut is attacked from the opposite direction, i.e. by injecting dissolved toxin into the insect body cavity. Toxin fed to insects kills rapidly. Lethality may or may not be increased by infection initiated by the spore, which ensures perpetuation of the species. Initially regarded only as an insect pathogen, rare in the wild, the species was found to be widespread but sparse in soil and has recently been found extensively in the phylloplane, opening up intriguing ecological questions about its role—or roles—in nature. However, it is the lethal toxins, on a par in potency with the most active organophosphate insecticides, that have attracted industry.

Industrial products based on *B. thuringiensis* are now well recognized weapons for pest control. These microbial insecticides are specific to herbivorous insects and, in the aquatic environment, to filter feeders. Thus they are safe for man and for virtually all beneficial animals and to plants, as well as being biodegradable, so fitting the yardsticks of modern environmentally friendly trends. The spores and crystals are readily produced by aerated liquid fermentation; they are easily harvested and, when properly formulated, have a long shelf life. The resulting products are easy to apply with conventional machinery, both from ground and air. Asporogenous strains can be used to make spore-free products that rely on the toxins alone. However, these apparently ideal microbial insecticides suffer constraints, which increase cost. Being a stomach poison, the toxin must be ingested to take effect, so coverage of the plant has to be correspondingly better than for a contact insecticide. Because the toxin crystals are particulate, they are less easy to apply than a soluble material, reaching only surface-feeding pests. Degradation in the environment is rather too rapid so that application often needs to be frequent. Present use levels in the order of 3000 tons/annum have not been won easily, but are the achievement of 40 years' research both basic and applied.

Each δ-endotoxin is the product of a single gene. This has enabled genes to be inserted into plants and expressed as toxin production in plant tissue, taking a prominent place among the growing number of factors that can be genetically engineered into plants to increase their protection against pests. Such systemic toxin reaches burrowing as well as surface-feeding pests. Filter feeding aquatic insects can be controlled by engineering toxin genes into some of their food microorganisms. Luckily insect resistance to the toxins usually does not develop readily, but continuous expression of the toxins in the insects' food raises the spectre of resistant insect pests.

The twelve chapters of the present book have been designed to cover comprehensively both research and development. Most emphasis has been given to recent advances, expounded by leading workers in the field. The stage is set by an overview, 'The Challenge of *Bacillus thuringiensis*', written by a worker well versed in the fundamental science and experienced in the development of microbial insecticides, particularly against forest pests in North America, the present largest market. Two chapters together cover our rapidly expanding knowledge of the biochemistry, structure, mode of action, mechanism of specificity, diversity and genetic determination of the δ-endotoxins. Then one chapter deals with transgenic lower organisms as toxin delivery systems, and another the genetic engineering of plants, both recent areas of study. From a more entomological approach, three chapters respectively elaborate control of Lepidoptera, the most destructive group of plant pests; Diptera, vectors of human disease; and Coleoptera, the most recent targets. Combining entomology and microbiology, a chapter investigates environmental aspects—safety and exciting ecological developments. A further important theme is the emotive issue of resistant insects, in theory and practice. The considerable value of *B. thuringiensis* to developing nations is carefully evaluated and the viability of its production on inexpensive organic wastes is described. A final chapter deals with the basic problems and methods of mass production using higher technology.

So wide a subject attracts a correspondingly wide readership. This book is the only recent comprehensive update of the whole subject. Of special interest to insect pathologists, specialists in bacterial toxins and entomologists, the book also concerns molecular biologists, bacteriologists, geneticists and ecologists from the aspects of research, university teaching and pest control, particularly in integrated control systems.

The future appears impressive, with progress likely to continue its exponential course. The ecology and molecular biology should expand with particular verve. As well as the δ-endotoxins, *B. thuringiensis* produces other toxins, metabolites, enzymes and antibiotics. These are beginning to establish an increasing interest in the species as a general industrial bacterium. Many strain searches are being conducted to find better δ-endotoxins for use in pest control. The resulting extensive culture collections will probably be screened for many additional materials.

H. D. Burges

1 The Challenge of *Bacillus thuringiensis*

KEES VAN FRANKENHUYZEN

Forest Pest Management Institute, Forestry Canada, PO Box 490, 1219 Queen Street East, Sault Ste. Marie, Ontario, Canada

INTRODUCTION

Over the first 70 years following its discovery at the start of this century, *Bacillus thuringiensis* (*B. t.*) captured the attention of relatively few microbiologists and entomologists. Now, 90 years later, it is being studied by scientists from a wide variety of disciplines, probing its secrets at molecular, physiological, and ecological levels. Today *B. t.* is not only the most successful commercial microbial insecticide with worldwide application for protection of crops, forests and human health, it is also starting to replace conventional chemical insecticides in several areas of application.

Although the foundation for this success was laid through many decades of research, it was the developments of the past decade that propelled *B. t.* into the foreground of commercial and scientific interest. During the 1980s, the convergence of new techniques, especially those afforded by recombinant DNA technology, and changing public and political attitudes towards pesticide use, precipitated a dramatic increase in *B. t.* research by industry, government and academia alike. Relatively low development costs, together with a high natural diversity of strains and toxins and good prospects for genetic manipulation, are providing the basis for the current worldwide endeavour to develop a new generation of more effective and environmentally acceptable insect control products.

But as challenges are overcome, new ones arise. Whereas in the 1980s the challenge of limited knowledge about toxin structure and mode of action was overcome to a large extent (although much remains to be learned), the challenge of the 1990s is to integrate that knowledge into an holistic understanding of how *B. t.* interacts with its natural environment to provide a basis for the development of use strategies that are effective and ecologically sound. To meet that challenge, a highly multidisciplinary approach will be required. The success of our collective effort to address

Bacillus thuringiensis, An Environmental Biopesticide: Theory and Practice. Edited by P. F. Entwistle, J. S. Cory, M. J. Bailey and S. Higgs
© 1993 John Wiley & Sons Ltd

that challenge will determine whether or not B. t.'s full potential will be realized during this decade.

HISTORICAL OVERVIEW: FROM DISCOVERY UNTIL 1980s

Discovery and early use

The discovery of B. t. and the early history of its development as a microbial insecticide have been reviewed in detail (e.g. Krieg, 1986; Briggs, 1986). In brief, the B. t. story began at the start of this century when a Japanese bacteriologist isolated a bacillus from diseased silkworm, Bombyx mori, larvae (Ishawata, 1901). About a decade later, a similar organism was isolated by Berliner from diseased larvae of the flour moth, Anagasta kuehniella, in Germany. Berliner named it Bacillus thuringiensis (Berliner, 1911, 1915).

First attempts to use B. t. for insect control took place in the late 1920s and early 1930s against the European corn borer, Ostrinia nubilalis, in south-eastern Europe as part of an international project funded by the USA. During the next two decades field testing continued against various lepidopteran pests in Europe and the USA. The first commercial product (Sporeine) was available in 1938 in France (Weiser, 1986), primarily for control of the flour moth (Jacobs, 1951). More extensive commercial production developed during the 1950s in several countries, including the USSR, Czechoslovakia, France and Germany (Weiser, 1986). In the USA, the work by Steinhaus (1951) stimulated interest in the commercialization of B. t., culminating in the production of Thuricide[R] (Bioferm Corporation) in 1957.

The availability of commercial products initiated a period of intermittent field testing throughout the 1960s in agriculture and forestry, generally with inconsistent results (Krieg, 1986; Kingshorn et al., 1961; Mott et al., 1961). Two developments in the 1960s were of particular significance in improving field efficacy and accelerated commercialization. First was the discovery of HD-1 (Dulmage, 1970), a kurstaki isolate that was 2–200 times more active against key agricultural pests than the thuringiensis isolates used in commercial products. HD-1 was adopted for commercial production and still forms the basis for most formulations used in agriculture and forestry. The second development was the establishment of an international system for standardizing the potency of commercial products. Products had originally been standardized on the basis of spore counts, which did not relate to total insecticidal activity (Menn, 1960; Krieg, 1965; Burges, 1967; Dulmage and Rhodes, 1971). Suggestions made by Bonnefoi et al. (1958) to express potency in biological units based on insect bioassays eventually resulted in the adoption of a European standard (E-61) (Burges, 1967), which was later replaced by HD-1 as the

North American standard (Dulmage *et al.*, 1971; Beegle *et al.* 1986). Insecticidal activity of HD-1 based products is now routinely expressed as the number of International Units (IU) per unit product as determined in bioassays against the cabbage looper, *Trichoplusia ni*.

Field development

The introduction of standardized formulations based on HD-1 generally improved the efficacy of field trials during the 1970s. Most of the field development during that decade took place in North America. Aerial spraying against the spruce budworm, *Choristoneura fumiferana*, and gypsy moth, *Lymantria dispar*, provided large single-user markets, which could accept a considerable degree of uncertainty in efficacy due to the absence of strictly defined economic damage thresholds. Field trials were conducted to develop application prescriptions in terms of hardware (aircraft type and spray atomizers), timing and frequency of treatment, and dosage and volume application rates (Morris *et al.*, 1975; Smirnoff and Morris, 1982; Grimble and Lewis, 1985). Although effectiveness of *B. t.* showed marked improvement during those trials, results remained inconsistent and treatment costs were much higher when compared with chemical insecticides. Cost-effectiveness started to improve in the late 1970s when commercial formulations became more concentrated, which permitted application of lower volumes (van Frankenhuyzen 1990a). By the end of the decade, performance of *B. t.* for spruce budworm control had improved to the point that it was considered an operational (but still more expensive) alternative to chemical insecticides (Morris, 1980). Subsequent operational use in the early 1980s precipitated further cost reductions (Irland and Rumpf, 1987; van Frankenhuyzen, 1990a) and paved the way for an unprecedented increase in the use of *B. t.* against forest insect pests during the rest of the decade (see 'Refinement of spray application'). Forestry use thus played a critical role in the commercialization of *B. t.* by establishing that it could indeed be developed as a commercially viable and effective alternative to conventional insecticides at a time when there was widespread uncertainty regarding its competitiveness.

Mode of action research

Research aimed at characterizing the pathogen and understanding its general mode of action lagged far behind commercial development. Although the existence of a parasporal body or 'Restkörper' in sporulated cells was noted by Berliner (1915) and Mattes (1927), it was not until the 1950s that it was identified as a protein crystal (Hannay, 1953; Hannay and Fitz-James, 1955). When Angus (1954) established that the crystal played a major role in pathogenicity, it became the centre of considerable attention all over the world. Over the next 20 years, research focused on crystal structure and its biochemistry, biogenesis during sporulation and general mode of action in lepidopteran larvae (see Norris, 1971; Cooksey, 1971; Fast, 1981). Knowledge available by the early 1980s can be highlighted as follows (Huber and Lüthy, 1981; Lüthy and Ebersold, 1981; Fast, 1981). The crystal is composed of a protoxin protein with a molecular weight of about 130 kDa and is solubilized at high pH in the larval gut; the actual toxin is a protease-resistant protein of about 60–70 kDa generated by enzymatic hydrolysis in the midgut; the toxin acts directly on the membranes of midgut epithelium cells, causing a gradual increase in membrane permeability followed by swelling and lysis of the cells; intoxication is associated with immediate feeding inhibition; leakage of ions from gut to haemolymph causes ionic imbalance in the haemolymph; in some species this ionic imbalance is sufficient to cause paralysis and death; in other species changes in gut and haemolymph conditions permit vegetative propagation of *B. t.* within the insect, resulting in a septicaemia, which contributes to or causes death.

As *B. t.* increased in commercial importance and different isolates were discovered, the need for a method to classify the isolates became apparent. An initial system to identify and classify crystalliferous bacteria on the basis of morphological and biochemical characteristics was introduced in the late 1950s by Heimpel and Angus (1958). That system was gradually replaced by a classification scheme based on serological analysis of vegetative cell flagellar H-antigens, combined with biochemical characteristics (de Barjac and Bonnefoi, 1962, 1968, 1973). Refinements were made by including electrophoretic patterns of esterases produced in vegetative cells (Norris, 1964) and serology of the crystals (Krywienczyk *et al.*, 1978), providing further subdivisions of some H-serotypes (de Barjac, 1981). This method proved reliable and reproducible as a taxonomic tool and brought a considerable degree of order to the classification of *B. t.* To date, 34 serovars are known, including 27 antigenic groups and seven subgroups (de Barjac and Frachon, 1990).

A topic of central interest that emerged during the 1960s and 1970s was the high degree of specificity among strains. That different strains exhibited different insecticidal activity spectra was already known in the 1950s (Angus, 1956; Hall and Dunn, 1958; Burgerjon and Grison, 1959). It quickly became apparent that insect species differed in their susceptibility to a particular strain while strains differed in their pathogenicity to a particular insect species. In the first major review on specificity, Burgerjon and Martouret (1971) emphasized the importance of the interplay between bacterial factors (crystal, spore and exotoxin) and insect-associated factors (gut pH, proteolytic processing) in producing the observed activity spectra. These authors focused in on the singular importance of crystals by observing that 'the activity spectrum of the spore-crystal complex seems to depend on the action of the crystals alone'. Supporting evidence that the diversity in strain specificity was due primarily to the diversity of delta-endotoxins came from the International Cooperative Program on the Spectra of Activity of *B. t.* (Dulmage *et al.*, 1981). Crystal serology demonstrated that an isolate can contain more than one crystal type and that a particular crystal type can be produced by more than one H-serotype (Krywienczyk *et al.*, 1978, 1981), while multiple-species bioassays indicated that activity spectra and crystal types were interrelated (Dulmage *et al.*, 1981). Much of the progress in molecular genetics and mode of action research during the 1980s can be traced back to the database and the culture collection that resulted from the International Cooperative Program.

THE 1980s: A DECADE OF CHANGE

Commercial use and constraints

Commercial production and worldwide use of *B. t.* was well established by the early 1980s. The market was dominated by products based on the HD-1 isolate of the *kurstaki* subspecies for control of lepidopteran pests in forestry and agriculture (Lüthy *et al.*, 1982; see Navon, Chapter 6). By far the most important market was for forest defoliator control in North America, where *B. t.* was starting to replace conventional insecticides, and which accounted for about 60% of worldwide sales by the mid-1980s (Burges and Daoust, 1986). New markets were opened by the discovery in 1976 of the *israelensis* subspecies (Goldberg and Margalit, 1977), which was followed by the first commercial products for control of blackfly and mosquito larvae in the early 1980s (see Becker and Margalit, Chapter 7).

Global sales of *B. t.* in the mid 1980s were estimated at about $50 million or less than 1% of the world insecticide market (Baldwin, 1987). Despite its widespread application, *B. t.* comprised only a small share of the market because of several factors that limited its use in crop protection. One limiting factor is its narrow spectrum of insecticidal activity. A high degree of target specificity makes *B. t.* a desirable alternative to broad-spectrum chemical insecticides from an environmental standpoint, but also limits its market potential. For example, several pest species of worldwide economic importance have little or no susceptibility to the HD-1 strain (such as *Spodoptera* spp.), or are not vulnerable because of their cryptic feeding habits (e.g. *Heliothis* spp. on cotton). A narrow activity spectrum is particularly a problem when dealing with a multi-species pest complex, as is often the case in agriculture. Another limiting factor is that efficacy tends to be more variable than efficacy of chemical insecticides because *B. t.* has to be ingested. In this situation, dose acquisition depends on proper timing of application and generally requires conditions that promote feeding (van Frankenhuyzen, 1990b). Inconsistent efficacy is further exacerbated by limited residual toxicity due to wash-off by rain (van Frankenhuyzen and Nystrom, 1989) and inactivation of both spore (Morris, 1983) and crystal (Pozsgay *et al.*, 1987; Puzstai *et al.*, 1991) by sunlight. Furthermore, transmission from diseased to healthy insects is poor or non-existent (Burges and Hurst, 1977). This often necessitates expensive repeat applications, particularly in the case of multivoltine species or species with long oviposition periods. However, with the growing recognition of *B. t.*'s potential as an environmentally acceptable insect control agent, the challenge of reducing those limitations became the main thrust of a worldwide research effort during the 1980s.

Changing commercial interest

The development of *B. t.* entered a new phase in the 1980s. Increasing restrictions on the use of existing chemical insecticides in many nations, pest resistance problems, and rapidly rising costs of developing new synthetics in an increasingly stringent regulatory environment stimulated the agrochemical industry to develop viable and environmentally acceptable alternatives. The demonstrated commercial success of *B. t. kurstaki* in forestry and the rapid commercialization of *B. t. israelensis* for vector control focused attention on *B. t.*

There was a growing optimism that the constraints of limited toxicity and inadequate delivery, as described above, could be solved. That optimism was fuelled by the discovery of strains with novel activity spectra, such as *israelensis* in the late 1970s and the *tenebrionis* subspecies with activity against Coleoptera in the mid 1980s (Krieg *et al.*, 1983; Herrnstadt *et al.*, 1986; see Keller and Langenbruch, Chapter 8). At the same time, the emergence of recombinant DNA technology and other genetic techniques provided unprecedented opportunities for manipulating toxicity and toxin delivery. Those conditions together set the stage for a surge in commercial and scientific interest in *B. t.* around the world. This is clearly illustrated by the number of agrochemical and newly established biotechnology companies that acquired or initiated significant *B. t.* research and development programmes during the 1980s (Table 1.1). The interplay between industry, academia and governments in pursuing the common objective of exploiting *B. t.*'s full potential, and the expansion of that potential by new technology, resulted in spectacular progress during the decade.

Much of the emphasis was on the challenge to reduce the limitations of inadequate toxicity and delivery (Table 1.1). By the end of the decade, that challenge had been overcome to a considerable extent. Rapidly advancing insights into the genetics of *B. t.* and into the molecular basis of its toxicity and specificity provided the underpinning for the

Table 1.1. The development of industrial interest in *Bacillus thuringiensis* during the 1980s

Year	Company	Unmodified strains	Improved strains	Resistant crops	Microbe-mediated delivery
< 1980	Abbott Laboratories	*			
	Biochem	*			
	Zoecon	*			
	Duphar	*			
> 1980	Abbott	*1	*		
	Solvay/Duphar	*2	*		
	Zoecon/Sandoz	*3	*	*	
	Novo	*4	*		
	Ciba-Geigy	*5	*		
	ICI	*6	*	*	
	Sumitomo	*	*		
	Dow Elanco	*	*		
	Ecogen	*7	*7		
	Mycogen	*8	*8		*
	Monsanto			*	*
	Rohm and Haas			*	
	Plant Genetic Systems			*	
	Agracetus			*	
	Calgene			*	
	Sungene Techn. Inc.			*	
	Agrigenetics				*
	Crop Genetics International				*

Registered products based on:

B. t. kurstaki:	*B. t. israelensis*:	*B. t. tenebrionis*:
[1] Dipel	Vectobac	
[2] Bactospeine, Futura, Florbac	Bactimos	
[3] Thuricide, Javelin	Teknar	Trident
[4] Foray, Biobit	Skeetal	Novodor
[5] Agree		
[6] Biodart		
[7] Cutlass, Condor, Foil		Foil
[8] MVP		M-One, MTrack

application of genetic engineering to develop better strains and novel delivery techniques, while concomitant improvements were made in the way *B. t.* was formulated and applied.

Molecular genetics

Two achievements in the early 1980s played a catalytic role in expanding our knowledge of *B. t.* genetics, namely the demonstration that crystal protein genes are located on plasmids and the cloning of those genes. Plasmids were only implicated in crystal formation until Gonzalez and co-workers established conclusively that the genes coding for the crystal proteins reside on large transmissible plasmids (Gonzalez et al., 1982) that can be cured (Gonzalez et al., 1981) and are readily exchanged between strains by means of a conjugation-like transfer (Gonzalez and Carlton, 1982). This knowledge provided a means of shuffling toxin genes between strains in an effort to decipher the relationship between gene composition and phenotypic toxicity, and to construct strains with novel insecticidal activity spectra (see 'Manipulation of toxicity'). Another critical achievement was the cloning of a B. t. kurstaki crystal toxin gene in Escherichia coli (Schnepf and Whiteley, 1981), which led to the cloning of additional toxin genes from other subspecies (e.g. Klier et al., 1982, 1985; McLinden et al., 1985; and see Whiteley and Schnepf, 1986). The availability of cloned genes opened the door to the sequencing of crystal protein genes, determination of their location in various subspecies (Kronstad et al., 1983; Klier et al., 1985), and investigations of the mechanisms regulating gene expression (Whiteley and Schnepf, 1986; Aronson et al., 1986; Höfte and Whiteley, 1989; Lereclus et al., 1989a).

The first crystal protein gene sequences were published in 1985 (Adang et al., 1985; McLinden et al., 1985; Schnepf et al., 1985; Shibano et al., 1985). They provided the basis for the construction of gene specific probes to screen strains by hybridization analysis for the presence of known nucleotide sequences (Kronstad and Whiteley, 1986; Préfontaine et al., 1987; Sanchis et al., 1988; Visser, 1989). This, combined with the development and application of more sensitive and selective techniques for isolating B. t. from the natural environment (Travers et al., 1987; Martin and Travers, 1989; Smith and Couche, 1991) and for characterizing crystal proteins from new isolates (Höfte et al., 1988), accelerated the discovery of new toxin genes. By 1989, 42 nucleotide sequences had been reported, permitting the distinction of 14 subclasses of genes based on sequence homology and toxicity spectra of the encoded proteins (Höfte and Whiteley, 1989; Lereclus et al., 1989a). Currently known gene types encode toxin proteins that are specific to either Lepidoptera (CryI), Lepidoptera and Diptera (CryII), Coleoptera (CryIII), or Diptera (CryIV), and toxins with broad cytolytic activity (Cyt) (see Lereclus et al., Chapter 2). Since 1989 additional variants of known gene types have been identified (Dardene et al., 1990; von Tersch et al., 1991) and new genes have been discovered that encode proteins with novel insecticidal activities, such as cryIE (Visser et al., 1990; Bossé et al., 1990), cryIF (Chambers et al., 1991), cryIIIB (Sick et al., 1990) and cryIIIC (Gawron-Burke et al., 1990). The discovery of a toxin type with dual activity against Lepidoptera and Coleoptera resulted in the recent creation of a fifth pathotype (CryV) (Tailor et al., 1992).

The increasing availability of toxin genes for genetic manipulation was paralleled by the development of new or improved gene transfer systems. A milestone was the use of the crown gall bacterium, *Agrobacterium tumefaciens*, to insert foreign genes into plant cells followed by recovery of whole plants from the transformed cells (Herrera-Estrella *et al.*, 1983). This, together with toxin gene sequence information, enabled expression of *B. t.* toxin genes in various crop species to reduce insect attack (see 'Transgenic plants', below, and Ely, Chapter 5). The development of high-velocity microprojectiles (Klein *et al.*, 1987) and electric discharge particle acceleration (McCabe *et al.*, 1988) to transfer foreign DNA into intact cells or tissues has permitted stable transformation of those crops that cannot be transformed by using *A. tumefaciens*, such as corn (Klein *et al.*, 1988) and rice (Christou *et al.*, 1991). Another milestone was the application of electroporation to reintroduce cloned toxin genes back into *B. t.*, which was reported by several groups (Bone and Ellar, 1989; Lereclus *et al.*, 1989b; Mahillon *et al.*, 1989; Masson *et al.*, 1989), and which opened the door for strain improvement by direct transformation (see 'Manipulation of toxicity'), below.

Mode of action

The availability of cloned toxin genes permitted bioassay of individual crystal proteins to study their spectrum of insecticidal activity without interference from other toxins or other pathogenicity factors. The

Table 1.2. Activity spectra of CryI toxins against Lepidoptera

Species	IA(a)	IA(b)	IA(c)	IB	IC	ID	IE	IF	Source
Spodoptera exigua	±	±	–	–	+	+	±	+	1, 2, 5, 8, 13, 14
S. littoralis	–	–	–	–	+	–	+		3, 6
Actebia fennica	–	–	–	–	–	–	–		7, 10
Trichoplusia ni	+	+	+		+	+	+	+	5, 15
Heliothis virescens	+	+	+	–	±	–	±	+	1, 3, 6, 8, 14
Ostrinia nubilalis	+	+	+				+	+	1, 4, 9, 14
Mamestra brassicae	+	+	–	–	+	–	–		3, 8
Pieris brassicae	+	+	+	+	+	–	–		3, 6
Manduca sexta	+	+	+	–	+	+	+		3, 6
Choristoneura fumiferana	+	+	+	+	+	+	+	+	7, 10, 13
Lymantria dispar	+	+	±	–	±	+	–	+	7, 10
Orgyia leucostigma	+	+	±	–	+	±	–	+	7, 10
Malacosoma disstria	+	+	+	+	+	+	+		7, 10
Bombyx mori	+	±	–	–	+	+	+		7, 10, 12
Plutella xylostella	+	+	+	+	+	–	–		11
Lambina fiscellaria	+	+	+	–	+	+	+		10

+ , active; ± , low activity; – , not active.
1, Gawron-Burke *et al.*, 1990; 2, Höfte *et al.*, 1990; 3, Höfte and Whiteley, 1989; 4, MacIntosh *et al.*, 1990a; 5, Moar *et al.*, 1990; 6, Van Rie *et al.*, 1990a; 7, van Frankenhuyzen *et al.*, 1991; 8, Visser *et al.*, 1990; 9, Von Tersch *et al.*, 1991; 10, van Frankenhuyzen *et al.*, unpublished data; 11, Ferré *et al.*, 1991; 12, Bossé *et al.*, 1990; 13, Masson *et al.*, 1992; 14, Chambers *et al.*, 1991; 15, Mycogen Corp., unpublished data.

Table 1.3. Activity spectra of CryII toxins against Lepidoptera and Diptera

Species	IIA	IIB	IIC	Source
Heliothis virescens	+	+	+	1,2,6
Lymantria dispar	+	+		1,2
Trichoplusia ni	+	+	+	2,6
Pieris brassicae	+			3
Manduca sexta	+	+	+	4,5,6
Aedes aegypti	+	−	−	1,2,4,5,6
Anopheles gambiae	+			3

+ , active; − , not active.
1, Donovan *et al.*, 1988; 2, Dankocsik *et al.*, 1990; 3, Nicholls *et al.*, 1989; 4, Widner and Whiteley, 1989; 5, Widner and Whiteley, 1990; 6, Wu *et al.*, 1991.

bioassay data available to date are summarized on a qualitative basis in Tables 1.2, 1.3 and 1.4 and clearly demonstrate how each toxin affects a specific spectrum of susceptible insects. Recent studies suggest that a major determinant of this specificity is the binding of toxins to putative receptor sites on midgut cells. Specific high-affinity binding to midgut brush border membrane vesicles of susceptible insect species was demonstrated for various insect–toxin combinations (Table 1.5). Binding studies also showed that a susceptible insect can possess binding sites for

Table 1.4. Activity spectra of CryIII toxins against Coleoptera

Species	IIIA	IIIB	IIIC	Source
Leptinotarsa decemlineata	+	+	+	1,2,3
Diabrotica undecimpunctata	−	−	+	1
Tenebrio molitor	+			2,4
Agelastica alni	+			2
Anthonomus grandis	+			3
Pyrrhalta luteola	+			3
P. viburni	+			4
Haltica tombacina	+			3
Otiorhynchus sulcatus	+			3
Attagenus unicolor	−			3
Gibbium psylloides	−			3
Tribolium castaneum	−			3
T. confusum	+			4
Phaedon cochleariae	−			5
Sitona spp.	+			6
Gasterophysa viridula	+			6

+ , active; − , not active.
1, Gawron-Burke *et al.*, 1990; 2, Krieg *et al.*, 1983; 3, Herrnstadt *et al.*, 1986; 4, Krieg *et al.*, 1987; 5, Carroll *et al.*, 1989; 6, Skøt *et al.*, 1990.

Table 1.5. Target species for which high-affinity saturable binding of CryI toxins to midgut brush border membranes has been demonstrated

Species	CryIA (a)	CryIA (b)	CryIA (c)	CryIB	CryIC	CryIE	Source
Manduca sexta	+	+	+		+	+	1, 2, 3
Heliothis virescens	+	+	+		+		1, 2
Spodoptera littoralis					+	+	2
Pieris brassicae		+		+			3
Plodia interpunctella		+			+		4
Lymantria dispar	+		+				5
Plutella xylostella		+		+	+		6

1, Van Rie *et al.*, 1989; 2, Van Rie *et al.*, 1990a; 3, Hofmann *et al.*, 1988; 4, Van Rie *et al.*, 1990b; 5, Wolfersberger, 1990; 6, Ferré *et al.*, 1991.

several toxins and that a toxin can recognize more than one binding site (Hofmann *et al.*, 1988; Van Rie *et al.*, 1989, 1990a). Such heterogeneity, combined with differences in affinity and concentration of binding sites, may to a large extent account for the observed diversity in toxicity spectra. However, additional factors can play a role as well, such as protoxin stability (Arvidson *et al.*, 1989), differential solubilization of crystals (Jaquet *et al.*, 1987; Lecadet and Martouret, 1987; Aronson *et al.*, 1991) and subsequent proteolytic processing (Haider *et al.*, 1988; Milne *et al.*, 1990). Although binding to midgut receptors may be essential for toxicity, binding characteristics only partially determine the level of toxicity or lytic activity (Ferré *et al.*, 1991; Van Rie *et al.*, 1990a; Wolfersberger, 1990). Toxicity *per se* appears to be a function of the ability of the toxin to form a pore in the cell membrane after binding to the receptor. Pore formation results in colloid-osmotic lysis of the cell, which is believed to be the main cytolytic mechanism that is common to all Cry toxins (Knowles and Ellar, 1987; Slatin *et al.*, 1990) (see Visser *et al.*, Chapter 3).

Different domains of the toxin molecule are responsible for the steps of receptor recognition and pore formation. CryI toxins consist of two structural domains following proteolytic processing, a conserved N-terminal half and a hypervariable C-terminal half (Choma *et al.*, 1990; Convents *et al.*, 1990, 1991). The C-terminal domain is involved in specificity (binding) and the N-terminal domain in toxicity (pore formation). In studies with a number of target insects and various toxins, the amino acid regions conferring toxicity were shown to be located in the C-terminal half of the active toxin core (Ge *et al.*, 1989, 1991; Schnepf *et al.*, 1990; Widner and Whiteley, 1990 and indications are that the same regions are also responsible for binding (Honée *et al.*, personal communication; Dean *et al.*, 1991). Evidence that the determinants for pore formation reside in the N-terminal of the activated toxin includes the highly conserved hydropathicity pattern in regions of low direct sequence homology (Chungjatupornchai *et al.*, 1988), a remarkable conservation of putative transmembrane helical segments in all Cry toxins (Hodgman and Ellar, 1990), and a reduction in toxicity by site-directed mutagenesis designed to disrupt one of those helices (Ahmad and Ellar, 1990; see also Visser *et al.*, Chapter 3). Spectacular progress in understanding the relationship between toxin structure and function was made recently by Li *et al.* (1991), who modelled the secondary and tertiary structure of a coleopteran-active toxin based on crystallographic data. The model confirms the multidomain toxin structure and undoubtedly will give rise to numerous experiments to prove the proposed functionalities of those domains (i.e. pore formation, receptor binding, and crystal formation or structural integrity). More work is also needed to test the authors' hypothesis that lepidopteran-active toxins are likely to fold in a similar three-domain structure because of the highly conserved interdomain and other internal core regions. Such knowledge of toxin structure, linked with ongoing efforts to identify receptors (Knowles *et al.*, 1991) and to elucidate their role in mediating cell lysis (Schwartz *et al.*, 1991), may ultimately lead to the design of proteins with improved toxicity and tailored specificity.

Manipulation of toxicity

The expansion in the insecticidal activity spectrum of *B. t.* with the discovery of dipteran- and coleopteran-active strains, and the availability of techniques for cloning, manipulation and transfer of toxin genes, as summarized above, raised the expectation that improved products directed against a broader array of pest species may well be a reality within the foreseeable future. Indeed, by the end of the decade that expectation had come true to a considerable extent.

One successful approach for constructing strains with broader or improved insecticidal activities was the use of non-recombinant genetic methods to rearrange protein composition of the crystal (e.g. Gonzalez *et al.*, 1982; Klier *et al.*, 1983). This approach involved a combination of plasmid curing to eliminate genes that contributed little to the activity against the target of interest, and conjugal plasmid transfer to bring in genes encoding toxins with desirable insecticidal properties from another parental background (Carlton *et al.*, 1990). Because the two processes are not viewed by regulatory agencies in the same light as recombinant DNA manipulations, application of that approach has enabled Ecogen to register several products based on newly constructed strains within 4 years of project initiation (Carlton *et al.*, 1990). Those products display either improved activity against specific target insects because of the production of more potent proteins (e.g. Condor®, for control of spruce budworm and gypsy moth; Cutlass®, for control of beet armyworm and other vegetable pests), or a broader activity spectrum by combining genes encoding proteins with different specificities (e.g. Foil ®, for control of European corn borer and Colorado potato beetle, *Leptinotarsa decemlineata*). These first generation products of genetically manipulated strains will serve as a stepping stone for registration of second generation products that are based on strains with (combinations of) toxin genes improved through the use of recombinant DNA techniques.

The feasibility of using recombinant DNA technology for constructing broader insecticidal activity spectra was demonstrated in several studies. For example, Honée et al. (1990) constructed a tandem gene that produced a fusion protein consisting of the N-terminal regions of two different protoxins, CryIA(b) with activity against tobacco budworm, *Heliothis virescens*, and large cabbage white butterfly, *Pieris brassicae*, and CryIC with activity against *Spodoptera exigua*. Expression of the fusion gene in *Escherichia coli* showed a biologically active protein with toxicity against all three targets. Crickmore et al. (1990) produced recombinant strains with broader activity spectra by using electroporation to insert toxin genes from one subspecies into another. With this technique they were able to confer coleopteran toxicity on both subspecies *israelensis* and *kurstaki* and to express lepidopteran toxicity in both subspecies *israelensis* and *tenebrionis*. They also reported a transiently expressed novel lepidopteran activity of subspecies *israelensis* when the *tenebrionis* toxin gene was introduced. A similar observation was reported earlier (Bassand et al., 1989). In that case, introduction (by conjugation) of the *kurstaki* HD-73 toxin gene into subspecies *tenebrionis* resulted in toxicity against Egyptian cotton leafworm, *Spodoptera littoralis*, an activity that neither parent strain possessed. Efficacy of the hybrid strain against several Lepidoptera and a coleopteran pest was demonstrated in the field. Thus, the possibility exists that new and unexpected cross specificities can be created through gene (product) interactions.

The goal of gene manipulation to increase the level of toxicity to a specific target species has proved to be more difficult to accomplish than obtaining a broader activity spectrum. Relatively minor increases of generally less than five-fold can be obtained, for example by the plasmid curing process mentioned above, or by random chemical mutagenesis (e.g. Bassand et al., 1989). Larger increases in toxicity have not been reported to date, with the exception of a recent report by Ge et al. (1991). Those authors observed a 30-fold increase in toxicity to *Heliothis virescens* of a chimeric protein in a study of reciprocal recombinant between the *cryIA(a)* and *cryIA(c)* genes. This observation suggests that it may be possible to improve toxicity through gene rearrangements.

Microbe-mediated delivery

One way to solve the constraints of inadequate delivery and limited residual toxicity is to insert *B. t.* toxin genes into microorganisms that are associated with the target insect habitat, with the idea that the transformed organisms will colonize and continue to synthesize sufficient quantities of toxin to prevent insect damage. The topic is discussed in detail by Gelernter and Schwab, in Chapter 4. That this approach is feasible was demonstrated by many studies, involving the insertion of specific lepidopteran-, coleopteran- and dipteran-active genes into bacteria that colonize plant leaf surface and roots externally, such as *Pseudomonas cepacia* (Stock *et al.*, 1990) and *Pseudomonas fluorescens* (Obukowicz *et al.*, 1986; McPherson *et al.*, 1988; Waalwijk *et al.*, 1991) or internally, such as *Clavibacter xyli* (Dimock *et al.*, 1988), *Rhizobium leguminosarum* (Skøt *et al.*, 1990) and *Bradyrhizobium* (Nambiar *et al.*, 1990). Other examples are the expression of mosquitocidal toxins in *Bacillus sphaericus* (Bar *et al.*, 1991) and in Cyanobacteria (Angsuthanasombat and Panyim, 1989, Tandeau de Marsac *et al.*, 1987) to improve persistence in the aquatic environment.

A main obstacle to taking this technology to the field is that it involves the release of living recombinant microorganisms, which evokes numerous concerns and regulatory restrictions. One concern is the stability of the introduced trait, i.e. how to confine the inserted gene to the recipient organism or how to minimize potential for horizontal gene transfer to other bacterial species (Tiedje *et al.*, 1989). Although several approaches have been developed to minimize this problem, for example by using transposase negative derivatives of Tn5 transposon (Obukowicz *et al.*, 1986) or suicide vectors where integration is dependent on recombination between homologous DNA sequences (Waalwijk *et al.*, 1991), organisms transformed this way have yet to undergo rigorous testing in the natural environment to determine their fate, their interaction with other microbial species, as well as their effectiveness in controlling the targeted pests.

An important stepping stone to registration of such living recombinant microorganisms is the use of recombinant DNA technology to express selected toxin genes in bacteria that are then killed to provide a non-living, cellular delivery system (Gaertner, 1990). Mycogen has used this approach to encapsulate several *B. t.* toxins within cells of *P. fluorescens*, under the trademark CellCap™. Cells are killed during stationary growth in a manner that stabilizes them to enhance field persistence of the toxin, thus providing a system for improved toxin delivery (Gaertner, 1990). Because of the non-living state of the cells, products produced this way have been field tested since 1985 without the concerns associated with outdoor testing of living genetically engineered microorganisms. Two

CellCap products are expected to receive registration in 1991, one for control of diamondback moth, *Plutella xylostella*, and several other major lepidopteran targets (cabbage looper, corn earworm, *Helicoverpa (Heliothis) zea*, European corn borer), and another for control of Colorado potato beetle. CellCap products for delivery of toxins with activity against *Spodoptera* or mosquitoes are being tested, and the use of alternative host cells for new applications is being explored.

Recent studies have demonstrated the possibility of expressing *B. t.* toxin genes in an insect baculovirus (Merryweather *et al.*, 1990; Martens *et al.*, 1990). Although the main objective of those studies was to improve a nuclear polyhedrosis virus (NPV) for insect control by using *B. t.* toxins to accelerate the speed of kill, a possible application is the development of virus-mediated toxin delivery systems thus combining the high host specificity of NPVs with the pathogenicity of *B. t.* Such high target specificity is a commercial disadvantage, but could be an advantage in situations where the effect on non-target Lepidoptera (e.g. endangered species) is of overriding concern. One example is the Karner blue, *Lycaeidus melissa samuelis*, a lupin feeding lycaenid with a distribution in Canada that is limited to areas in southern Ontario where *B. t.* is used for control of the gypsy moth.

Transgenic plants

An alternative approach to overcome the problems associated with toxin delivery to the target insect is the expression of the *B. t.* toxin gene in the plant itself, as discussed in detail by Ely in Chapter 5. The availability of the *A. tumefaciens*-mediated DNA transfer system in the early 1980s was followed quickly by the first reports of insertion of *B. t.* toxin genes into tobacco (Adang *et al.*, 1987; Barton *et al.*, 1987, Vaeck *et al.*, 1987) and tomato (Fischhoff *et al.*, 1987). Recent reports extended the list of *B. t.*-transformed crops to include cotton (Perlak *et al.*, 1990), potato (Leemans *et al.*, 1990) and poplar (McCown *et al.*, 1991) and that list is growing continuously. Field evaluation of transgenic plants have been underway since 1987 and have demonstrated successful protection of tobacco (Leemans *et al.*, 1990) and tomato (Delannay *et al.*, 1989) against tobacco budworm, *Manduca sexta*, and *H. virescens*. Protection against damage by less susceptible species, such as the tomato fruit worm (*H. zea*) and tomato pinworm, *Keiferia lycopersicella*, was not adequate because of low expression levels in the plants (Delannay *et al.*, 1989). Significant progress was made recently by modification of the *B. t.* coding sequence (Perlak *et al.*, 1991) which resulted in a 50- to 100-fold increase in toxin protein expression in cotton plants. That was sufficient to provide total protection against *S. exigua*, an insect that is 100-fold less susceptible than *T. ni* (Perlak *et al.*, 1990). The first genetically engineered seed for insect-resistant crops is expected to reach the market place within a few years (Meeusen and Warren, 1989). However, implementation of transgenic crops in agricultural production is now poised on the brink of the possible development of insect resistance and the design of use strategies that will minimize that risk (see Marrone and MacIntosh, Chapter 10).

Refinement of spray application

During the 1980s significant progress towards improving toxin delivery was achieved by refining the way B. t. is formulated and applied. Most of that progress took place within the context of aerial forestry applications in North America, where large-scale defoliator control programmes precipitated the need for reducing application costs. The technology for application of B. t. is basically the same as for synthetic insecticides and is based on controlled droplet application. The main principle of this concept, which was developed in England in the late 1970s (Johnstone, 1978), is that the optimum spray droplet size is determined by the nature of the target and other relevant application considerations. This contrasts with traditional application in which the spray droplet size is determined by the spray volume selected. Controlled droplet application thus involves the use of a minimum volume rate at a chosen droplet size that will effect control, and tends toward ultra-low volume (ULV) application (Johnstone, 1978). The concept was applied to aerial forestry spraying by the Cranfield Institute of Technology and the UK Forestry Commission in Scotland in trials with fenitrothion to control the pine beauty moth, *Panolis flammea* (Holden and Bevan, 1978), and a few years later by the New Brunswick Spray Efficacy Research Group to optimize spruce budworm control operations in Canada (Picot *et al.*, 1985, 1986).

Various studies have provided overwhelming evidence that the numerically most abundant droplets on coniferous foliage are $< 100 \mu m$ in diameter (Himel, 1969; Spillman, 1984; Picot *et al.*, 1986). The use of such small droplets requires efficient atomization of the spray formulation by using, for example, rotary atomizers. The availability in the early 1980s of non-volatile, low-viscosity, high-potency formulations that were designed for undiluted application permitted direct translation of the controlled droplet concept to the application of B. t. for the control of spruce budworm, and made undoubtedly the most important contribution to reducing the constraints of high cost and unreliable efficacy. Those improvements, together with a concurrent shift in political climate that favoured the use of biologicals, resulted in the acceptance of B. t. as a fully operational insecticide for control of spruce budworm and other defoliators by the mid-1980s. In most Canadian jurisdictions, B. t. is now the only insecticide allowed for aerial forestry applications. As a result, the area treated with B. t. increased from $< 5\%$ of the total

Table 1.6. Operational use of *Bacillus thuringiensis* for control of spruce budworm, *Choristoneura fumiferana*, in eastern Canada in relation to the total area infested (millions of hectares with moderate to severe defoliation) and total area treated with insecticides (in millions of hectares)

Year	Total area defoliated	Total area treated	% of treated area sprayed with *B. t.*
1980	36.2	1.9	4
1981	37.0	2.9	2
1982	18.2	3.0	2
1983	23.8	2.8	3
1984	16.8	1.5	20
1985	20.2	1.5	45
1986	12.3	0.8	45
1987	8.4	0.9	51
1988	6.2	0.7	63
1989	7.4	0.8	39
1990	8.5	1.1	63

area sprayed in the early 1980s to >60% by the end of the decade (Table 1.6).

The trend of the 1970s to increase product potency continued in the 1980s. The potency of products registered for forestry applications in North America increased from 8.4 billion international units (BIU)/l in 1980 to 12.7 BIU/l in 1984, 16.9 BIU/l in 1988, and 33.0 BIU/l in 1990. The major benefit of those formulations is that they can be applied undiluted, which permits a dramatic reduction in application volume (Irland and Rumpf, 1987; van Frankenhuyzen, 1990a). In the early 1980s, formulations containing up to 8.8 BIU/l were routinely applied in diluted form at 20–30 BIU in 4.7–9.5 l/ha, as compared with application of the same dosage rate in 1.6–2.4 l/ha by the mid to late 1980s. The use of such low spray volumes increased spray plane efficiency and reduced handling time and costs, thus narrowing the cost differential between *B. t.* and chemical insecticide. Towards the end of the decade a single application of *B. t.* (at 30 BIU l/ha) was about 1.5 times more expensive than two applications of fenitrothion (at 210 grams of active ingredient/ha) (van Frankenhuyzen, 1990a).

Further improvements in the cost-effectiveness of aerial *B. t.* applications were achieved in recent years in New Brunswick, Canada. Refinement of spray atomization and the availability of formulations containing 33.0 BIU/l enabled a further reduction in application volume to as low as 0.45 l/ha. Refinement of atomization was based on the demonstration by Van Vliet and Picot (1987) that the droplet volume generated by rotary atomizers in the size range that is physically optimal for spruce budworm–insecticide contact (15–55 μm, Picot *et al.*, 1985, 1986) can be maximized by reducing the flow rate through each atomizer to 2.0 l/min and by increasing the cage rotation to 14 000 rpm. Trials in New Brunswick between 1987 and 1989 demonstrated that by using such low flow rates effective dosage application rates could be reduced from 30 to 15 BIU/ha (E. G. Kettela, personal communication; Carter, 1990). With the registration of a formulation containing 33 BIU/l in 1990, the reduced dosage could be applied in 0.45 l/ha, which makes it possible to use two applications of *B. t.* at only a 25% higher cost than two applications of fenitrothion (Carter and Lavigne, 1991).

The switch to applying undiluted high-potency formulations in the mid 1980s also played a key role in reducing the problem of unreliable efficacy. Field trials in the early 1980s established that a deposit of about 0.5 droplets per needle of a 12.7 BIU/l formulation was required to obtain a 50% reduction in spruce budworm defoliation (Lambert, 1987) and that the required droplet density decreased as product potency increased (Wiesner and Kettela, 1987). Because of the higher amount of active ingredient delivered to the target foliage in each spray droplet, the switch from diluted to undiluted application generally increased efficacy and its reliability in spruce budworm control programmes (van Frankenhuyzen, 1990a). However, inconsistency in efficacy from year to year remains a problem (Carter, 1990).

Commercial use at the end of the decade

Despite the increase in forestry use, the current worldwide market for *B. t.* products is dominated by agricultural applications. By the end of the decade, global *B. t.* sales were estimated to be in the $50–80 million range (Carlton *et al.*, 1990). Exact estimates are difficult to obtain and do not include significant commercial production in the former USSR (no estimates available) and China (about 730 000 kg in 1989, Tianjian *et al.*, 1990) and the growing production in developing countries (see Salama and Morris, Chapter 11). Applications of *B. t. kurstaki* in agriculture accounted for about 60% of total sales in 1990. The largest market is protection of vegetable crops in North America and elsewhere against a complex of species including *P. xylostella*, small white butterfly, *Pieris rapae*, *P. brassicae*, *T. ni*, cabbage moth, *Mamestra brassicae*, and *S. exigua*. Other uses include protection of soybean (velvet bean caterpillar, *Anticarsia gemmatalis*), tobacco (*M. sexta*, *H. virescens*), corn (*O. nubilalis*, mostly on seed corn) and cotton (early season control of *Heliothis* spp.). Products based on Coleoptera-active strains, particularly for control of the Colorado potato beetle, *L. decemlineata*, have just entered the market (see Keller and Langenbruch, Chapter 8). More details regarding agricultural applications of *B. t.* can be found in Navon, Chapter 6.

Forestry applications of *B. t. kurstaki* accounted for about 20% of 1990 sales, again mostly in North America. The main target species are spruce budworm in Canada and gypsy moth in the USA. Between 1985 and 1990, a cumulative total of almost five million hectares was treated with *B. t.* for control of those species. About 60% of all budworm and gypsy moth control programmes is currently conducted with *B. t.* It is also used for aerial control of the Eastern hemlock looper, *Lambdina fiscellaria* (West *et al.*, 1989), tent caterpillars, *Malacosoma* spp., and blackheaded budworm, *Acleris* spp. Forestry applications elsewhere include control of *Lymantria* spp., pine processing moth, *Thaumetopoea pityocampa*, larch budmoth, *Zeiraphera diniana*, winter moth, *Operophtera brumata*, and green oak tortrix, *Tortrix viridana*, on roughly 20 000 to 100 000 ha annually in central Europe (Weiser, 1986; Svestka, 1989), and control of the Siberian silkworm, *Dendrolimus sibericus*, in the former USSR (Weiser, 1986).

Vector control with *B. t. israelensis* accounted for the remaining 20% of sales in 1990, mostly in North America and West Africa. The largest single use market is the Onchocerciasis Control Programme in West Africa, which was initiated by the World Health Organization in 1974 (Akpoboua *et al.*, 1989). Thousands of river kilometres were treated frequently with organophosphates to reduce populations of *Simulium damnosum*, the vector of the *Onchocerca* filaria. *B. t. israelensis* was introduced shortly after the first signs of organophosphate resistance became apparent in 1980. Its use increased rapidly from 8000 litres in 1981 in one country to 700 000 litres in 11 countries in 1989 (Akpoboua *et al.*, 1989). The second major use is nuisance control of *Aedes* and *Culex* mosquito larvae, such as in the Upper Rhine Valley in Germany and many municipalities in North America and elsewhere, as discussed by Becker and Margalit, Chapter 7.

THE FUTURE CHALLENGE

During the past 10 years interest in *B. t.* has gained tremendous momentum: commercial potential is stimulating scientific interest while the resulting knowledge is expanding that potential. Progress in both areas has been exponential and indicates clearly that it may be possible to solve the constraints that are currently limiting the use of *B. t.* Formulations have been improved and spray application techniques have been refined. New spray products based on more potent strains or strains with improved toxicity and altered specificity are just entering the market. Their operational use during the next few years will show whether or not those improvements translate into superior product performance. If so, user confidence and acceptance will be boosted while providing the revenue for the development of a second generation of improved spray products. The knowledge base for making further improvements is growing rapidly through a continuing and concerted effort to better understand the mode of action at the molecular and physiological levels. In addition, microbe-mediated and crop-mediated delivery strategies are, at least from a technological perspective, close to operational implementation. Thus, the stage appears to be set for the realization of *B. t.*'s true potential as a bioinsecticide and for substantial market expansion.

However, current technological advances are outstripping the ecological knowledge that is required for their deployment as environmentally acceptable insect control strategies. Ecological risks (Tiedje *et al.*, 1989) need to be assessed, new regulatory requirements have to be met, and public acceptance has to be gained before transgenic plants and living recombinant microorganisms can be put to use. But even the use of non-recombinant techniques is giving rise to public concerns, such as the potentially greater impact on non-target organisms of products with broader insecticidal activity. It is becoming increasingly clear that successful application of those new technologies as effective and acceptable insect control strategies requires a broad understanding of the pathogen, including its mode of action at the molecular level as well as its interactions at the community level. The need for and the complexity of that understanding and the range of multidisciplinary inputs required are probably best illustrated by the many questions surrounding the issue of *B. t.* resistance, an issue that will undoubtedly take centre stage in the 1990s (see Marrone and MacIntosh, Chapter 10).

Resistance to *B. t.* has been documented in both laboratory and field populations of several target species, which is reviewed in detail by Marrone and MacIntosh in Chapter 10. It is now clear that resistance can emerge if selection pressure is sufficient, that the expression of *B. t.* toxin genes in plants and associated microorganisms will alter that selection pressure, and that strategies for managing resistance need to be devised. The development of those strategies will require an unprecedented level of interaction between molecular biologists, insect physiologists, microbiologists, ecologists, entomologists, evolutionary biologists and population geneticists, as exemplified by the following selection of the many questions that need to be answered.

The molecular and genetic mechanisms of resistance Mechanisms of resistance and their genetic basis need to be understood before use strategies can be designed that will minimize the development of resistance. Most genetic models are based on the assumptions that resistance is monogenic and functionally recessive (Tabashnik, 1989). Research into the mechanisms of *B. t.* resistance has just begun and current understanding is limited. The data available so far suggest little or no alteration of normal physiological mechanisms, such as enhanced proteolytic processing (Johnson *et al.*, 1990), and favour the explanation that resistance involves a target site alteration, such as a change in toxin membrane receptors (Van Rie *et al.*, 1990b; Ferré *et al.*, 1991). Genetic studies indicate that resistance is controlled by a single major factor which is a partially recessive trait, at least in the case of Indian mealmoth, *Plodia interpunctella* (McGaughey, 1985; McGaughey and Beeman, 1988). However, resistance of *H. virescens* to a toxin expressed in *Pseudomonas* was under the control of several genetic factors (Sims and Stone, 1991) and involved multiple changes in receptor binding parameters and modification of postbinding events (MacIntosh *et al.*, 1991). Resistance mechanisms, therefore, need to be studied in a variety of species before generalizations can be made.

Field studies of resistance   There is an urgent need to study the development of resistance and its mechanisms under field conditions because field populations can respond differently to selection with *B. t.* than laboratory populations (Tabashnik *et al.*, 1991). For example, in laboratory populations limited genetic diversity and limited environmental stress may favour selection of polygenic resistance as opposed to a single major mechanism in the field (Roush and McKenzie, 1987). In field populations, interaction with plant allelochemicals can alter selection pressures (Brattsten *et al.*, 1986). Furthermore, selection of resistance in a no-choice (laboratory) situation may result in different fitness tradeoffs than when the insects do have a choice (e.g. Gould and Anderson, 1991; Gould *et al.*, 1991). Experimental studies of the proposed resistance management strategies (Brunke and Meeusen, 1991; and see Marrone and MacIntosh, Chapter 10) and of the genetic models underlying those strategies should be based on resistances that are genetically similar to those that develop in the field. Field studies are also needed to investigate insect behaviour and life tables on transgenic crops before widespread use (Gould, 1988), and to assess fitness differences between adapted and unadapted genotypes (Gould and Anderson, 1991; Brewer, 1991).

The interaction between B. t. *and plant allelochemicals and implications for other trophic levels*   Plant defensive chemicals can modify the development of resistance by altering selection pressures (Brattsten *et al.*, 1986; Brattsten, 1988). Potentiation of *B. t.* by allelochemicals could increase selection for cross resistance, particularly when allelochemicals have a profound effect on fitness of the pest (Raffa, 1989), whereas antagonism between the two may reduce the risk of resistance by providing conflicting selection pressures. Current knowledge on how *B. t.* interacts with plant defensive chemicals is limited to a few studies (Lüthy *et al.*,

1985; Krischik *et al.*, 1988; Felton and Dahlman, 1984; Ludlum *et al.*, 1991; MacIntosh *et al.*, 1990b; see also Navon, Chapter 6), and they suggest that the direction of the interaction depends on how well the target insect is adapted to the host plant. Thus, selection pressure imposed by increased exposure to *B. t.* may be different for a generalist herbivore such as *T. ni* than for a specialist herbivore such as *M. sexta* (Barbosa, 1988), and for crops that rely on qualitative, toxic defences (e.g. nicotine in tobacco) versus quantitative, digestibility reducing defences (e.g. tannins in cotton). Prediction of the selection response and choice of the optimal management strategy will therefore depend on the particular pest-crop system that is targeted for control. Another consideration is that plant allelochemicals do not only affect fitness of the herbivorous insect, but also of its natural enemies (Barbosa, 1988). Insect adaptation to plant defences involves a balance between the benefits of overcoming the defence and the costs associated with doing that. These include physiological costs, for example due to increased metabolic expenditure, as well as ecological costs, for example due *inter alia* to altered predation or parasitism (Price *et al.*, 1980). This means that adaptation to *B. t.* toxins as a new plant defence in transgenic crops will be influenced not only by the interaction between the toxins and existing plant defences but also by the effect of that interaction on the pest's natural enemies. Because these multiple trophic level interactions are poorly understood, evolutionary responses of pest species to insect-resistant transgenic crops may follow unexpected trajectories.

The ecological niche of B. t. *and the evolutionary significance of its toxins*   Understanding the pathogen's ecological niche and the evolutionary significance of its insect-killing capability provides the overall evolutionary context for predicting insect response to increased *B. t.* exposure. Such understanding is still in its infancy (see also Meadows, Chapter 9). The detection of *B. t.* in soils from all over the world (DeLucca *et al.*, 1981; Ohba and Aizawa, 1986), and the apparent lack of association between its incidence and insect activity, resulted in the speculation that *B. t.* is essentially a soil microbe that possesses incidental insecticidal activity (Martin and Travers, 1989). An alternative hypothesis was advanced recently by Smith and Couche (1991), who suggested that *B. t.* is a natural component of the phylloplane microflora and has evolved in a symbiotic or mutualistic association with plants to provide protection against herbivores.

   The above examples clearly illustrate the level of interdisciplinary interaction that will be required for realizing *B. t.*'s full potential as an insect control agent. The *B. t.* research community seems particularly well suited to successfully meet this challenge as it already comprises a broad range of scientific disciplines with a high degree of interaction. That interaction should be stimulated to the greatest possible extent. One way to achieve that is to establish government–industry partnerships to provide funding for interdisciplinary research at the precompetitive level. An example is the consortium-based approach in the USA to fund research into resistance mechanisms. Another key component for success is to have frequent meetings to enhance interdisciplinary communication and to provide a platform for interaction with regulatory agencies and public interest groups in order to create a social and political environment that will accept the new technologies as ecologically sound ways to protect our crops and our health.

REFERENCES

Adang, M. J., Staver, M. J., Rocheleau, T. A., Leighton, J., Barker, R. F. and Thompson, D. V. (1985). 'Characterized full-length and truncated plasmid clones of the crystal protein of *Bacillus thuringiensis* subsp. *kurstaki* HD-73 and their toxicity to *Manduca sexta'*. *Gene*, **36**, 289–300.

Adang, M. J., Firoozabady, E., Klein, J., DeBoer, D., Sekar, V., Kemp, J. D., Murray, E., Rocheleau, T. A., Rashka, K., Staffeld, G., Stock, C., Sutton, D. and Merlo, D. J. (1987). 'Expression of a *Bacillus thuringiensis* insecticidal crystal protein gene in tobacco plants'. In *Molecular Strategies for Crop Protection*, pp. 345–353, Alan R. Liss.

Ahmad, W. and Ellar, D. J. (1990). 'Directed mutagenesis of selected regions of a *Bacillus thuringiensis* entomocidal protein'. *FEMS Microbiol. Lett.*, **68**, 97–104.

Akpoboua, L. K. B., Guillet, P., Kurtak, D. C. and Pangalet, P. (1989). 'Le role du *Bacillus thuringiensis* H14 dans la lutte contre *Simulium damnosum* (Diptera: Simulidae), vecteur de l'onchocercose en Afrique occidentale'. *Naturaliste Can.* (*Rev. Ecol. Syst.*), **116**, 167–174.

Angsuthanasombat, C. and Panyim, S. (1989). 'Biosynthesis of 130-kilodalton mosquito larvicide in the cyanobacterium *Agmenellum quadruplicatum* PR-6' *Appl. Env. Microbiol.*, **55**, 2428–2430.

Angus, T. A. (1954). 'A bacterial toxin paralysing silkworm larvae'. *Nature (London)*, **173**, 545–546.

Angus, T. A. (1956). 'General characteristics of certain insect pathogens related to *Bacillus cereus'*. *Can. J. Microbiol.*, **2**, 111–121.

Aronson, A. I., Beckman, W. and Dunn, P. (1986). '*Bacillus thuringiensis* and related insect pathogens'. *Microbiol. Rev.*, **50**, 1–24.

Aronson, A. I., Han, E. S., McGaughey, W. and Johnson, D. (1991). 'The solubility of inclusion proteins from *Bacillus thuringiensis* is dependent upon protoxin composition and is a factor in toxicity to insects'. *Appl. Env. Microbiol.*, **57**, 981–986.

Arvidson, H., Dunn, P. E., Strand, S. and Aronson, A. I. (1989). 'Specificity of *Bacillus thuringiensis* for lepidopteran larvae: factors involved in vivo and in the structure of a purified protoxin. *Mol. Microbiol.*, **3**, 1533–1543.

Baldwin, B. (1987). 'Commercialization of microbially produced pesticides'. *Intern. Industr. Biotechn.*, **7**, 290–293.

Bar, E., Lieman-Hurwitz, J., Rahamim, E., Kenyan, A. and Sandler, N. (1991).

'Cloning and expression of *Bacillus thuringiensis israelensis* delta-endotoxin DNA in *B. sphaericus*'. *J. Invert. Pathol.*, **57**, 149–158.

Barbosa, P. (1988). 'Natural enemies and herbivore–plant interactions: influence of plant allelochemicals and host specificity'. In *Novel Aspects of Insect–Plant Interactions* (Eds. P. Barbosa and D. K. Letourneau), pp. 201–229, John Wiley & Sons, New York.

Barton, K. A., Whiteley, H. R. and Yang, N. (1987). '*Bacillus thuringiensis* delta-endotoxin expressed in transgenic *Nicotiana tabacum* provides resistance to lepidopteran insects'. *Plant Physiol.*, **85**, 1103–1109.

Bassand, D., Jellis, C. L. and Piot, J. C. (1989). 'Application des techniques d'exchange et de génie génétiques a l'amélioration des propriétés insecticides de *Bacillus thuringiensis*'. *C. R. Acad. Agric. Fr.*, **75.**, 127–134.

Becker, N. (1990). 'Microbial control of mosquitoes and black flies'. In *Proceedings Vth International Colloquium on Invertebrate Pathology*, pp. 84–89, Adelaide, Australia.

Beegle, C. C., Couch, T. L., Alls, R. T., Versoi, P. L. and Lee, B. L. (1986). 'Standardization of HD-1-S-1980: US standard for assay of lepidopterous-active *Bacillus thuringiensis*'. *Bull. Entomol. Soc. Am.*, **32**, 44–45.

Berliner, E. (1911). 'Über die Schlaffsucht der Mehlmottenraupe'. *Z. Gesamte Getreidewesen (Berlin)*, **3**, 63–70.

Berliner, E. (1915). Über die Schlaffsucht der Mehlmottenraupe'. *Z. Ang. Entomol.*, **2**, 29–56.

Bone, E. J. and Ellar, D. J. (1989). 'Transformation of *Bacillus thuringiensis* by electroporation'. *FEMS Microbiol. Lett.*, **58**, 171–178.

Bonnefoi, A., Burgerjon, A. and Grison, P. (1958). 'Titrage biologique des preparations de spores de *Bacillus thuringiensis*'. *C. R. Acad. Sci.*, **27**, 1418–1420.

Bossé, M., Masson, L. and Brousseau, R. (1990). 'Nucleotide sequence of a novel crystal protein gene isolated from *Bacillus thuringiensis* subspecies *kenyae*'. *Nucl. Acids Res.*, **18**, 7443.

Brattsten, L. B. (1988). 'Potential role of plant allelochemicals in the development of insecticide resistance'. In *Novel Aspects of Insect–Plant Interactions* (Eds. P. Barbosa and D. K. Letourneau), pp. 313–348, John Wiley & Sons, New York.

Brattsten, L. B., Holyoke, C. W., Leeper, J. R. and Raffa, K. F. (1986). 'Insecticide resistance: challenge to pest management and basic research'. *Science*, **231**, 1255–1260.

Brewer, G. J. (1991). 'Resistance to *Bacillus thuringiensis* in the sunflower moth (Lepidoptera: Pyralidae)'. *Env. Entomol.*, **20**, 316–322.

Briggs, J. D. (1986). 'Pioneering and advanced phases of commercial use of *Bacillus thuringiensis* in North America'. In *Mitteilungen aus der Biologischen Bundesanstalt für Land- und Forstwirtschaft Berlin-Dahlem* (Eds. A. Krieg and A. M. Huger), Heft 233, pp. 25–36, Paul Parey, Berlin.

Brunke, K. J. and Meeusen, R. L. (1991). 'Insect control with genetically engineered crops'. *TibTech*, **9**, 197–200.

Burgerjon, A. and Grison, P. (1959). 'Sensibilité de differents lépidoptères à la souche "anduze" de *Bacillus thuringiensis* Berliner'. *Entomophaga*, **4**, 205–209.

Burgerjon, A. and Martouret, D. (1971). 'Determination and significance of the host spectrum of *Bacillus thuringiensis*'. In *Microbial Control of Insects and Mites* (Eds. H. D. Burges and N. W. Hussey), pp. 305–325, Academic Press, London.

Burges, H. D. (1967). 'The standardization of products based on *Bacillus thuringiensis*'. In *Insect Pathology and Microbial Control* (Ed. P. van der Laan), pp. 306–314, North Holland Publ. Co., Amsterdam.

Burges, H. D. and Daoust, R. A. (1986). 'Current status of the use of bacteria as biocontrol agents'. In *Fundamental and Applied Aspects of Invertebrate Pathology* (Eds. R. A. Samson, J. M. Vlak, and D. Peters), pp. 514–517, Soc. Invert. Pathol., Wageningen.

Burges, H. D. and Hurst, J. A. (1977). 'Ecology of *Bacillus thuringiensis* in storage moths'. *J. Invert. Pathol.*, **30**, 131–139.

Carlton, B. C., Gawron-Burke, C. and Johnson, T. B. (1990). 'Exploiting the genetic diversity of *Bacillus thuringiensis* for the creation of new bioinsecticides'. In *Proceedings Vth International Colloquium on Invertebrate Pathology and Microbial Control*, pp. 18–22, Adelaide, Australia.

Carroll, J., Li, J. and Ellar, D. J. (1989). 'Proteolytic processing of a coleopteran-specific delta-endotoxin produced by *Bacillus thuringiensis* var. *tenebrionis*'. *Biochem. J.*, **261**, 99–105.

Carter, N. (1990). 'Protection spraying against spruce budworm in New Brunswick in 1989'. New Brunswick Dep. Natural Res. and Energy, Fredericton, NB.

Carter, N. and Lavigne, D. (1991). 'Protection spraying against spruce budworm in New Brunswick in 1990', New Brunswick Dep. Natural Res. and Energy, Fredericton, NB.

Chambers, J. A., Jelen, A., Gilbert, M. P., Jany, C. S., Johnson, T. B. and Gawron-Burke, C. (1991). 'Isolation and characterization of a novel insecticidal crystal protein gene from *Bacillus thuringiensis* subsp. *aizawai*'. *J. Bacteriol.*, **173**, 3966–3976.

Choma, C. T., Surewicz, W. K., Carey, P. R., Pozsgay, M., Raynor, T. and Kaplan, H. (1990). 'Unusual proteolysis of the protoxin and toxin from *Bacillus thuringiensis*, structural implications'. *Eur. J. Biochem.*, **189**, 523–527.

Christou, P., Ford, T. L. and Kofron, M. (1991). 'Production of transgenic rice plants from agronomically important indica and japonica varieties via electric discharge particle acceleration of exogenous DNA into immature zygotic embryos'. *Bio/Technology*, **9**, 957–962.

Chungjatupornchai, W., Höfte, H., Seurinck, J., Angsuthanasombat, C. and Vaeck, M. (1988). 'Common features of *Bacillus thuringiensis* toxins specific for Diptera and Lepidoptera'. *Eur. J. Biochem.*, **173**, 9–16.

Convents, D., Houssier, C., Lasters, I. and Lauwereys, M. (1990). 'The *Bacillus thuringiensis* delta-endotoxin: evidence for a two domain structure of the minimal toxic fragment'. *J. Biol. Chem.*, **265**, 1369–1375.

Convents, D., Cherlet, M., van Damme, J., Lasters, I. and Lauwereys, M. (1991). 'Two structural domains as a general fold of the toxic fragment of the *Bacillus thuringiensis* delta-endotoxins'. *Eur. J. Biochem.*, **195**, 631–635.

Cooksey, K. E. (1971). 'The protein crystal toxin of *Bacillus thuringiensis*': biochemistry and mode of action'. In *Microbial Control of Insects and Mites* (Eds. H. D. Burges and N. W. Hussey), pp. 247–274, Academic Press, London.

Crickmore, N., Nicholls, C., Earp, D. J., Hodgman, T. C. and Ellar, D. J. (1990). 'The construction of *Bacillus thuringiensis* strains expressing novel entomocidal delta–endotoxin combinations'. *Biochem. J.*, **270**, 133–136.

Dankocsik, C., Donovan, W. P. and Jany, C. S. (1990). 'Activation of a cryptic crystal protein gene of *Bacillus thuringiensis* subspecies *kurstaki* by gene fusion and determination of the crystal protein insecticidal specificity'. *Mol. Microbiol.*, **4**, 2087–2094.

Dardene, F., Seurink, J., Lambert, B. and Peferoen, M. (1990). 'Nucleotide sequence and deduced amino acid sequence of a CryIA(c) variant from *Bacillus thuringiensis*'. *Nucl. Acids Res.*, **18**, 5546.

Dean, D. H., Almond, B. D., Ge, A. Z., Lee, M. K., Cheng, C. I., Liebig, B. and Milne, R. (1991). 'Overview of specificity and receptor binding domains on CryIA delta-endotoxins', in *Abstracts XXIV Annual Meeting Soc. Invert. Pathol.*, Flagstaff, Arizona, p. 73.

de Barjac, H. (1981). 'Identification of H-serotypes of *Bacillus thuringiensis'*. In *Microbial Control of Pests and Plant Diseases 1970–1980* (Ed. H. D. Burges), pp. 35–43, Academic Press, London.

de Barjac, H. and Bonnefoi, A. (1962). 'Essai de classification biochemique et serologique de 24 souches de Bacillus du type *B. thuringiensis*. *Entomophaga*, **7**, 5–31.

de Barjac, H. and Bonnefoi, A. (1968). 'A classification of strains of *Bacillus thuringiensis* with a key to their differentiation'. *J. Invert. Pathol.'*, **11**, 335–347.

de Barjac, H. and Bonnefoi, A. (1973). 'Mise au point sur la classification des *Bacillus thuringiensis'*. *Entomophaga*, **18**, 5–17.

de Barjac, H. and Frachon, E. (1990). 'Classification of *Bacillus thuringiensis* strains'. *Entomophaga*, **35**, 233–240.

Delannay, X., Lavallee, B. J., Proksch, R. K., Fuchs, R. L., Sims, S. R., Greenplate, J. T., Marrone, P. M., Dodson, R. B., Augustine, J. J., Layton, J. G. and Fischhoff, D. A. (1989). 'Field performance of transgenic tomato plants expressing the *Bacillus thuringiensis* var. *kurstaki* insect control agents'. *Bio/Technology*, **7**, 1265–1269.

DeLucca, A. J., Simonson, J. G. and Larson, A. D. (1981). '*Bacillus thuringiensis* distribution in soils of the United States'. *Can. J. Microbiol.*, **27**, 865–870.

Dimock, M. B., Beach, R. M. and Carlson, P. S. (1988). 'Endophytic bacteria for the delivery of crop protection agents'. In *Biotechnology, Biological Pesticides and Novel Plant-Pest Resistance for Insect Pest Management* (Eds. D. W. Roberts and R. S. Granados), pp. 88–92, Boyce Thompson Institute, Ithaca, New York.

Donovan, W. P., Dankocsik, C. D., Gilbert, M. P., Gawron-Burke, M. C., Groat, R. G. and Carlton, B. C. (1988). 'Amino acid sequence and entomocidal activity of the P2 crystal protein'. *J. Biol. Chem.*, **263**, 561–567.

Dulmage, H. T. (1970). 'Insecticidal activity of HD-1, a new isolate of *Bacillus thuringiensis* var. *alesti'*. *J. Invert. Pathol.*, **15**, 232–239.

Dulmage, H. T. and Rhodes, R. A. (1971). 'Production of pathogens in artificial media'. In *Microbial Control of Insects and Mites* (Eds. H. D. Burges and N. W. Hussey), pp. 507–540, Academic Press, London.

Dulmage, H. T., Boening, O. P., Rehnborg, C. S. and Hansen, G. D. (1971). 'A proposed standardized bioassay for formulations of *Bacillus thuringiensis* based on the International Unit'. *J. Invert. Pathol.*, **18**, 240–246.

Dulmage, H. T. and Cooperators (1981). 'Insecticidal activity of isolates of *Bacillus thuringiensis* and their potential for pest control'. In *Microbial Control of Pests and Plant Diseases 1970–1980* (Ed. H. D. Burges), pp. 193–222, Academic Press, London.

Fast, P. G. (1981). 'The crystal toxin of *Bacillus thuringiensis'*. In *Microbial Control of Pests and Plant Diseases 1970–1980* (Ed. H. D. Burges), pp. 221–248, Academic Press, London.

Felton, G. W. and Dahlman, D. L. (1984). 'Allelochemical induced stress: effects of l-canavanine on the pathogenicity of *Bacillus thuringiensis* in *Manduca sexta'*. *J. Invert. Pathol.*, **44**, 187–191.

Ferré, J., Real, M. D., van Rie, J., Jansens, S. and Peferoen, M. (1991). 'Resistance to the *Bacillus thuringiensis* bioinsecticide in a field population of *Plutella xylostella* is due to a change in a midgut membrane receptor'. *Proc. Natl Acad. Sci. USA*, **88**, 5119–5123.

Fischhoff, D. A., Bowdish, K. S., Perlak, F. J., Marrone, P. G., McCormick, S. M., Niedermeyer, J. G., Dean, D. A., Kusano-Kretzmer, K., Mayer, E. J., Rochester,

D. E., Rogers, S. G. and Fraley, R. T. (1987). 'Insect tolerant transgenic tomato plants'. *Bio/Technology*, **5**, 807–813.

Gaertner, F. (1990). 'Cellular delivery systems for insecticidal proteins: living and non-living microorganisms'. In *Controlled Delivery of Crop Protection Agents* (Ed. R. M. Wilkins), pp. 245–257, Taylor & Francis, London.

Gawron-Burke, C. M., Chambers, J., Jelen, A., Donovan, W., Rupar, M., Jany, C., Slaney, A., Baum, J., English, L. and Johnson, T. (1990). 'Molecular biology and genetics of *Bacillus thuringiensis*'. In *Proceedings Vth International Colloquium on Invertebrate Pathology and Microbial Control*', pp. 456–460, Adelaide, Australia.

Ge, A. Z., Rivers, D., Milne, R. and Dean, D. H. (1991). 'Functional domains of *Bacillus thuringiensis* insecticidal crystal proteins: refinement of the *Heliothis virescens* and *Trichoplusia ni* specificity domains on CryIA(c)'. *J. Biol. Chem.*, **266**, 17954–17958.

Ge, A. Z., Shivarova, N. I. and Dean, D. H. (1989). 'Location of the *Bombyx mori* specificity domain on a *Bacillus thuringiensis* delta-endotoxin protein'. *Proc. Natl. Acad. Sci. USA*, **86**, 4037–4041.

Goldberg, L. J. and Margalit, J. (1977). 'A bacterial spore demonstrating rapid larvicidal activity against *Anopheles sergentii*, *Uranotaenia unguiculata*, *Culex univitattus*, *Aedes aegypti*, and *Culex pipiens*'. *Mosq. News*, **37**, 355–358.

Gonzalez, J. M. and Carlton, B. C. (1982). 'Plasmid transfer in *Bacillus thuringiensis*'. In *Genetic Exchange: A Celebration and a New Generation* (Eds. U. N. Streips, S. H. Goodall, W. R. Guild and G. A. Wilson), pp. 85–95, Marcel Dekker, New York.

Gonzalez, J. M., Dulmage, H. T. and Carlton, B. C. (1981). 'Correlation between specific plasmids and delta-endotoxin production in *Bacillus thuringiensis*'. *Plasmid*, **11**, 28–38.

Gonzalez, J. M., Brown, B. J. and Carlton, B. C. (1982). 'Transfer of *Bacillus thuringiensis* plasmids coding for delta-endotoxin among strains of *B. thuringiensis* and *B. cereus*'. *Proc. Natl Acad. Sci. USA*, **79**, 6951–6955.

Gould, F. (1988). 'Evolutionary biology and genetically engineered crops'. *BioScience*, **38**, 26–33.

Gould, F. and Anderson, A. (1991). 'Effects of *Bacillus thuringiensis* and HD-73 delta-endotoxin on growth, behaviour, and fitness of susceptible and toxin-adapted strains of *Heliothis virescens* (Lepidoptera: Noctuidae)'. *Env. Entomol.*, **201**, 30–38.

Gould, F., Anderson, A., Landis, D. and van Mellaert, H. (1991). 'Feeding behaviour and growth of *Heliothis virescens* larvae on diets containing *Bacillus thuringiensis* formulations or endotoxins'. *Ent. Exp. Appl.*, **58**, 199–210.

Grimble, D. G. and Lewis, F. B. (1985). *Proceedings of a Symposium: Microbial Control of Spruce Budworms and Gypsy Moths*, Windsor Locks, Conn. USDA For. Serv. GTR-NE-100, Broomall, Penns.

Haider, M. Z., Knowles, B. H., and Ellar, D. J. (1988). 'Specificity of *Bacillus thuringiensis* var. *colmeri* insecticidal delta-endotoxin is determined by differential proteolytic processing of the protoxin by larval gut proteases'. *Eur. J. Biochem.*, **156**, 531–540.

Hall, I. M. and Dunn, P. (1958). 'Susceptibility of some insect pests to infection by *Bacillus thuringiensis* in laboratory tests'. *J. Econ. Entomol.*, **51**, 296–298.

Hannay, C. L. (1953). 'Crystalline inclusions in aerobic spore-forming bacteria'. *Nature (London)*, **172**, 1004.

Hannay, C. L. and Fitz-James, P. (1955). 'The protein crystals of *Bacillus thuringiensis* Berliner'. *Can. J. Microbiol.*, **1**, 694–709.

Heimpel, A. M. and Angus, T. A. (1958). 'The taxonomy of insect pathogens related to *Bacillus cereus'. Can. J. Microbiol.*, **4**, 531–541.

Herrera-Estrella, L., Depicker, A., van Montagu, M. and Schell, J. (1983). 'Expression of chimaeric genes transferred into plant cells using a Ti-plasmid-derived vector'. *Nature London*, **303**, 209–213.

Herrnstadt, C., Soares, G. C., Wilcox, E. R. and Edwards, D. L. (1986). 'A new strain of *Bacillus thuringiensis* with activity against coleopteran insects'. *Bio/Technology*, **4**, 305–308.

Himel, C. M. (1969). 'The optimum size for insecticide droplets'. *J. Econ. Entomol.*, **62**, 919–925.

Hodgman, T. C. and Ellar, D. J. (1990). 'Models for the structure and function of the *Bacillus thuringiensis* delta-endotoxins determined by compilational analysis'. *J. DNA Sequencing and Mapping*, **1**, 97–106.

Hofmann, C., Vanderbruggen, H., Höfte, H., van Rie, J., Jansens, S. and Van Mellaert, H. (1988). 'Specificity of *Bacillus thuringiensis* delta-endotoxins is correlated with the presence of high-affinity binding sites in the brush border membrane of target insect midguts'. *Proc. Natl Acad. Sci. USA*, **85**, 7844–7848.

Höfte, H. and Whiteley, H. R. (1989). 'Insecticidal crystal proteins of *Bacillus thuringiensis'. Microbiol. Rev.*, **53**, 242–255.

Höfte, H., van Rie, J., Jansens, S., Van Houten, A., Vanderbruggen, H. and Vaeck, M. (1988). 'Monoclonal antibody analysis and insecticidal spectrum of three types of lepidopteran-specific insecticidal crystal proteins of *Bacillus thuringiensis'. App. Env. Microbiol.*, **54**, 2010–2017.

Höfte, H., Soetaert, P., Jansens, S. and Peferoen, M. (1990). 'Nucleotide sequence and deduced amino acid sequence of a new Lepidoptera-specific crystal protein gene from *Bacillus thuringiensis'. Nucl. Acids Res.* **18**, 5545.

Holden, A. V. and Bevan, D. (1978). 'Control of pine beauty moth by fenitrothion'. *Forestry Commission Report*, Farnham, Surrey.

Honée, G., Vriezen, W. and Visser, B. (1990). 'A translation fusion product of two different insecticidal crystal protein genes of *Bacillus thuringiensis* exhibits an enlarged insecticidal spectrum'. *Appl. Env. Microbiol.*, **56**, 823–825.

Huber, H. E. and Lüthy, P. (1981). '*Bacillus thuringiensis* delta-endotoxin: composition and activation'. In *Pathogenesis of Invertebrate Microbial Diseases* (Ed. E. W. Davidson), pp. 209–234, Allanheld, Osmun Publ. Totowa, N.J.

Irland, L. C. and Rumpf, T. A. (1987). 'Cost trends for *Bacillus thuringiensis* in the Maine spruce budworm control program'. *Bull. Ent. Soc. Am.*, **33**, 86–90.

Ishawata, S. (1901). 'On a kind of severe flacherie (sotto disease)'. *Dainihon Sanshi Kaiho*, **114**, 1–5.

Jacobs, S. E. (1951). 'Bacterial control of the flour moth, *Ephestia kuehniella'. Proc. Soc. Appl. Bacteriol.*, **13**, 83–91.

Jaquet, F., Hütter, R. and Lüthy, P. (1987). 'Specificity of *Bacillus thuringiensis* delta-endotoxin'. *Appl. Env. Microbiol.*, **53**, 500–504.

Johnson, D. E., Brookhart, G. L., Kramer, K. J., Barnett, B. D. and McGaughey, W. H. (1990). 'Resistance to *Bacillus thuringiensis* by the Indian meal moth, *Plodia interpunctella*: comparison of midgut proteinases from susceptible and resistant larvae'. *J. Invert. Pathol.*, **55**, 235–244.

Johnstone, D. R. (1978). 'The influence of physical and meteorological factors on the deposition and drift of spray droplets of controlled size'. In *Proceedings of a Symposium on Controlled Drop Application*, April 1978, University of Reading/British Crop Protection Council Monograph, **22**, 43–57.

Kingshorn, J. M., Fisher, R. A., Angus, T. A. and Heimpel, A. M. (1961). 'Aerial spray trials against the blackheaded budworm in British Columbia. *Can. Dept. For. Bimonthly Progress Report*, **17**, 3–4.

Klein, T. M., Wolf, E. D. and Allen, N. (1987). 'High-velocity microprojectiles for delivering nucleic acids into living cells'. *Nature (London)*, **327**, 70–73.

Klein, T. M., Gradziel, T., Fromm, M. E. and Sanford, J. C. (1988). 'Factors influencing gene delivery into *Zea mays* cells by high velocity microprojectiles'. *Bio/Technology*, **6**, 559–563.

Klier, A., Fargette, F., Ribier, J., Hoch, J., Aronson, A. I. and Minnich, S. A. (1982). 'Cloning and localization of the lepidopteran protoxin gene of *Bacillus thuringiensis* subsp. *kurstaki*'. *Proc. Natl Acad. Sci. USA*, **79**, 6065–6069.

Klier, A., Bourgouin, C. and Rappoport, G. (1983). 'Mating between *Bacillus subtilis* and *Bacillus thuringiensis* and transfer of cloned crystal genes'. *Mol. Gen. Genet.*, **191**, 257–262.

Klier, A., Lereclus, D., Ribier, J., Bourgouin, C., Menou, G., Lecadet, M.-M. and Rappoport, G. (1985). 'Cloning and expression in *Escherichia coli* of the crystal protein gene from *Bacillus thuringiensis* strain aizawai 7-29 and comparison of the structural organization of genes from different serotypes'. In *Molecular Biology of Microbial Differentiation* (Eds. J. A. Hoch and P. Setlow), pp. 217–224, American Society for Microbiology, Washington, DC.

Knowles, B. and Ellar, D. J. (1987). 'Colloid-osmotic lysis is a general feature of the mechanisms of action of *Bacillus thuringiensis* delta-endotoxins with different insect specificity'. *Biochim. Biophys. Acta*, **924**, 509–518.

Knowles, B. H., Knight, P. J. K. and Ellar, D. J. (1991). 'N-acetyl galactosamine is part of the receptor in insect gut epithelia that recognizes an insecticidal protein from *Bacillus thuringiensis*'. *Proc. R. Soc. Lond. B.*, **245**, 31–35.

Krieg, A. (1965). 'Über die vivo-titration von insektenpathogenen, speziell von *Bacillus thuringiensis*'. *Entomophaga*, **10**, 3–20.

Krieg, A. (1986). 'Die Entdeckung des *Bacillus thuringiensis* durch Dr. Ernst Berliner: ein Meilenstein der Insecktenpathologie und mikrobiologischen Bekämpfung von Schadinsekten'. In *Mitteilungen aus der Biologischen Bundesanstalt für Land- und Forstwirtschaft Berlin-Dahlem* (Eds. A. Krieg and A. M. Huger), Heft 233, pp. 11–24, Paul Parey, Berlin.

Krieg, A., Huger, A. M., Langenbruch, G. A. and Schnetter, W. (1983). '*Bacillus thuringiensis* var. *tenebrionis*: Ein neuer, gegenüber Larven von Coleopteren wirksamer Pathotyp'. *Z. Ang. Entomol.*, **96**, 500–508.

Krieg, A., Huger, A. M. and Schnetter, W. (1987). '"*Bacillus thuringiensis* var. *san diego*" Stamm M-7 ist identisch mit dem zuvor in Deutschland isolierten käferwirksamen *B. thuringiensis* subsp. *tenebrionis* Stamm BI 256–82'. *J. Appl. Entomol.* **104**, 417–424.

Krishik, V. A., Barbosa, P. and Reichelderfer, C. F. (1988). 'Three trophic level interactions: allelochemicals, *Manduca sexta*, and *Bacillus thuringiensis* var. *kurstaki* Berliner'. *Env. Entomol.*, **17**, 476–482.

Kronstad, J. W. and Whiteley, H. R. (1986). 'Three classes of homologous *Bacillus thuringiensis* crystal-protein genes'. *Gene*, **43**, 29–40.

Kronstad, J. W., Schnepf, H. E. and Whiteley, H. R. (1983). 'Diversity of locations for *Bacillus thuringiensis* crystal protein genes'. *J. Bacteriol.*, **154**, 419–428.

Krywienczyk, J., Dulmage, H. T. and Fast, P. G. (1978). 'Occurrence of two serologically distinct groups within *Bacillus thuringiensis* serotype 3a, b var. *kurstaki*'. *J. Invert. Pathol.*, **31**, 372–375.

Krywienczyk, J., Dulmage, H. T., Hall, I. M., Beegle, C. C., Arakawa, K. Y. and Fast, P. G. (1981). 'Occurrence of Kurstaki k-1 crystal activity in *Bacillus thuringiensis* subsp. *thuringiensis* serovar (H1)'. *J. Invert. Pathol.*, **37**, 62–65.

Lambert, M. (1987). 'Quantification of spray deposits in experimental and operational aerial spraying operations'. In *Proceedings of a Symposium on the Aerial Application of Pesticides in Forestry* (Ed. G. W. Green), pp. 125–129, AFA-TN-18, NRC No. 29197.

Lecadet, M. M. and Martouret, D. (1987). 'Host specificity of the *Bacillus thuringiensis* delta-endotoxins toward lepidopteran species: *Spodoptera littoralis* and *Pieris brassicae*'. *J. Invert. Pathol.*, **49**, 37–48.

Leemans, J., Reynaerts, A., Höfte, H., Peferoen, M., Van Mellaert, H. and Joos, H. (1990). 'Insecticidal crystal protein genes from *Bacillus thuringiensis* and their use in transgenic crops'. In *New Directions in Biological Control: Alternatives for Suppressing Agricultural Pests and Diseases*, pp. 573–581, Alan R. Liss.

Lereclus, D., Bourgouin, C., Lecadet, M. M., Klier, A. and Rapoport, G. (1989a). 'Role, structure, and molecular organization of the genes coding for the parasporal delta-endotoxins of *Bacillus thuringiensis*'. In *Regulation of Procaryotic Development* (Eds. I. Smith, R. A. Slepecky, and P. Setlow), pp. 255–276, American Society for Microbiology, Washington, DC.

Lereclus, D., Arantes, O., Chauffaux, J. and Lecadet, M. M. (1989b). 'Transformation and expression of a cloned delta-endotoxin gene in *Bacillus thuringiensis*'. *FEMS Microbiol. Lett.*, **60**, 211–218.

Li, J., Carroll, J. and Ellar, D. J. (1991). 'Crystal structure of insecticidal delta-endotoxin from *Bacillus thuringiensis* at 2.5 A resolution'. *Nature (London)*, **353**, 815–821.

Ludlum, C. T., Felton, G. W. and Duffy, S. S. (1991). 'Plant defenses: chlorogenic acid and polyphenol oxidase enhance toxicity of *Bacillus thuringiensis* to *Heliothis zea*'. *J. Chem. Ecol.*, **17**, 217–237.

Lüthy, P. and Ebersold, H. R. (1981). '*Bacillus thuringiensis* delta-endotoxin: histopathology and molecular mode of action'. In *Pathogenesis of Invertebrate Microbial Diseases* (Ed. E. W. Davidson), pp. 235–267, Allanheld, Osmun Publ., Totowa, N.J.

Lüthy, P., Cordier, J. L. and Fischer, H. M. (1982). '*Bacillus thuringiensis* as a bacterial insecticide: basic considerations and application'. In *Microbial and Viral Pesticides* (Ed. E. Kurstak), pp. 35–74. Marcel Dekker, New York.

Lüthy, P., Hofmann, C. and Jaquet, F. (1985). 'Inactivation of delta-endotoxin of *Bacillus thuringiensis* by tannin'. *FEMS Microbiol. Lett.*, **28**, 31–33.

MacIntosh, S. C., Stone, T. B., Sims, S. R., Hunst, P. L., Greenplate, J. T., Marrone, P. G., Perlak, F. J., Fischhoff, D. A. and Fuchs, R. L. (1990a). 'Specificity and efficacy of purified *Bacillus thuringiensis* proteins against agronomically important insects'. *J. Invert. Pathol.*, **56**, 258–266.

MacIntosh, S. C., Kishore, G. M., Perlak, F. J., Marrone, P. M., Stone, T. B., Sims, S. R. and Fuchs, R. L. (1990b). 'Potentiation of *Bacillus thuringiensis* insecticidal activity by serine protease inhibitors'. *J. Agric. Food Chem.*, **38**, 1145–1152.

MacIntosh, S. C., Stone, T. B., Jokerst, R. S. and Fuchs, R. L. (1991). 'Binding of *Bacillus thuringiensis* proteins to a laboratory-selected line of *Heliothis virescens*'. *Proc. Natl Acad. Sci.*, **88**, 8930–8933.

Mahillon, J., Chungjatupornchai, W., Decock, J., Diedrickx, S., Michiels, F., Peferoen, M. and Joos, H. (1989). 'Transformation of *Bacillus thuringiensis* by electroporation'. *FEMS Microbiol. Lett.*, **60**, 205–210.

Martens, J. W. M., Honée, G., Zuidema, D., van Lent, J. W. M., Visser, B. and Vlak, J. M. (1990). 'Insecticidal activity of a bacterial crystal protein gene expressed by a recombinant baculovirus in insect cells'. *Appl. Env. Microbiol.*, **56**, 2764–2770.

Martin, P. A. W. and Travers, R. S. (1989). 'Worldwide abundance and distribution of *Bacillus thuringiensis* isolates'. *Appl. Env. Microbiol.*, **55**, 2437–2442.

Masson, L., Préfontaine, G. and Brousseau, R. (1989). 'Transformation of *Bacillus thuringiensis* vegetative cells by electroporation'. *FEMS Microbiol. Lett.*, **60**, 273–512.

Masson, L., Moar, W. J., van Frankenhuyzen, K., Bossé, M. and Brousseau, R. (1992). 'Insecticidal properties of a novel crystal protein gene isolated from *Bacillus thuringiensis* subsp. *kenyae'*. *Appl. Env. Microbiol.*, **58**, 642–646.

Mattes, O. (1927). 'Parasitaire Krankheiten der Mehlmottenlarven und Versuche über ihre Verwendbarkeit als biologischen Bekämpfungsmittel'. *Sitzber. Ges. Beförd. Ges. Naturwiss. Marburg*, **62**, 382–417.

McCabe, D. E., Swain, W. F., Martinell, B. J. and Christou, P. (1988). 'Stable transformation of soybean by particle acceleration'. *Bio/Technology*, **6**, 923–926.

McCown, B. H., McCabe, D. E., Russell, D. R., Robinson, D. J., Barton, K. A. and Raffa, K. F. (1991). 'Stable transformation of *Populus* and incorporation of pest resistance by electric discharge particle acceleration'. *Plant Cell Reports*, **9**, 590–594.

McGaughey, W. H. (1985). 'Insect resistance to the biological insecticide *Bacillus thuringiensis'*. *Science*, **229**, 193–195.

McGaughey, W. H. and Beeman, R. W. (1988). 'Resistance to *Bacillus thuringiensis* in colonies of Indian meal moth and Almond moth (Lepidoptera: Pyralidae)'. *J. Econ. Entomol.*, **81**, 28–33.

McLinden, J. H., Sabourin, J. R., Clark, B. D., Gensler, D. R., Workman, W. E. and Dean, D. H. (1985). 'Cloning and expression of an insecticidal k-73 type crystal protein gene from *Bacillus thuringiensis* var. *kurstaki* into *Escherichia coli'*. *Appl. Env. Microbiol.*, **50**, 623–628.

McPherson, S. A., Perlak, F. J., Fuchs, R. L., Marrone, P. G., Lavrik, P. B. and Fischhoff, D. A. (1988). 'Characterization of the coleopteran-specific protein gene of *Bacillus thuringiensis* var. *tenebrionis'*. *Bio/Technology*, **6**, 61–66.

Meeusen, R. L. and Warren, G. (1989). 'Insect control with genetically engineered crops'. *Ann. Rev. Entomol.*, **34**, 373–381.

Menn, J. J. (1960). 'Bioassay of a microbial insecticide containing spores of *Bacillus thuringiensis'*. *J. Insect Pathol.*, **2**, 134–138.

Merryweather, A. T., Weyer, U., Harris, M. P. G., Hirst, M., Booth, T. and Possee, R. D. (1990). 'Construction of genetically engineered baculovirus insecticides containing the *Bacillus thuringiensis* subsp. *kurstaki* HD-73 delta endotoxin'. *J. Gen. Virol.*, **71**, 1535–1544.

Milne, R., Ge, A. Z., Rivers, D. and Dean, D. H. (1990). 'Specificity of insecticidal crystal proteins'. In *Analytical Chemistry of Bacillus thuringiensis* (Eds. L. A. Hickle and W. L. Fitch), American Chemical Society Symposium Series 432, Washington, DC.

Moar, W. J., Masson, L., Brousseau, R. and Trumble, J. T. (1960). 'Toxicity to *Spodoptera exigua* and *Trichoplusia ni* of individual P1 protoxins and sporulated cultures of *Bacillus thuringiensis* subsp. *kurstaki* HD-1 and NRD-12'. *Appl. Env. Microbiol.*, **56**, 2480–2483.

Morris, O. N. (1980). 'Report of the 1980 cooperative *Bacillus thuringiensis* spray trials'. *Information Report FPM-X-40*, Can. For. Serv., Saulte Ste. Marie, Ontario.

Morris, O. N. (1983). 'Protection of *Bacillus thuringiensis* from inactivation by sunlight'. *Can. Entomol.*, **109**, 1239–1248.

Morris, O. N., Angus, T. A. and Smirnoff, W. A. (1975). 'Field trials of *Bacillus thuringiensis* against the spruce budworm'. In *Aerial Control of Forest Insects in Canada* (Ed. M. L. Prebble), pp. 129–133, Can. Dept. Environm., Ottawa.

Mott, D. G., Angus, T. A., Heimpel, A. M. and Fisher, R. A. (1961). 'Aerial application of Thuricide against the spruce budworm in New Brunswick'. *Can. Dept. For. Bimonthly Progress Report*, **17**, 2.

Nambiar, P. T. C., Ma, S. W. and Iyer, V. N. (1990). 'Limiting an insect infestation of nitrogen-fixing root nodules of the pigeon pea by engineering the expression of an entomocidal gene in its root nodules'. *Appl. Env. Microbiol.*, **56**, 2866–2869.
Nicholls, C. N., Ahmad, W. and Ellar, D. (1989). 'Evidence for two different types of insecticidal P2 toxins with dual specificity in *Bacillus thuringiensis* subspecies'. *J. Bacteriol.*, **171**, 5141–5147.
Norris, J. R. (1964). 'The classification of *Bacillus thuringiensis*'. *J. Appl. Bacteriol.*, **27**, 439–447.
Norris, J. R. (1971). 'The protein crystal toxin of *Bacillus thuringiensis*: biosynthesis and physical structure'. In *Microbial Control of Insects and Mites* (Eds. H. D. Burges and N. W. Hussey), pp. 229–246, Academic Press, London.
Obukowicz, M. G., Perlak, F. J., Kusano-Kretzmer, K., Mayer, E. J., Bolten, S. L. and Watrud, L. S. (1986). 'Tn5-mediated integration of the delta-endotoxin gene from *Bacillus thuringiensis* into the chromosome of root-colonizing Pseudomonads'. *J. Bacteriol.*, **168**, 982–989.
Ohba, M. and Aizawa, K. (1986). 'Distribution of *Bacillus thuringiensis* in soils of Japan'. *J. Invert. Pathol.*, **47**, 277–282.
Perlak, F. J., Deaton, R. W., Armstrong, T. A., Fuchs, R. L., Sims, S. R., Greenplate, J. T. and Fischhoff, D. A. (1990). 'Insect resistant cotton plants'. *Bio/Technology*, **8**, 939–943.
Perlak, F. J., Fuchs, R. L., Dean, D. A., McPherson, S. L. and Fischhoff, D. A. (1991). 'Modification of the coding sequence enhances plant expression of insect control proteins'. *Proc. Natl Acad. Sci. USA*, **88**, 3324–3328.
Picot, J. J. C., Bontemps, X. and Kristmanson, D. D. (1985). 'Measuring spray atomizer droplet spectrum down to 0.5 μm size'. *Trans. ASAE*, **28**, 1367–1370.
Picot, J. J. C., Kristmanson, D. D. and Basak-Brown, N. (1986). 'Canopy deposit and off-target drift in forestry aerial spraying: the effects of operational parameters'. *Trans. ASAE*, **29**, 90–96.
Pozsgay, M., Fast, P. G., Kaplan, H. and Carey, P. (1987). 'The effect of sunlight on the protein crystals from *Bacillus thuringiensis* subsp. *kurstaki* HD-1 and NRD-12; a Raman spectroscopic study'. *J. Invert. Pathol.*, **50**, 246–253.
Préfontaine, G., Fast, P. G., Lau, P. C. K., Hefford, M. A., Hanna, Z. and Brousseau, R. (1987). 'Use of oligonucleotide probes to study the relatedness of delta-endotoxin genes among *Bacillus thuringiensis* subspecies and strains'. *Appl. Env. Microbiol.*, **53**, 2808–2814.
Price, P. W., Bouton, C. E., Gross, P., McPheron, B. A., Thompson, J. N. and Weis, A. E. (1980). 'Interactions among three trophic levels: influence of plants on interactions between insect herbivores and natural enemies'. *Ann. Rev. Ecol. Syst.*, **11**, 41–65.
Pusztai, M., Fast, P., Gringorten, L., Kaplan, H., Lessard, T. and Carey, P. R. (1991). 'The mechanism of sunlight-mediated inactivation of *Bacillus thuringiensis* crystals'. *Biochem. J.*, **273**, 43–47.
Raffa, K. F. (1989). 'Genetic engineering of trees to enhance resistance to insects. Evaluating the risks of biotype evolution and secondary pest outbreak'. *BioScience*, **39**, 524–534.
Roush, R. T. and McKenzie, J. A. (1987). 'Ecological genetics of insecticide and acaricide resistance'. *Ann. Rev. Entomol.*, **32**, 361–380.
Sanchis, V., Lereclus, D., Menou, G., Chaufaux, J. and Lecadet, M.-M. (1988). 'Multiplicity of delta-endotoxin genes with different insecticidal specificities in *Bacillus thuringiensis aizawai* 7.29'. *Mol. Microbiol.*, **2**, 393–404.
Schnepf, H. E. and Whiteley, H. R. (1981). 'Cloning and expression of the *Bacillus*

thuringiensis crystal protein gene in *Escherichia coli'. Proc. Natl Acad. Sci. USA*, **78**, 2893–2897.

Schnepf, H. E., Wong, H. C. and Whiteley, H. R. (1985). 'The amino acid sequence of a crystal protein from *Bacillus thuringiensis* deduced from the DNA base sequence'. *J. Biol. Chem.*, **260**, 6264–6272.

Schnepf, H. E., Tomczak, K., Ortega, J. P. and Whiteley, H. R. (1990). 'Specificity-determining regions of a lepidopteran-specific insecticidal protein produced by *Bacillus thuringiensis'. J. Biol. Chem.*, **265**, 20923–20930.

Schwartz, J. L., Garneau, L., Masson, L. and Brousseau, R. (1991). 'Early response of cultured lepidopteran cells to exposure to delta-endotoxin from *Bacillus thuringiensis*: involvement of calcium and anionic channels'. *Biochm. Biophys. Acta*, **1065**, 250–260.

Shibano, Y., Yamagata, A., Nakamura, N., Iizuka, T. and Takanami, M. (1985). 'Nucleotide sequence coding for the insecticidal fragment of the *Bacillus thuringiensis* crystal protein gene'. *Gene*, **34**, 243–251.

Sick, A., Gaertner, F. and Wong, A. (1990). 'Nucleotide sequence of a coleopteran-active toxin gene from a new isolate of *Bacillus thuringiensis* subsp. *tolworthi'. Nucl. Acids Res.*, **18**, 1305.

Sims, S. R. and Stone, T. B. (1991). 'Genetic basis of tobacco budworm resistance to an engineered *Pseudomonas fluorescens* expressing the delta-endotoxin of *Bacillus thuringiensis kurstaki'. J. Invert. Pathol.*, **57**, 206–210.

Skøt, L., Harrison, S. P., Nath, A., Mytton, L. R. and Clifford, B. C. (1990). 'Expression of insecticidal activity in *Rhizobium* containing the delta-endotoxin gene cloned from *Bacillus thuringiensis* subsp. *tenebrionis'. Plant Soil*, **127**, 285–295.

Slatin, S. L., Abrams, C. K. and English, L. (1990). 'Delta-endotoxins form cation-selective channels in planar lipid bilayers'. *Biochem. Biophys. Res. Commun.*, **169**, 765–772.

Smirnoff, W. A. and Morris, O. N. (1982). 'Field development of *Bacillus thuringiensis* in eastern Canada'. In *Biological Control Programmes against Insects and Weeds in Canada 1969–1980* (Eds. J. S. Kelleher and M. A. Hulme), pp. 238–247, Commonwealth Agricultural Bureaux.

Smith, R. A. and Couche, G. A. (1991). 'The phylloplane as a source of *Bacillus thuringiensis* variants'. *Appl. Env. Microbiol.*, **57**, 311–315.

Spillman, J. (1984). 'Spray impaction, retention, and adhesion: an introduction to basic characteristics'. *Pest. Sci.*, **15**, 97–106.

Steinhaus, E. A. (1951). 'Possible use of *Bacillus thuringiensis* as an aid in the biological control of the alfalfa caterpillar'. *Hilgardia*, **20**, 359–381.

Stock, C. A., McLoughlin, T. J., Klein, J. A. and Adang, M. J. (1990). 'Expression of a *Bacillus thuringiensis* crystal protein gene in *Pseudomonas cepacia* 526'. *Can. J. Microbiol.*, **36**, 879–884.

Svestka, M. (1989). 'Results and problems of aerial application of pesticides and biopreparations in Czechoslovak forestry'. *Lesnictvi*, **35**, 1003–1014.

Tabashnik, B. E. (1989). 'Managing resistance with multiple pesticide tactics: theory, evidence, and recommendations'. *J. Econ. Entomol.*, **82**, 1263–1269.

Tabashnik, B. E., Finson, N. and Johnson, M. W. (1991). 'Managing resistance to *Bacillus thuringiensis*: lessons from the diamondback moth (Lepidoptera: Plutellidae)'. *J. Econ. Entomol.*, **84**, 49–55.

Tailor, R., Tippett, J., Gibb, G., Pells, S., Pike, D., Jordan, L. and Ely, S. (1992). 'Identification and characterisation of a novel *Bacillus thuringiensis* δ-endotoxin entomocidal to coleopteran and lepidopteran larvae'. *Mol. Microbiol.*, **6** (9) (in press).

Tandeau de Marsac, N., de la Torre, F. and Szulmajster, J. (1987). 'Expression of the larvicidal gene of Bacillus sphaericus 1593M in the cyanobacterium Anacystis nidulans R2'. Mol. Gen. Genet., **214**, 42–47.

Tianjian, X., Bingao, H., Liansen, Z. and Gixin, W. (1990). 'Commercial production and application of Bt insecticide in China'. In Proceedings Vth International Colloquium on Invertebrate Pathology, p. 16, Adelaide, Australia.

Tiedje, J. M., Colwell, R. K., Grossman, Y. L., Hodson, R. E., Lenski, R. E., Mack, R. N. and Regal, P. J. (1989). 'The planned introduction of genetically engineered organisms: ecological considerations and recommendations'. Ecology, **70**, 298–315.

Travers, R. S., Martin, P. A. and Reichelderfer, C. F. (1987). 'Selective process for efficient isolation of soil Bacillus sp.'. Appl. Env. Microbiol., **53**, 1263–1266.

Vaeck, M., Reynaerts, A., Höfte, H., Jansens, S., de Beuckeleer, M., Dean, C., Zabeau, M., van Montagu, M. and Leemans, J. (1987). 'Transgenic plants protected from insect attack'. Nature (London), **328**, 33–37.

van Frankenhuyzen, K. (1990a). 'Development and current status of Bacillus thuringiensis for control of defoliating forest insects'. Forestry Chronicle, **56**, 498–507.

van Frankenhuyzen, K. (1990b). 'Effect of temperature and exposure time on toxicity of Bacillus thuringiensis spray deposits to spruce budworm, Choristoneura fumiferana (Lepidoptera: Tortricidae)'. Can. Entomol., **122**, 69–75.

van Frankenhuyzen, K. and Nystrom, C. (1989). 'Residual toxicity of a high-potency formulation of Bacillus thuringiensis to spruce budworm (Lepidoptera: Tortricidae)'. J. Econ. Entomol., **82**, 868–872.

van Frankenhuyzen, K., Gringorten, J. L., Milne, R. E., Gauthier, D., Puzstai, M., Brousseau, R. and Masson, L. (1991). 'Specificity of activated CryIA proteins from Bacillus thuringiensis subsp. kurstaki HD-1 for defoliating forest Lepidoptera'. Appl. Env. Microbiol., **57**, 1650–1655.

Van Rie, J., Jansens, S., Höfte, H., Degheele, D. and Van Mellaert, H. (1989). 'Specificity of Bacillus thuringiensis delta-endotoxins: importance of specific receptors on the brush border membrane of the midgut of target insects'. Eur. J. Biochem., **186**, 239–247.

Van Rie, J., Jansens, S., Höfte, H., Degheele, D. and Van Mellaert, H. (1990a). 'Receptors on the brush border membrane of the insect midgut as determinants of the specificity of Bacillus thuringiensis delta-endotoxins'. Appl. Env. Microbiol., **56**, 1378–1385.

Van Rie, J., McGaughey, W. H., Johnson, D. E., Barnett, D. and Van Mellaert, H. (1990b). 'Mechanism of insect resistance to the microbial insecticide Bacillus thuringiensis'. Science, **247**, 72–74.

Van Vliet, M. and Picot, J. J. (1987). 'Drop spectrum characterization for the Micronair AU4000 aerial spray atomizer'. Atom. and Spray Techn., **3**, 123–134.

Visser, B. (1989). 'A screening for the presence of four different crystal protein gene types in 25 Bacillus thuringiensis strains'. FEMS Microbiol. Lett., **58**, 121–124.

Visser, B., Munsterman, E., Stoker, A. and Dirkse, W. G. (1990). 'A novel Bacillus thuringiensis gene encoding a Spodoptera exigua specific crystal protein'. J. Bacteriol., **172**, 6783–6788.

Von Tersch, M. A., Robbins, H. L., Jany, C. S. and Johnson, T. B. (1991). 'Insecticidal toxins from Bacillus thuringiensis subsp. kenyae: gene cloning and characterization and comparison with Bacillus thuringiensis subsp. kurstaki CryIA(c) toxins'. Appl. Env. Microbiol., **57**, 349–358.

Waalwijk, C., Dullemans, A. and Maat, C. (1991). 'Construction of a bioinsecticidal rhizosphere isolate of Pseudomonas fluorescens'. FEMS Microbiol. Lett., **77**, 257–264.

Weiser, J. (1986). 'Impact of *Bacillus thuringiensis* on applied entomology in eastern Europe and in the Soviet Union'. In *Mitteilungen aus der Biologischen Bundesanstalt für Land- und Forstwirtschaft Berlin-Dahlem* (Eds. A. Krieg and A. M. Huger), Heft 233, pp. 37–50, Paul Parey, Berlin.

West, R. J., Raske, A. G. and Sundaram, A. (1989). 'Efficacy of oil-based formulations of *Bacillus thuringiensis* against hemlock looper, *Lambdina fiscellaria fiscellaria* (Lepidoptera: Geometridae)'. *Can. Entomol.*, **121**, 55–63.

Whiteley, H. R. and Schnepf, H. E. (1986). 'The molecular biology of parasporal crystal body formation in *Bacillus thuringiensis*'. *Ann. Rev. Microbiol.*, **40**, 549–576.

Widner, W. R. and Whiteley, H. R. (1989). 'Two highly related insecticidal crystal proteins of *Bacillus thuringiensis* subsp. *kurstaki* possess different host range specificities'. *J. Bacteriol.*, **171**, 965–974.

Widner, W. R. and Whiteley, H. R. (1990). 'Location of the dipteran specificity region in a lepidopteran-dipteran crystal protein from *Bacillus thuringiensis*'. *J. Bacteriol.*, **172**, 2826–2832.

Wiesner, C. J. and Kettela, E. G. (1987). 'Dipel deposit-efficacy studies and bioassay validation'. *New Brunswick Research and Productivity Council Report C/86/110*, Fredericton, New Brunswick.

Wolfersberger, M. G. (1990). 'The toxicity of two *Bacillus thuringiensis* delta-endotoxins to gypsy moth larvae is inversely related to the affinity of binding sites on midgut brush border membranes for the toxins'. *Experientia*, **46**, 475–477.

Wu, D., Cao, X. L., Bai, Y. Y. and Aronson, A. I. (1991). 'Sequence of an operon containing a novel delta-endotoxin gene from *Bacillus thuringiensis*'. *FEMS Microbiol. Lett.*, **81**, 31–36.

2 Diversity of *Bacillus thuringiensis* Toxins and Genes

DIDIER LERECLUS, ARMELLE DELÉCLUSE AND
MARGUERITE-M. LECADET
*Unité de Biochimie Microbienne, Institut Pasteur,
25 rue du Docteur Roux, 75724 Paris Cedex 15*, France

DIVERSITY OF *BACILLUS THURINGIENSIS* STRAINS AND TOXINS

The entomopathogenic properties of *Bacillus thuringiensis* (*B. t.*) are highly variable. In recent years this has resulted in extensive searches for new strains with different target spectra. Several thousand natural strains have been isolated from various geographical areas and from different sources, including grain dust, soil, insects and plants (Martin and Travers, 1989; Smith and Couche, 1991). These isolates can be classified into about 30 serotypes based on biochemical properties and flagellar antigens or H-antigens (de Barjac and Frachon, 1990). However, this classification does not reflect the pathotype of the bacteria, which is essentially defined by the δ-endotoxins that make up the characteristic crystalline inclusion of the *B. t.* strains. These insecticidal proteins are synthesized after stage II of sporulation (t_2–t_3) and accumulate in the mother cell as a crystal which can account for up to 25% of the dry weight of the sporulated cells (Ribier and Lecadet, 1973).

Most *B. t.* strains can synthesize more than one crystal, which may itself be formed by different, although related, δ-endotoxins. Depending on their δ-endotoxin composition, the crystals have various forms and a partial correlation between structure and protein composition of the crystals has been established (Table 2.1). Since each type of δ-endotoxin is active against a limited range of insects, this correlation can be extended to the target insects. Thus, microscopic examination of the crystals produced by a given strain can provide preliminary indications about its insect specificity.

Analysis of the protein composition of the crystals, by polyacrylamide gel electrophoresis, also gives useful information. The electrophoretic patterns of δ-endotoxin crystal preparations are characteristic of strains

Bacillus thuringiensis, An Environmental Biopesticide: Theory and Practice. Edited by P. F. Entwistle, J. S. Cory, M. J. Bailey and S. Higgs

38

Table 2.1. Structure, composition and activity spectra of the crystals of *Bacillus thuringiensis*

Crystal components		Susceptible insects[b]	B. thuringiensis strains (examples)	Serotype	Structure	References
Class[a]	Size (kDa)					
I	130–140	L	*thuringiensis berliner*	1	Bipyramidal	Lecadet (1970)
			kurstaki HD-1, KT0	3		Aronson et al. (1986)
			entomocidus 6.01	6		Unpublished data
			aizawai 7.29	7		
	135	D, L	*aizawai* IC 1	7		Haider and Ellar (1987)
II	71	D, L	*kurstaki* HD-1	3	Cuboidal	Yamamoto and McLaughlin (1981)
	71	L	*kurstaki* HD-1	3	?	Widner and Whiteley (1989)
III	66–73	C	*tenebrionis* *san diego*	8	Flat rhomboid	Krieg et al. (1983, 1987) Herrnstadt et al. (1986)
IV	125–145	D	*morrisoni* PG14	8	Composite (spherical and rectangular)	Ibarra and Federici (1986)
	68	D	*israelensis*	14		Charles and de Barjac (1982)
Cyt	26–28	Non-specific (cytolytic activity)				
?	81	C, L	*kurstaki* JHCC 4835	3	Bipyramidal	Blenk et al. (1989)
?	40–45	?	*thompsoni*	12	Pyramidal	Unpublished data

[a] According to the classification proposed by Höfte and Whiteley (1989).
[b] C, D and L indicate Coleoptera, Diptera and Lepidoptera, respectively.

active against Lepidoptera (*kurstaki* KT0 and *aizawai* 7.29), both Lepidoptera and Diptera (*thuringiensis berliner* 1715 and *kurstaki* HD-1), Coleoptera (*tenebrionis*) and Diptera (*israelensis*), shown in Figure 2.1. Several types of crystal protein, each with a defined activity spectrum, can be distinguished by electrophoretic analysis.

The cloning of numerous δ-endotoxin genes and knowledge of their nucleotide sequence allow the use of monoclonal antibodies (Höfte *et al.*, 1988) and of specific DNA probes (Préfontaine *et al.*, 1987) to determine the δ-endotoxin content of a given strain and, thereby, its putative insecticidal specificities.

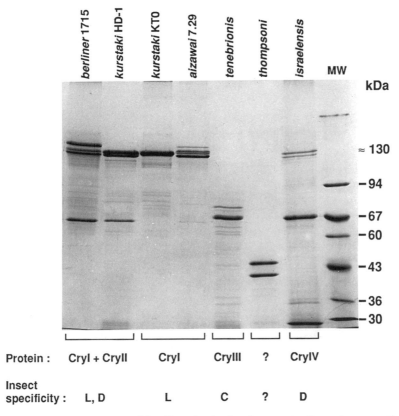

Figure 2.1. Protein patterns of *Bacillus thuringiensis* crystals. Crystal preparations isolated from various *B. thuringiensis* strains (indicated at the top of the figure) were analysed by sodium dodecyl sulphate polyacrylamide gel electrophoresis (acrylamide, 10.5%) and coomassie blue staining. MW, molecular weight markers; kDa, kilodaltons. CryI, II, III and IV refer to the classification of Cry proteins described in Tables 2.1 and 2.3. C, D and L indicate Coleoptera, Diptera and Lepidoptera, respectively

Recently, Höfte and Whiteley (1989) proposed a system of nomenclature and classification of the δ-endotoxins according to their insecticidal properties and molecular relationships. Four major classes of δ-endotoxins (CryI, II, III and IV) and a cytolysin (Cyt), found in the crystals of the mosquito active strains, have been described to date. The different δ-endotoxins belonging to each of the four Cry classes contain homologous domains and presumably constitute a single family of proteins (Höfte and Whiteley, 1989; Lereclus et al., 1989b). Generally these proteins are active against Lepidoptera (CryI), both Lepidoptera and Diptera (CryII), Coleoptera (CryIII) and Diptera (CryIV), although there are exceptions.

Recent extensive screening programmes have revealed numerous strains producing δ-endotoxins which do not fall in these four categories. An example of such atypical strains is an isolate belonging to the B. t. subspecies thompsoni (Figure 2.1), which produces small cuboidal crystals composed of two polypeptides of 40 and 45 kDa. The insecticidal specificity of these polypeptides, if any, has not yet been determined. Moreover, new isolates displaying previously unknown toxic activities have also been described, for example, B. t. strains toxic for nematodes (Edwards et al., 1988). Similarly, a nematocidal activity has been recently associated with a 130 kDa δ-endotoxin produced by the mosquitocidal strain B. t. israelensis HD-567 (Schwab and Halloran, 1989). Novel strains producing an 81 kDa δ-endotoxin active against both lepidopteran and coleopteran species have also been isolated (Blenk et al., 1989).

In addition to the large variety of insecticidal crystal proteins (δ-endotoxins and cytolysin), some B. t. strains synthesize heat-stable toxins designated β-exotoxins (Lecadet and de Barjac, 1981). These compounds are nucleotide analogues whose insecticidal toxicity may be due to inhibition of RNA polymerase by competition with ATP (Sebesta and Horska, 1970). Although the β-exotoxins are not insect specific they presumably contribute to the overall toxicity of a B. t. strain.

LOCALIZATION AND MOLECULAR ORGANIZATION OF THE TOXIN GENES

Plasmids

A feature common to many B. t. subspecies is the presence of a complex array of plasmids. The number and the size of these plasmids (1.4 MDa to 180 MDa) vary considerably between strains (González and Carlton, 1980; Stahly et al., 1978) but are independent of serotype and pathotype.

Hybridization experiments indicate that the plasmids fall into two size groups: small plasmids (< 10 MDa) with some degree of relatedness and with no known function and large plasmids (> 30 MDa) sharing homologous DNA sequences (Lereclus *et al.*, 1982). The two groups appear to be unrelated.

Despite their diversity and the apparent absence of any essential function, the *B. t.* plasmids are generally inherited with remarkable stability, suggesting that maintenance mechanisms and very efficient replication functions could operate.

It is tempting to relate the two size groups to recent observations of Gruss and Ehrlich (1989) on plasmid replication in Gram-positive species. According to these investigators, a class of small plasmids (< 10 MDa, generally with a high copy-number) replicate by a rolling-circle mechanism involving accumulation of single-stranded DNA (ssDNA) intermediates. A second group of Gram-positive replicons, including larger plasmids, replicate by a theta mechanism and do not accumulate ssDNA: they are structurally stable and present at low copy-number (Bruand *et al.*, 1991; Jannière *et al.*, 1990). Recent observations reported by McDowell and Mann (1991), illustrate the former through the characterization of the small plasmid pHD2 from *B. t. kurstaki*. Sequence data analysis suggests that this plasmid belongs to the family of plasmids replicating via a ssDNA intermediate.

The larger low copy-number *B. t.* plasmids are very stably maintained in the cell and may correspond to the latter group. This is consistent with recent observations of Baum and Gilbert (1991) and of Lereclus and Arantes (1992) who characterized the replication regions of large plasmids and of the highly stable plasmid pHT1030, respectively. No ssDNA intermediate was found, and it is suggested that pHT1030 belongs to a new class of Gram-positive replicons, since only host proteins are required to ensure its replication.

Most *B. t.* plasmids are cryptic. The main function that has been assigned to them is the production of entomocidal toxins. Evidence for a correlation between crystal protein synthesis and the presence of a given plasmid was provided by curing experiments leading to the loss of large plasmids (González *et al.*, 1981). These investigators further demonstrated the transfer of such plasmids and others between *B. t.* strains through a conjugation-like process (González *et al.*, 1982). It was thus possible to restore the ability to produce insecticidal crystals to cured strains, by reintroducing appropriate plasmids. Direct evidence for this role of plasmids came from cloning the toxin structural genes, which were then used as probes to determine their localization in most of the known strains (reviewed in Lereclus *et al.*, 1989a; Whiteley and Schnepf, 1986). These genes are normally found on large plasmids (40–150 MDa) in many subspecies.

The possible chromosomal location of *cry* genes in some subspecies (*entomocidus* type, *aizawai* 7.29, *dendrolimus*, *wuhanensis*) has already been discussed and reviewed (Aronson *et al.*, 1986; Lereclus *et al.*, 1989b).

In fact, the possibility that such genes could be carried by very large plasmids (>150 MDa), which cannot be discriminated from the chromosome, cannot be ruled out.

The presence of one or several toxin genes on the same or different replicons, including very large plasmids or the chromosome, is now an established feature of the δ-endotoxin genes. There are examples of strains (*kurstaki* HD-1, *israelensis* and *aizawai* 7.29) harbouring at least five separate crystal protein genes. The *aizawai* strain contains: a *crylA(b)* gene present in duplicate on two different replicons, two other genes, *crylC* and *crylD*, located 3 kb apart, in the same orientation and presumed 'chromosomal', and a *crylA(a)* gene (Sanchis *et al.*, 1988). A second striking example is the *israelensis* strain in which four δ-endotoxin genes, plus a gene encoding a cytolysin, are located on the same 72 MDa plasmid. In contrast, a unique *crylA(c)* gene has been identified on a 50 MDa plasmid in strain *kurstaki* HD-73.

Information is now available on the location of the genetic determinants responsible for the synthesis of the heat stable β-exotoxin. In five lepidopteran-active strains belonging to three different H-serotypes, Levinson *et al.* (1990) were able to establish a correlation between β-exotoxin production with large plasmids (75 or 110 MDa) carrying a δ-endotoxin gene. This was demonstrated after curing experiments and transferring the plasmid to a Cry negative (Cry⁻), β-exotoxin-non-producing host.

It has also been shown that thuricin, a bacteriocin produced by the *B. t.* strain HD-2, recently isolated and characterized by Favret and Yousten (1989), correlates with the presence of a 150 MDa plasmid. This plasmid could encode either a structural or a regulatory gene for thuricin production. This bacteriocin was active against 48 of 56 strains examined and against some Gram-positive species. In partially purified preparations it appeared to be associated with a phospholipase A activity.

It has been suggested that certain plasmids, particularly small plasmids, affect the ultraviolet (UV) sensitivity of spores (Benoit *et al.*, 1990). The presence of plasmids may decrease spore resistance by increasing the amount of dipicolinic acids or by altering the patterns of small acid-soluble proteins present.

In strain *kurstaki* HD-1 several apparently cryptic plasmids (of 110, 29 and 4.9 MDa) were designated as possibly involved in a complex mechanism conditionally temperature-regulating crystal protein synthesis (Minnich and Aronson, 1984). In particular, the 4.9 MDa plasmid was presumed to regulate the amount of protoxin produced in *Bacillus cereus* transcipients.

Structural organization of plasmids harbouring crystal protein genes

It is currently accepted that several plasmid genes encoding δ-endotoxins are parts of composite structures which include several mobile genetic elements. This was first demonstrated in four lepidopteran-active strains by a variety of approaches (Kronstad and Whiteley, 1984; Lereclus *et al.*, 1984). As shown in Figure 2.2 the δ-endotoxin gene is flanked by two sets of insertion sequences, IS231 and IS232, and the transposon Tn4430. The characteristics of these transposable elements are summarized in Table 2.2.

Tn4430, which was the first transposon isolated from bacilli, encodes a transposase (TnpA) related to those of the Tn3 family (Lereclus *et al.*, 1986). However, the site-specific recombinase (TnpI) which ensures cointegrate resolution is not a resolvase but belongs to the integrase family (Mahillon and Lereclus, 1988). Thus, Tn4430 defines a novel type of Tn3 related transposable element (Sherratt, 1989).

Three IS231 elements were first identified in lepidopteran-active strains of *B. t.*, in which they constitute a family of isoelements (IS231, A, B, C . . .) (Mahillon *et al.*, 1985, 1987). It has been demonstrated (Hallet *et al.*, 1991), that IS231A is functional and can transpose in *Escherichia coli*, generating target duplication at the insertion site. Comparison of sequences at diverse sites of insertion indicated that the different target sites are similar to the extremities of Tn4430 which are a preferential insertion site for IS231. Five new IS231 elements are now being characterized (R. Rezsohazy, personal communication). IS231D and E were isolated from *B. t. finitimus*, and IS231F, V and W were isolated from the 72 MDa plasmid of the strain *israelensis*. As shown in Figure 2.2, IS231W was previously mapped in the vicinity of the *cytA* and *cryIVD* genes (Adams *et al.*, 1989).

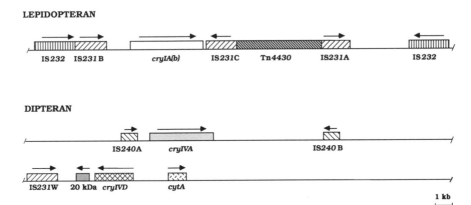

Figure 2.2. Structural organization of plasmid toxin genes and transposable elements. Arrows above the IS231, IS232 and IS240 elements indicate their relative orientation. Arrows above the *cry* genes and the 20 kDa polypeptide gene indicate the direction of transcription. The structural organizations shown are typical of *cry* genes found on large plasmids of lepidopteran- and dipteran-specific strains

Table 2.2. Characteristics of the transposable elements identified in *Bacillus thuringiensis*

Element	Length (bp)	Target duplication (bp)	Terminal inverted repeats (IR)		Encoded polypeptides		Mode of transposition[b]
			Size (bp)	Related IR	Size (aa)	Related mobile elements	
IS231	1656	11	20	Gram (−): IS4	478	Gram (−): IS4	C
IS232	2184	0, 4 or 6	28/37[a]	NI	431 / 250	Gram (−): IS21	C
IS240	865	NI	16	Gram (−): IS26 family / Gram (+): IS431 ISS1, IS6100	235	Gram (−): IS26 family / Gram (+): IS431 ISS1, IS6100	NI
Tn4430	4149	5	38	Gram (−): Tn3 family / Gram (+): Tn917	TnpI : 284 / TnpA : 987	Integrase family / Gram (−): Tn3 family / Gram (+): Tn917	R

[a] Number of identical base pairs (bp) in each IR/total length of the IR.
[b] R: replicative transposition (cointegrate formation); C : conservative transposition ('cut and paste' mechanism). NI: not identified; aa: amino acid.

Interestingly, IS231V and W are flanked by invert repeats of 12 bp similar to the extremities of Tn4430.

Another set of insertion elements found in lepidopteran-active strains have been characterized and named IS232 (Menou et al., 1990). This insertion sequence encodes two polypeptides homologous to the IstA and IstB proteins of the Gram-negative insertion sequence IS21. This element could transpose in E. coli after insertion of a genetic marker upstream of the two open reading frames.

The crylVA gene, in the dipteran-active B. t. israelensis, was flanked by two repeated sequences in opposite orientations (Bourgouin et al., 1988). They have been shown to be true insertion sequences designated IS240A and IS240B (Delécluse et al., 1989). The transposase of IS240 is homologous to the transposases of the widely spread insertion sequences belonging to the IS26 family. Sequences related to IS240 have also been identified in strains morrisoni PG14 and darmstadiensis 73 E10-16 which are active against mosquito larvae (Delécluse et al., 1990).

Finally, it should be mentioned that smaller cryptic plasmids, which are not involved in toxin production, have been shown to harbour Tn4430 and/or copies of IS231 (Lereclus et al., 1988; Mahillon et al., 1988; Mahillon and Seurinck, 1988). Thus, in B. t., mobile genetic elements are widely distributed among different and unrelated replicons.

Conjugation

An important technical breakthrough in genetic exchange in B. t. was the transference of plasmids during cell mating. Despite the absence of knowledge about the physiology of such conjugation phenomena, this procedure has been widely used. The large plasmids that harbour the crystal protein genes were identified by this approach (González et al., 1982). Due to the high frequency of plasmid transfer between B. t. subspecies grown in mixed cultures it was possible, in some cases, to detect DNA exchange without the help of a selective marker.

Self-transmissible plasmids are large plasmids. However, smaller plasmids may transfer simultaneously with them (González and Carlton, 1984; Lereclus et al., 1983). This was illustrated by Reddy et al. (1987) who characterised four self-transmissible plasmids pXO13, pXO14, pXO15 and pXO16, isolated from four different B. t. subspecies; they promote transfer of smaller plasmids (pBC16) into a variety of recipients including Bacillus cereus and Bacillus anthracis. Two small cryptic plasmids from B. t. israelensis have been shown to be mobilizable in mating experiments (Andrup et al., 1991). A DNA sequence from one of these plasmids (pTX14-3) displayed homology with a gene (mob2) encoding a putative mobilization protein found in plasmids of several Gram-positive species, including B. t. (Josson et al., 1990). The israelensis sequence was designated mob14-3.

Two other conjugative B. t. plasmids, pXO11 and pXO12 (Battisti et al., 1985), have been shown to mediate transfer of the pXO1 and pXO2 B. anthracis plasmids and it was suggested that such transfer could occur by conduction via the formation of cointegrate structures (Green et al., 1989). Interestingly, the transposon Tn4430 appeared to be involved in this process acting, presumably, in the mobilization of pXO1 and pXO2. The same study also indicated, that in plasmid pXO12, Tn4430 may be located close to sequences involved in mediating conjugal transfer; pXO12 also harbours a crystal protein gene. Furthermore, Tn4430 was isolated after a mating between Streptococcus faecalis and B. t. involving the conjugative plasmid pAMβ1; the plasmid pAMβ1::Tn4430 was thus obtained (Lereclus et al., 1983).

There is also an example of plasmid transfer from Bacillus subtilis to B. t., thereby introducing a cloned δ-endotoxin gene into a Cry⁻strain, or B. t. israelensis (Klier et al., 1983). More surprisingly, a bifunctional plasmid, carrying the oriT region of RK2 (Guiney and Yakobson, 1983), was successfully transferred between the distantly related species E. coli and B. t., via forced cell to cell contact (Trieu-Cuot et al., 1987). These procedures provide additional tools for the manipulation and transference of crystal protein genes.

Novel strategies are available to exploit the conjugative transfer of plasmids so as to isolate improved strains with new combinations of genes. It has been suggested (Carlton et al., 1990) that this natural process of genetic exchange can be used in conjunction with plasmid curing to select strains combining more than one desirable insecticidal specificity.

Finally, this approach is consistent with natural phenomena occurring within certain ecological niches, particularly within insect larvae themselves. To simulate such conditions, Jarrett and Stephenson (1990) infected lepidopteran larvae with two or more B. t. strains differing in their plasmid content. They were able to show that transfer of plasmids occurred within larvae at rates similar to those obtained in broth cultures.

Thus conjugative transfer between B. t. strains or related species occurs in nature resulting in new subspecies with various plasmid arrays. The intermolecular mobility of the cry genes, presumably mediated by transposable elements, and the evolution of the plasmid content of strains, may largely explain the diverse and complex activity spectra observed in B. t.

THE CRYSTAL PROTEIN GENES AND PRODUCTS

Eighteen different crystal protein genes, isolated from various B. t. strains, have been cloned and sequenced. The determination of both the amino acid sequence similarities and the insecticidal activities of the encoded polypeptides has allowed the classification of these genes into five classes (Höfte and Whiteley, 1989). As described below, despite the molecular relatedness observed between some of these genes, the insect specificities of their products vary.

Crystal protein gene relationships

The *cryI* gene class, the type most frequently identified, contains six subclasses of genes (*cryIA–F*) as shown in Table 2.3. The encoded polypeptides with molecular weights of 130–140 kDa are active against lepidopteran larvae and are related in amino acid sequence: they display greater than 55% identity (Höfte and Whiteley, 1989). On the basis of sequence similarities, the *cryIA* gene subclass has itself been divided into three subgroups (*cryIA(a)*, *(b)* and *(c)*); the amino acid sequences of the corresponding CryIA proteins show more than 80% amino acid identity within each subgroup. The carboxy-terminal halves of all CryI encoded proteins are highly conserved; the amino-terminal part is more variable. However, as summarized in Figure 2.3, five domains in the N-terminal part are conserved in all of the lepidopteran-active CryI toxins (Sanchis *et al.*, 1989). As will be shown below, this N-terminal variable region represents the active fragment. In contrast, the conserved C-terminal region is not required for toxicity: it may be involved in crystallization.

Many *B. t.* strains produce 71 kDa proteins in addition to the CryI polypeptides. Some of these 71 kDa proteins are toxic for both lepidopteran and dipteran larvae. In the strain *kurstaki* HD-1, the gene encoding this polypeptide belongs to the *cryII* class, and is referred to as the *cryIIA* gene (Donovan *et al.*, 1988b; Widner and Whiteley, 1989). The *cryIIA* gene is the distal gene of an operon which is comprised of three open reading frames

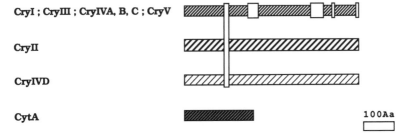

Figure 2.3. Toxic fractions of the major classes of crystal proteins. The horizontal open lines represent the toxic fractions of the crystal proteins. The white boxes indicate the domains highly conserved between all the different δ-endotoxins, as previously described (Höfte and Whiteley, 1989; Lereclus *et al.*, 1989b; Sanchis *et al.*, 1989). The regions variable between all the toxins are symbolized by hatched boxes. Aa; amino acid

Table 2.3. *Bacillus thuringiensis* crystal protein genes and product specificity

Gene type[a]	Predicted molecular weight (kDa)	Host range[b]	Examples of target insects	References[c]
cryIA(a)	132.2	L	*Manduca sexta, Bombyx mori, Pieris brassicae, Plutella xylostella, Ostrinia nubilalis*	Schnepf *et al.* (1985); Ge *et al.* (1989); Masson *et al.* (1989); Moar *et al.* (1990)
cryIA(b)	131	L	*Pieris brassicae, Manduca sexta, Heliothis virescens*	Wabiko *et al.* (1986); Höfte *et al.* (1986); Hofmann *et al.* (1988); Moar *et al.* (1990)
cryIA(c)	130	L/D	*Pieris brassicae, Aedes aegypti*	Haider and Ellar (1989)
	133.3	L	*Heliothis virescens, Manduca sexta, Trichoplusia ni, Ostrinia nubilalis, Pieris brassicae*	Adang *et al.* (1985); Höfte *et al.* (1988); Von Tersch *et al.* (1991)
cryIB	138	L	*Pieris brassicae*	Brizzard and Whiteley (1988); Höfte *et al.* (1988)
cryIC	134.8	L	*Spodoptera littoralis, Spodoptera exigua, Mamestra brassicae, Pieris brassicae*	Sanchis *et al.* (1988, 1989); Visser *et al.* (1988); Honée *et al.* (1988)
cryID	132.5	L	*Manduca sexta, Spodoptera exigua*	Sanchis *et al.* (1988); Visser *et al.* (1988); Höfte *et al.* (1990)
cryIE	130	L	*Manduca sexta, Spodoptera littoralis, Spodoptera exigua*	Van Rie *et al.* (1990); Bossé *et al.* (1990); Visser *et al.* (1990)
cryIF	133.6	L	*Ostrinia nubilalis, Heliothis virescens, Spodoptera exigua*	Chambers *et al.* (1991)
cryIIA	70.9	L/D	*Heliothis virescens, Lymantria dispar, Manduca sexta, Aedes aegypti*	Donovan *et al.* (1988b); Widner and Whiteley (1989); Ahmad *et al.* (1989)
cryIIB	70.8	L	*Manduca sexta, Lymantria dispar, Heliothis virescens, Trichoplusia ni*	Widner and Whiteley (1989); Dankocsik *et al.* (1990)

cryIIC	69.5	L	Manduca sexta, Lymantria dispar, Trichoplusia ni	Wu et al. (1991)
cryIIIA	73.1	C	Leptinotarsa decemlineata, Phaedon cochleariae	Donovan et al. (1988c); Herrnstadt et al. (1987); Höfte et al. (1987); Jahn et al. (1987); McPherson et al. (1988); Sekar et al. (1987)
cryIIIB	74.2	C	Leptinotarsa decemlineata	Sick et al. (1990)
cryIVA	134.4	D	Aedes aegypti, Anopheles stephensi, Culex pipiens	Bourgouin et al. (1988); Ward and Ellar (1987, 1988); Sen et al. (1988)
cryIVB	127.8	D	Aedes aegypti, Anopheles stephensi	Bourgouin et al (1986); Ward and Ellar (1988); Sen et al. (1988); Yamamoto et al. (1988); Sekar and Carlton (1985); Tungpradubkul et al. (1988); Angsuthanasombat et al. (1987)
cryIVC	77.8	D	Aedes aegypti	Thorne et al. (1986)
cryIVD	72.4	D	Aedes aegypti, Culex pipiens	Donovan et al. (1988a); Chang et al. (1990)
cryVA	81.2	L/C	Diabrotica spp., Leptinotarsa decemlineata, Ostrinia nubilalis	Blenk et al. (1989); Tailor et al. (1992)
cytA	27.4	non-specific		Ward et al. (1984); Ward and Ellar (1986); Waalwijk et al. (1985); Delécluse et al. (1991)

[a] According to Höfte and Whiteley (1989).
[b] L: lepidopteran, D: dipteran, C: coleopteran.
[c] References correspond both to the isolation of the gene and determination of gene product activity.

(Widner and Whiteley, 1989); the role of the two proximal genes remains unknown. A *crylIA*-related DNA fragment, now designated *crylIB*, has also been isolated (Dankocsik *et al.*, 1990; Widner and Whiteley, 1989). Unlike *crylIA*, *crylIB* is poorly expressed in *B. t.* cells (Dankocsik *et al.*, 1990). Recently, a third *crylI*-type gene, *crylIC*, has been cloned and sequenced (Wu *et al.*, 1991). This gene, isolated from a novel *B. t.* strain, is also part of an operon which presents similarities to that described for the *crylIA* gene. The CrylIA, B and C proteins share about 80–90% amino acid identity, but are dissimilar to the other Cry proteins, except in the first N-terminal conserved domain (Figure 2.3).

A third class of toxin genes encodes 73 kDa coleopteran-specific proteins (CrylII). Two genes belonging to this class have been characterized: *crylIIA* (Donovan *et al.*, 1988c; Herrnstadt *et al.*, 1987; Höfte *et al.*, 1987; Jahn *et al.*, 1987; McPherson *et al.*, 1988, Sekar *et al.*, 1987) and *crylIIB* (Sick *et al.*, 1990), which are distantly related (only 67% DNA identity). Recently, a novel strain displaying a different coleopteran-specificity from that described for CrylIIA and B (see below) has been isolated (Rupar *et al.*, 1991). Interestingly, the CrylIIA and B proteins are homologous to the amino-terminal half of the CryI protoxins through the five conserved domains (Figure 2.3), but lack the region corresponding to the carboxy-terminal part of these molecules.

The *crylV* class of crystal protein genes was isolated from *B. t. israelensis* which is active against dipteran larvae. This class contains four genes (*crylVA, B, C* and *D*) that encode polypeptides with predicted molecular masses of 135, 128, 74 and 72 kDa. Amino acid sequence comparisons of the CrylVA and CrylVB proteins revealed that these two 130 kDa type toxins possess a common carboxy-terminal part and differing amino-terminal halves (with 40% similarity). The *crylVC* gene, previously referred to as ORF1, is located upstream from a second gene, ORF2, and the two genes could form an operon (Thorne *et al.*, 1986). The amino acid sequence analysis indicated that the CrylVC and ORF2 products taken together are similar to the CrylVA and CrylVB proteins with CrylVC and ORF2 corresponding respectively to the variable region and carboxy-terminal part of these two polypeptides (Delécluse *et al.*, 1988). Therefore it is probable that *crylVC* and ORF2 are descended from an ancestral 130 kDa type protein gene. The 72 kDa polypeptide, which is a major component of the *B. t. israelensis* crystals, is also encoded by a *crylV* type gene, *crylVD*, despite its limited similarity with *crylVA, B* and *C* genes. The only amino acids conserved between the CrylVD protein and the other crystal proteins are restricted to a short region corresponding to the first N-terminal domain (Figure 2.3). The only polypeptide found in *B. t. israelensis* crystals is a polypeptide of 28 kDa, encoded by the *cytA* gene. The CytA protein shows no sequence homology to the other crystal polypeptides and has therefore been assigned to a class of proteins which does not belong to the endotoxin family.

Finally, a novel gene encoding a 81 kDa protein has recently been isolated (Blenk *et al.*, 1989). The encoded polypeptide displays similarities with the five conserved domains present in CryI, III and IV proteins (Figure 2.3); in addition, the carboxy-terminal part of both CryI and CryIV proteins contain similarity with the end of the 81 kDa protein, although the latter appears to be truncated. Moreover, as presented below, this protein presents a dual specificity, active against both lepidopteran and coleopteran larvae. This is, to our knowledge, the first example of dual specificity to these insect orders. Hence, a novel gene class (*cryV*) has been defined for this gene (Tailor *et al.*, 1992).

Four groups of *B. t.* crystal proteins can therefore be distinguished by amino acid sequence analysis; these, represented in Figure 2.3, are defined as follows: (a) a group of proteins including CryI, CryIII, CryIVA, CryIVB, and CryIVC proteins as well as the recently described 81 kDa protein with five domains conserved in each toxic fraction, (b) a 71 kDa type protein class, composed of all the CryII polypeptides, (c) a 72 kDa type protein corresponding to the CryIVD toxin and (d) CytA protein.

Toxic fragment of crystal proteins

The toxins cited above are in fact protoxins which are activated by insect midgut proteases to yield toxic fragments. The CytA protein is an exception as it is active in its native form, although a 25 kDa proteolytically derived fragment can be obtained (Armstrong *et al.*, 1985; Chilcott and Ellar, 1988; Thomas and Ellar, 1983). The CryIVD polypeptide has been shown to be converted into an active fragment of 30 kDa by proteases (Chilcott and Ellar, 1988; Pfannenstiel *et al.*, 1986). It is not known where the active toxic fragment maps in the native protein. In contrast, several studies have reported the identification of the active moiety of the 130 kDa protein type.

The toxic domain of the 130 kDa proteins is localized in the amino-terminal half of the protoxins. This has been demonstrated for lepidopteran-active CryI protoxins including CryIA(a) (Schnepf and Whiteley, 1985), CryIA(b) (Höfte *et al.*, 1986; Wabiko *et al.*, 1986), CryIA(c) (Adang *et al.*, 1985), CryIC (Sanchis *et al.*, 1989) and CryIE (Botterman *et al.*, 1989) by deletion analysis of the corresponding genes. Similar experiments with the *cryIVB* gene also revealed that the carboxy-terminal part of the CryIVB protein, active against dipteran larvae, is not essential for toxicity (Chungjatupornchai *et al.*, 1988; Delécluse *et al.*, 1988; Paointara *et al.*, 1988). The determination of N-terminal amino acid sequence of the trypsin-activated toxin together with deletion analysis showed that several amino acids from the N-terminus of the protoxins could be removed without affecting toxicity. The toxic fragment appears therefore to be about 60–65 kDa and corresponds to the protoxin truncated both at its amino- and carboxy-termini. Compilation of results obtained by several groups shows that the minimal toxic core of the CryI toxin is delineated by residues 29 and 607 (Höfte *et al.*, 1986; Nagamatsu *et al.*, 1984), and for CryIVB is contained between residues 39 and 662 (Delécluse *et al.*, 1988; Pao-intara *et al.*, 1988). The CryIII toxin lacks a region corresponding to the C-terminal part of the CryI proteins. It could therefore be considered as a naturally truncated version of a 130 kDa type protoxin. It can be extensively cleaved at the N-terminal end (up to amino acid 159) without loss of activity (Carroll *et al.*, 1989). Interestingly, sequences containing potential tryptic cleavage sites are found at positions approximating to the ends of the active fragments. The sequence present at the C-terminal end corresponds to one of the five domains (block 5) conserved in all the 130 kDa Cry proteins; in contrast, the sequence determined for the N-terminus of the toxic fragment is not found in all toxins, and suggests that N-terminal processing varies between toxins.

Insecticidal specificity of crystal gene products

Most *B. t.* strains simultaneously produce more than one crystal protein. The determination of the toxicity spectrum of individual proteins, which is investigated by analysing cloned gene products, has contributed an understanding of the diverse activities of *B. t.* strains.

Despite their homologies, there are considerable differences between the activity spectra of the eight CryI toxins. As shown in Table 2.3, most CryI protoxins are active, to varying extents, against several lepidopteran species. Among the eight *cryI* genes described, only *cryIC* and *E*, and at a lesser extent *cryID* and *F*, encode polypeptides which, despite their distant relatedness, are toxic to *Spodoptera* species. The CryIA proteins do not have this activity. The three lepidopteran-specific CryIA proteins, which are structurally closely related, show differences in insect toxicity spectrum, although there is some overlap. Even when active against the same insect, the specific activities of different CryIA proteins can differ. For example, the CryIA(a) protein was found to be 400 times more active against silkworm, *Bombyx mori*, than was the CryIA(c) protein. CryIA(c) was, in contrast, nine times more active against cabbage looper, *Trichoplusia ni*, than CryIA(a), while both have similar activities against tobacco hornworm, *Manduca sexta* (Ge *et al.*, 1989); several other reports have also documented the specificities of CryIA proteins (Höfte *et al.*, 1988; Moar *et al.*, 1990; van Frankenhuyzen *et al.*, 1991). The basis of this specificity has been investigated, and it is apparently related to a small domain in the variable amino-terminal region of CryIA protoxin. Ge *et al.*

(1989) defined the *B. mori* specificity domain of the CryIA(a) protein to a variable region, between amino acids 332 and 450. Interestingly, one of the CryIA(b) proteins is toxic to both lepidopteran and dipteran larvae (Table 2.3). Despite this dual activity, the protein isolated from *B. t. aizawai* IC1 has been classified as CryIA(b) because its amino acid sequence differs from that of a monospecific lepidopteran CryIA(b) polypeptide by only three amino acids (Haider and Ellar, 1988). This toxin, with dual specificity, appears to undergo differential proteolysis and hence activation, depending on the origin of the activating gut proteases (lepidopteran or dipteran). The structural basis for this differential conversion was shown to be related to the three amino acids that distinguish the dual-specific protein from the monospecific toxin (Haider and Ellar, 1989).

The host range specificity of the CryII type proteins, which are always found in strains that also produce CryI polypeptides, has also been determined. The first cloned protein identified, CryIIA, is toxic to the lepidopteran larvae tobacco budworm, *Heliothis virescens*, gypsy moth, *Lymantria dispar* and *M. sexta*, and the dipteran larvae *Aedes aegypti*. In contrast, CryIIB and CryIIC toxins are active only against lepidopteran species, despite their relative similarity to the CryIIA protein. A short region of 76 amino acids in CryIIA toxin has been identified as being important for mosquitocidal activity (Widner and Whiteley, 1990); this region differs in CryIIB and CryIIC at only 18 amino acid positions. It therefore appears that a small number of changes can substantially alter the specificities of these toxins. Interestingly, the segment of CryIIA specifying toxicity to the dipteran larvae *A. aegypti* overlaps the *B. mori* segment from CryIA(a) previously described. The toxic determinants therefore could be located in a similar region, even for non-related toxins.

Two Coleoptera-specific toxins, CryIIIA and B, have been found and analysed. CryIIIA is active against Colorado potato beetle, *Leptinotarsa decemlineata*, and *Phaedon cochleariae*. Little information is available regarding the toxicity of the CryIIIB protein. In contrast, a recent report (Rupar *et al.*, 1991) concerning novel coleopteran active *B. t.* strains refers to a toxin not previously described. It is more toxic than the CryIIIA protein for *L. decemlineata* larvae and is lethal for larvae of *Diabrotica undecimpunctata*, whereas the CryIIIA is not.

The crystals from *B. t. israelensis* active against dipteran larvae are composite, and contain three different types of polypeptides: (a) the 130 kDa type: CryIVA, CryIVB or CryIVC, (b) the 72 kDa protein designated CryIVD and (c) a cytolytic factor of 28 kDa referred to as CytA. This diversity has complicated the identification of the protein(s) responsible for mosquitocidal activity. Analysis of the cloned gene products has clarified the situation. All CryIV proteins are involved to varying degrees in their toxicity to mosquito larvae, but each of these polypeptides possesses a different spectrum of activity: the three 130 kDa type proteins, CryIVA, B and C, do not present the same specific activity towards the three mosquito larvae species generally tested (*Ae. aegypti*, *Anopheles stephensi*, and *Culex pipiens*). CryIVB protein is toxic for *A. aegypti* and *Anopheles stephensi* but inactive against *C. pipiens* (Délécluse *et al.*, 1988). CryIVA although less active than the CryIVB, is toxic for all three mosquito species (Bourgouin *et al.*, 1988). The CryIVC polypeptide is only slightly toxic for *A. aegypti* (Thorne *et al.*, 1986) but is totally inactive against *Anopheles stephensi* and *C. pipiens*. The CryIVD toxin which is not related to the CryIVA, B or C proteins exhibits activity towards *Ae. aegypti* (Donovan *et al.*, 1988a), and recently toxicity to *C. pipiens* has been reported for this polypeptide (Chang *et al.*, 1990). However, this toxicity has not been compared with those of the 130 kDa type proteins. Interestingly, it has been shown that none of these four polypeptides were individually as toxic as the native crystals, against any of the mosquito species tested (Délécluse, unpublished results). The high activity of the intact crystals is probably the result of synergy between several crystal components. Synergistic effects have been reported between the CryIVB and CryIVC toxins; the simultaneous presence of these two polypeptides results in mosquitocidal activity to *C. pipiens* larvae, whereas each of these polypeptides are non-toxic when tested alone (Délécluse *et al.*, 1988). Recently, a similar effect was detected between the CryIVA and CryIVB toxins; the presence of both polypeptides enhances the mosquitocidal activity against all the three mosquito species cited above. The level of activity observed when both polypeptides were present was nevertheless lower than that of the native crystals (Délécluse, unpublished results).

These results suggest that the activity of the native crystals is due to the presence of several polypeptides acting synergistically. The CytA protein, which is the major component of the crystals, is not homologous to the other crystal proteins (see above). The role of this polypeptide in mosquitocidal activity has long been the subject of controversy. Recent reports revealed that the CytA polypeptide is not essential for toxicity at least against the three species tested (Délécluse *et al.*, 1991). However, this protein is responsible for the *in vitro* cytolytic activity of the crystals (Armstrong *et al.*, 1985; Thomas and Ellar, 1983); its cytolytic activity is non-specific and directed against a wide variety of mammalian and insect cells. The role of the CytA polypeptide thus remains open to debate.

Finally, the 81 kDa polypeptide active against both lepidopteran and coleopteran larvae is the first toxin described with dual specificity on these insect orders. The domain(s) of the protein responsible for this specificity remain(s) unknown. Based on this particular activity, we suggest that the gene encoding the 81 kDa polypeptide be designated as *cryVA* (Table 2.3).

In view of these results, it appears that a large variety of toxins with different insect specificities contribute to the high insecticidal activity of the *B. t.* strains. A more detailed knowledge of the nature of the polypeptide segments involved in the specificity could allow the construction of chimeric toxins with increased insecticidal activities and/or enlarged spectra of activity.

REGULATION OF CRYSTAL SYNTHESIS

The very high level of crystal protein synthesis in *B. t.* and its rigorous coordination with sporulation constitute a remarkable and interesting system for the investigation of genetic regulation in sporulating Gram-positive bacteria.

At least three distinct mechanisms of transcriptional or posttranscriptional molecular regulation have been found to be devoted to the synthesis of crystals in *B. t.* Moreover, the level of toxin production in a given strain is probably dependent on the *cry* gene copy-number.

Temporal and spatial control of *cry* gene expression

Crystal protein genes of *B. t.* provide a typical example of sporulation genes specifically expressed in the mother cell compartment. Electron microscopic and kinetic studies of sporulation in *B. t.* indicate that the parasporal inclusion appears in the mother cell at t_2 (after that the asymmetric septation of the cell is triggered), and increases in size until t_{12} (t_0 is the onset of sporulation and t_n indicates the number of hours after t_0) (Ribier and Lecadet, 1973). Studies of sporulation negative (Spo$^-$) mutants clearly show that the synthesis of the crystal is developmentally regulated and is dependent on sporulation (Ribier and Lecadet, 1981). Spo$^-$ mutants blocked at t_0 failed to produce crystals, whereas those blocked after t_2 can produce crystals. The production of δ-endotoxins from a cloned *cry* gene in *Bacillus subtilis* was sporulation specific (Klier *et al.*, 1982).

Extensive studies of the sporulation of *B. subtilis* have given much detailed understanding of the complex mechanisms which temporally and spatially control the different steps of this differentiation process (reviewed in Losick and Kroos, 1989). The control of gene expression is, at least in part, exerted at the transcriptional level by the successive activation of sigma (σ) factors which bind the core RNA polymerase (E) to direct the RNA polymerase (Eσ) to specific promoters (Moran, 1989; Stragier and Losick, 1990). Among the four sigma factors (σ^E, σ^F, σ^G and σ^K) produced exclusively in sporulating cells, only σ^E and σ^K are able to function in the mother cell compartment after t_2. σ^E appears at t_2 and disappears about 3 hours later and is present in both forespore and mother cell compartments (Carlson and Haldenwang, 1989). The synthesis of σ^K results from a DNA rearrangement in the mother cell chromosome. It is only functional in the mother cell compartment, and there only between about t_3 and t_7 (Stragier et al., 1989).

The crylA(a) gene is transcribed in B. t. from two overlapping promoters (BtI and BtII) used sequentially (Wong et al., 1983). BtI is active between about t_2 and t_6 of sporulation, and BtII from about t_5 onwards. Similar regions (containing both types of promoters) are located upstream of all the lepidopteran-specific cryI genes and upstream of the cyt gene. To date, only one type of promoter (BtI or BtII) has been found upstream of the other cry genes (reviewed in Lereclus et al., 1989b).

Brown and Whiteley (1988, 1990) have isolated the two sigma factors that specifically direct the transcription of cry genes from BtI or BtII. $E\sigma^{35}$ can transcribe from the promoter BtI, which is active from early to mid sporulation, and $E\sigma^{28}$ activates transcription from the promoter BtII, which functions from mid to late sporulation. The genes encoding σ^{35} and σ^{28} have been cloned and sequenced, and their deduced amino acid sequences show 88 and 85% identity with σ^E and σ^K, respectively (Adams et al., 1991). Moreover, the respective B. t. sigma factors can complement B. subtilis mutants defective in σ^E or σ^K. Similarly, the B. t. core RNA polymerase reconstituted with either the σ^{35} or σ^{28} polypeptides directs transcription from B. subtilis promoters specifically recognized by $E\sigma^E$ and $E\sigma^K$, respectively (Adams et al., 1991). These results are in perfect agreement with the fact that in B. t. and B. subtilis both types of sigma factors are functional in the same cellular compartment at the same time of sporulation. Surprisingly, unlike σ^K, the functional σ^{28} gene does not result from a mother cell-specific chromosomal rearrangement of two complementary truncated genes.

mRNA stability and termination of transcription

One of the hypotheses proposed to account for the high level of δ-endotoxin production in B. t. is that the crystal protein mRNA is very stable. This was suggested by results of Glatron and Rapoport (1972) which indicated that the mRNAs coding for the crystal proteins had a half-life of 10 minutes. Since this observation, it has been shown that the putative transcriptional terminator of a toxin gene (a stem-loop structure) acts as a positive retroregulator and increases the mRNA half-life (Wong and Chang, 1986). It is notable that the potential terminator sequences found downstream of the different cry genes are frequently highly conserved. The fusion of such a putative terminator with the 3' end of heterologous genes enhances the half-life of their transcripts and consequently their expression level. When placed between a promoter and a reporter gene the efficiency of termination measured was about 50% (Hess and Graham, 1990).

These results suggest that the positive retroregulation of the cry genes should not be directly attributed to termination but rather to the presence of the stem-loop structure at the 3' end of the mRNA which protects it from exonuclease degradation.

Post-transcriptional effect

McLean and Whiteley (1987) have shown that a 0.8 kb DNA fragment located on the 72 MDa plasmid of *B. t. israelensis* (Figure 2.2) enhanced the expression level of CytA in *E. coli*. This positive effect is presumably post-transcriptional, since the amount of cytolysin-specific mRNA remains the same in the presence and in absence of the 0.8 kb sequence. More recently, it was shown that this DNA fragment encodes a 20 kDa polypeptide which is directly responsible for the enhancement of cytolysin production in *E. coli* cells (Adams et al., 1989). The analysis of the effect of the 20 kDa protein on fusions between *lacZ* and a DNA fragment carrying the promoter region and the first 30 codons of *cytA* indicates that the protein is not involved in the initiation of translation and therefore acts at some later stage of CytA synthesis. The 20 kDa protein mRNA appears 2 hours earlier than the CytA mRNA in *B. t.* and the two proteins are synthesized concurrently beginning at about t_2. Moreover the 20 kDa protein is found in solubilized preparations of *B. t. israelensis* crystals and can be coimmunoprecipitated with CytA by a single antibody directed against either polypeptide (Visick and Whiteley, 1991). These results suggest a protein–protein interaction in which the 20 kDa protein might act as a molecular chaperone to direct folding of CytA into a stable conformation or to protect it from proteolytic degradation.

Visick and Whiteley (1991) have shown that the presence of the 20 kDa protein in *E. coli* also increases the expression level of a cloned *crylVD* gene. However, there is no precise indication of the protein specificity of the 20 kDa protein nor its function in crystal production in *B. t.* Several results indicate that high amounts of crystal can be obtained in *B. t.* strains harbouring a cloned *cry* gene in the absence of the 20 kDa protein gene. Similarly, phase bright inclusions were obtained in *B. subtilis* and *B. t.* when the *cytA* gene was cloned alone (Bone and Ellar, 1989; Ward et al., 1986). Nevertheless it is possible that the Bacilli strains used as recipients in these experiments possess a gene coding for a polypeptide with a similar function to that of the 20 kDa protein.

The *cry* gene copy-number and differential expression

That the level of *cry* gene expression depends on its copy-number, at least in part, is evident. However, the copy-number of each *cry* gene in a given strain has to be considered if the overall toxicity of the strain is to be understood. In addition to the mechanisms of regulation involved at transcriptional and post-transcriptional levels, the copy-number of the plasmids carrying the cry genes is probably a major determinant of the relative production of each type of δ-endotoxin, and therefore in the insecticidal activity spectrum of the strain. It is important that the relative cry gene copy-number should be considered when constructing recombinant B. t. strains harbouring several δ-endotoxin genes.

B. t. kurstaki HD-1 contains the three homologous genes cryIA(a), (b) and (c); however, the corresponding δ-endotoxins are not found in equal amounts in the crystals produced by this strain (Masson et al., 1990; Yamamoto et al., 1988a). As these genes presumably possess similar promoter regions, this result implies the involvement of additional regulatory systems in the synthesis of the crystal. The copy-number of the plasmids carrying these genes could be partly responsible for their differential expression.

It was shown that in the wild strain thuringiensis 407, which harbours two different cryIA genes, the cryIA(a) gene is poorly expressed in comparison with the other. In sharp contrast with this situation is the high level of CryIA(a) production obtained when the gene is cloned alone in a B. t. Cry⁻ strain using the low copy-number plasmid pHT3101 (Lereclus et al., 1989a). The presumed promoter region of the cryIA(a) gene is not modified in the recombinant. Its very low expression in the parent strain may therefore be due to a difference in the copy-number of the two cry genes present in this strain and subsequent titration effects (e.g. competition for specific σ factors) creating conditions unfavourable for cryIA(a) expression.

Similarly, when a cryIA(c) gene is cloned in a strain harbouring other cry genes, significantly less CryIA(c) protein is produced than when cloned into an acrystalliferous strain of B. t. (Baum et al., 1990). When the cry genes are cloned on high copy-number plasmids and transferred into B. subtilis or Bacillus megaterium, they reduce the ability of the cells to form spores (Donovan et al., 1988a,b; Shivakumar et al., 1989). This effect is probably the result of increasing the dosage of sporulation specific promoters and is consistent with the findings described above for B. t.

The effect of plasmid copy-number on δ-endotoxin production was recently tested by cloning a cryIIIA gene into each of a series of vectors varying only by their copy-number. The results indicated that low δ-endotoxin production is obtained with a low copy-number plasmid (four copies per equivalent chromosome), while a high level of δ-endotoxin production is reached when the cryIIIA gene is cloned into a vector having about 15 copies per equivalent chromosome (Arantes and Lereclus, 1991).

In conclusion to this section it is tempting to predict that other mechanisms of regulation exist in B. t. to control the synthesis of the crystal. The recently available techniques of transformation in this bacterium may allow their characterization.

CONCLUSIONS AND PERSPECTIVES

The survey of the natural entomopathogenic properties of *B. t.*, presented in this review, reveals a striking example of bacterial genetic variability. The large family of δ-endotoxins found reflects the mobility and the flexibility of the insecticidal toxin genes, which is most probably dependent on their association with accessory genetic elements such as plasmids and transposable elements.

Höfte and Whiteley (1989) have proposed a classification system for the crystal protein genes established both on insect specificity and the primary structure of the proteins. Although the correlation between these two criteria is not always strictly observed (e.g. a CryIA protein displays a double activity against Lepidoptera and Diptera, and reciprocally CryIIB and C proteins are only toxic for Lepidoptera, see Tables 2.1 and 2.3), it appears that this system of nomenclature provides a useful framework to classify and designate the novel types of *B. t.* crystal protein toxins discovered by the current extensive programmes of screening. According to this classification, we suggest that the toxins active against both Lepidoptera and Coleoptera (e.g. the 81 kDa polypeptide indicated in Tables 2.1 and 2.3) define a CryV group of δ-endotoxins; the numerous crystal proteins, whose insect targets are as yet unknown, could represent new Cry classes (VI, VII, etc.), once their host specificities and amino acid sequences are determined.

The current molecular analyses of the mechanism of action of the δ-endotoxins could lead to the determination of the toxin domains which direct the specific interactions with the receptors of insects. The precise identification of the polypeptide sequences required for host specificity and toxicity could improve the present nomenclature of the *cry* genes or define a secondary system of classification based on the structure–function relationship of the δ-endotoxins (i.e. on the specificity determinants of the molecules). It is not unreasonable to assume that in the future an accurate and unambiguous classification of the δ-endotoxins will be largely dependent on the rigorous and concerted control of the numerous data concerning this aspect of *B. t.* research.

The broad diversity of *cry* genes, their cloning and the characterization of their products, the modification of their insecticidal specificities by *in vitro* mutagenesis and the understanding of the regulation of their expression, promise to contribute to the prospects of constructing genetically engineered *B. t.* strains with improved properties for pest control.

This requires the development of an efficient host–vector system. The recent emergence of electroporation procedures now makes the first part of this work feasible. Indeed, most *B. t.* strains can be easily transformed by plasmid DNA using electroporation procedures or transduction

(Lecadet *et al.*, 1992). The second part of the project has been addressed by using replication regions of *B. t.* resident plasmids to construct convenient shuttle vectors for cloning experiments (Baum *et al.*, 1990; Lereclus *et al.*, 1989a). These plasmids appear to be both segregationally and structurally stable and have allowed the cloning of several *cry* genes in *B. t.* and the subsequent production of crystals in the transformant clones. Moreover, the scope for genetic manipulation was increased recently when it was shown that *in vivo* recombination can occur in *B. t.* between homologous sequences carried by a non-replicative or thermosensitive plasmid and a resident plasmid (Delécluse *et al.*, 1991; Lereclus *et al.*, 1992).

Specific plasmid vectors and *in vivo* recombination mediated by homologous DNA sequences are two potentially valuable ways of introducing DNA into *B. t.* Recombinant strains harbouring various combinations of insecticidal crystal protein genes and displaying broader but defined activity spectra could thereby be constructed.

REFERENCES

Adams, L. F., Visick, J. E. and Whiteley, H. R. (1989). 'A 20-kilodalton protein is required for efficient production of the *Bacillus thuringiensis* subsp. *israelensis* 27-kilodalton crystal protein in *Escherichia coli*'. *J. Bacteriol.*, **171**, 521–530.

Adams, L. F., Brown, K. L. and Whiteley, H. R. (1991). 'Molecular cloning and characterization of two genes encoding sigma factors that direct transcription from a *Bacillus thuringiensis* crystal protein gene promoter'. *J. Bacteriol.*, **173**, 3846–3854.

Adang, M. J., Staver, M. J., Rocheleau, T. A., Leighton, J., Barker, R. F. and Thompson, D. V. (1985). 'Characterized full-length and truncated plasmid clones of the crystal protein of *Bacillus thuringiensis* subsp. *kurstaki* HD-73 and their toxicity to *Manduca sexta*'. *Gene*, **36**, 289–300.

Ahmad, W., Nicholls, C. and Ellar, D. J. (1989). 'Cloning and expression of an entomocidal protein gene from *Bacillus thuringiensis galleriae* toxic to both lepidoptera and diptera'. *FEMS Microbiol. Lett.*, **59**, 197–202.

Andrup, L., Bolander, G., Boe, L., Madsen, S. M., Nielsen, T. T. and Wassermann, K. (1991). 'Identification of a gene (*mob14-3*) encoding a mobilization protein from the *Bacillus thuringiensis* subsp. *israelensis* plasmid pTX14-3'. *Nucl. Acids Res.*, **19**, 2780.

Angsuthanasombat, C., Chungjatupornchai, W., Kertbundit, S., Luxananil, P., Settasatian, C., Wilairat, P. and Panyim, S. (1987). 'Cloning and expression of 130-kd mosquito larvicidal δ-endotoxin gene of *Bacillus thuringiensis* var. *israelensis* in *Escherichia coli*'. *Mol. Gen. Genet.*, **208**, 384–389.

Arantes, O. and Lereclus, D. (1991). 'Construction of cloning vectors for *Bacillus thuringiensis*'. *Gene*, **108**, 115–119.

Armstrong, J. L., Rohrman, G. F. and Beaudreau, G. S. (1985). 'Delta endotoxin of *Bacillus thuringiensis* subsp. *israelensis*'. *J. Bacteriol.*, **161**, 39–46.

Aronson, A. I., Beckman, W. and Dunn, P. (1986). 'Bacillus thuringiensis and related insect pathogens'. Microbiol. Rev., 50, 1–24.

Battisti, L., Green, B. D. and Thorne, C. B. (1985). 'Mating system for transfer of plasmids among Bacillus anthracis, Bacillus cereus and Bacillus thuringiensis'. J. Bacteriol., 162, 543–550.

Baum, J. A. and Gilbert, M. P. (1991). 'Characterization and comparative sequence analysis of replication origins from three large Bacillus thuringiensis plasmids'. J. Bacteriol., 173, 5280–5289.

Baum, J. A., Coyle, D. M., Gilbert, M. P., Jany, C. S. and Gawron-Burke, C. (1990). 'Novel cloning vectors for Bacillus thuringiensis'. Appl. Env. Microbiol., 56, 3420–3428.

Benoit, T. G., Wilson, G. R., Bull, D. L. and Aronson, A. I. (1990). 'Plasmid associated sensitivity of Bacillus thuringiensis to UV light'. Appl. Env. Microbiol., 56, 2282–2286.

Blenk, R. G., Ely, S., Tailor, R. H. and Tippett, J. M. (1989). 'Bacterial genes'. International Patent Application, PCT/GB90/00706.

Bone, E. J. and Ellar, D. J. (1989). 'Transformation of Bacillus thuringiensis by electroporation'. FEMS Microbiol. Lett., 58, 171–178.

Bossé, M., Masson, L. and Brousseau, R. (1990). 'Nucleotide sequence of a novel crystal protein gene isolated from Bacillus thuringiensis subspecies kenyae'. Nucl. Acids Res., 18, 7443.

Botterman, J., Peferoen, M., Höfte, H. and Joos, H. (1989). 'Plants transformed with a DNA sequence from Bacillus thuringiensis lethal to lepidoptera'. European Patent Application, 0 358 557 A2.

Bourgouin, C., Klier, A. and Rapoport, G. (1986). 'Characterization of the genes encoding the haemolytic toxin and the mosquitocidal delta-endotoxin of Bacillus thuringiensis israelensis'. Mol. Gen. Genet., 205, 390–397.

Bourgouin, C., Delécluse, A., Ribier, J., Klier, A. and Rapoport, G. (1988). 'A Bacillus thuringiensis subsp. israelensis gene encoding a 125-kilodalton larvicidal polypeptide is associated with inverted repeat sequences'. J. Bacteriol., 170, 3575–3583.

Brizzard, B. L. and Whiteley, H. R. (1988). 'Nucleotide sequence of an additional crystal protein gene cloned from Bacillus thuringiensis subsp. thuringiensis'. Nucl. Acids Res., 16, 2723–2724.

Brown, K. L. and Whiteley, H. R. (1988). 'Isolation of a Bacillus thuringiensis RNA polymerase capable of transcribing crystal protein genes'. Proc. Natl Acad. Sci. USA, 85, 4166–4170.

Brown, K. L. and Whiteley, H. R. (1990). 'Isolation of the second Bacillus thuringiensis RNA polymerase that transcribes from a crystal protein gene promoter'. J. Bacteriol., 172, 6682–6688.

Bruand, C., Ehrlich, S. D. and Jannière, L. (1991). 'Unidirectional theta replication in the Streptococcus faecalis plasmid pAMβ1'. EMBO J., 10, 2171–2177.

Carlson, H. C. and Haldenwang, W. G. (1989). 'The σ^E subunit of Bacillus subtilis RNA polymerase is present in both forespore and mother cell compartments'. J. Bacteriol., 171, 2216–2218.

Carlton, B. C., Gawron-Burke, C. and Johnson, T. B. (1990). 'Exploiting the genetic diversity of Bacillus thuringiensis for the creation of new bioinsecticides'. Proceedings of the Vth International Colloquium on Invertebrate Pathology and Microbial Control, pp. 18–22, Adelaide, Australia.

Carroll, J., Li, J. and Ellar, D. J. (1989). 'Proteolytic processing of a coleopteran-specific δ-endotoxin produced by Bacillus thuringiensis var. tenebrionis'. Biochem. J., 261, 99–105.

Chambers, J. A., Jelen, A., Gilbert, M. P., Jany, C. S., Johnson, T. B. and Gawron-Burke, C. (1991). 'Isolation and characterization of a novel insecticidal crystal protein gene from *Bacillus thuringiensis* subsp. *aizawai*'. *J. Bacteriol.*, **173**, 3966–3976.

Chang, C., Dai, S.-M., Frutos, R., Federici, B. A. and Gill, S. S. (1990). 'Expression and toxicity of the 72 kDa toxin of *Bacillus thuringiensis* subsp. *morrisoni* (PG-14)'. *Proceedings of the Vth International Colloquium on Invertebrate Pathology and Microbial Control*, p. 293, Adelaide, Australia.

Charles, J. F. and de Barjac, H. (1982). 'Sporulation et cristallogénèse de *Bacillus thuringiensis* var. *israelensis* en microscopie électronique'. *Ann. Microbiol. (Inst. Pasteur)*, **133A**, 425–442.

Chilcott, C. N. and Ellar, D. J. (1988). 'Comparative toxicity of *Bacillus thuringiensis* var. *israelensis* crystal proteins *in vivo* and *in vitro*'. *J. Gen. Microbiol.*, **134**, 2551–2558.

Chungjatupornchai, W., Höfte, H., Seurinck, J., Angsuthanasombat, C. and Vaeck, M. (1988). 'Common features of *Bacillus thuringiensis* toxins specific for Diptera and Lepidoptera'. *Eur. J. Biochem.*, **173**, 9–16.

Dankocsik, C., Donovan, W. P. and Jany, C. S. (1990). 'Activation of a cryptic crystal protein gene of *Bacillus thuringiensis* subspecies *kurstaki* by gene fusion and determination of the crystal protein insecticidal specificity'. *Mol. Microbiol.*, **4**, 2087–2094.

de Barjac, H. and Frachon, E. (1990). 'Classification of *Bacillus thuringiensis* strains'. *Entomophaga*, **35**, 233–240.

Delécluse, A., Bourgouin, C., Klier, A. and Rapoport, G. (1988). 'Specificity of action on mosquito larvae of *Bacillus thuringiensis israelensis* toxins encoded by two different genes'. *Mol. Gen. Genet.*, **214**, 42–47.

Delécluse, A., Bourgouin, C., Klier, A. and Rapoport, G. (1989). 'Nucleotide sequence and characterization of a new insertion element, IS*240*, from *Bacillus thuringiensis israelensis*'. *Plasmid*, **21**, 71–78.

Delécluse, A., Bourgouin, C., Menou, G., Lereclus, D., Klier, A. and Rapoport, G. (1990). 'IS*240* associated with the *cryIVA* gene from *Bacillus thuringiensis israelensis* belongs to a family of Gram(+) and Gram(–) IS elements'. In *Genetics and Biotechnology of Bacilli* (Eds. M. M. Zukowski, A. T. Ganesan and J. A. Hoch), Vol. 3, pp. 181–190, Academic Press, San Diego, CA.

Delécluse, A., Charles, J.-F., Klier, A. and Rapoport, G. (1991). 'Deletion by *in vivo* recombination shows that the 28-kilodalton cytolytic polypeptide from *Bacillus thuringiensis* subsp. *israelensis* is not essential for mosquitocidal activity'. *J. Bacteriol.*, **173**, 3374–3381.

Donovan, W. P., Dankocsik, C. and Gilbert, M. P. (1988a). 'Molecular characterization of a gene encoding a 72 kilodalton mosquito-toxic crystal protein from *Bacillus thuringiensis* subsp. *israelensis*'. *J. Bacteriol.*, **170**, 4732–4738.

Donovan, W. P., Dankocsik, C. C., Gilbert, M. P., Gawron-Burke, M. C., Groat, R. G. and Carlton, B. C. (1988b). 'Amino acid sequence and entomocidal activity of the P2 crystal protein. An insect toxin from *Bacillus thuringiensis* var. *kurstaki*'. *J. Biol. Chem.*, **263**, 561–567.

Donovan, W. P., González, J. M., Jr., Gilbert, M. P. and Dankocsik, C. (1988c). 'Isolation and characterisation of EG2158, a new strain of *Bacillus thuringiensis* toxic to coleopteran larvae, and nucleotide sequence of the toxin gene'. *Mol. Gen. Genet.*, **214**, 365–372.

Edwards, D. L., Payne, J. and Soares, G. G. (1988). 'Novel isolates of *Bacillus thuringiensis* having activity against nematodes'. *European Patent Application*, EP 0 303 426 A2.

Favret, M. E. and Yousten, A. A. (1989). 'Thuricin: the bacteriocin produced by *Bacillus thuringiensis*'. *J. Invert. Pathol.*, **53**, 206–216.

Ge, A. Z., Shivarova, N. I. and Dean, D. H. (1989). 'Location of the *Bombyx mori* specificity domain on a *Bacillus thuringiensis* δ-endotoxin protein'. *Proc. Natl Acad. Sci. USA*, **86**, 4037–4041.

Glatron, M. F. and Rapoport, G. (1972). 'Biosynthesis of the parasporal inclusion of *Bacillus thuringiensis*: half-life of its corresponding messenger RNA'. *Biochimie*, **54**, 1291–1301.

González, J. M., Jr, and Carlton, B. C. (1980). 'Patterns of plasmid DNA in crystalliferous and acrystalliferous strains of *Bacillus thuringiensis*'. *Plasmid*, **3**, 92–98.

González, J. M., Jr, and Carlton, B. C. (1984). 'A large transmissible plasmid is required for crystal toxin production in *Bacillus thuringiensis* variety *israelensis*'. *Plasmid*, **11**, 28–38.

González, J. M., Jr, Dulmage, H. T. and Carlton, B. C. (1981). 'Correlation between specific plasmids and δ-endotoxin production in *Bacillus thuringiensis*'. *Plasmid*, **5**, 351–365.

González, J. M., Jr, Brown, B. J. and Carlton, B. C. (1982). 'Transfer of *Bacillus thuringiensis* plasmids coding for delta-endotoxin among strains of *B. thuringiensis* and *B. cereus*'. *Proc. Natl Acad. Sci. USA*, **79**, 6951–6955.

Green, B. D., Battisti, L. and Thorne, C. B. (1989). 'Involvement of Tn*4430* in transfer of *Bacillus anthracis* plasmids mediated by *Bacillus thuringiensis* plasmid pXO12'. *J. Bacteriol.*, **171**, 104–113.

Gruss, A. and Ehrlich, S. D. (1989). 'The family of highly interrelated single-stranded deoxyribonucleic acid plasmids'. *Microbiol. Rev.*, **53**, 231–241.

Guiney, D. G. and Yakobson, E. (1983). 'Location and nucleotide sequence of the transfer origin of the broad host range plasmid RK2'. *Proc. Natl Acad. Sci., USA*, **80**, 3595–3598.

Haider, M. Z. and Ellar, D. J. (1987). 'Characterization of the toxicity and cytopathic specificity of a cloned *Bacillus thuringiensis* crystal protein using insect cell culture'. *Mol. Microbiol.*, **1**, 59–66.

Haider, M. Z. and Ellar, D. J. (1988). 'Nucleotide sequence of a *Bacillus thuringiensis aizawai* IC1 entomocidal crystal protein gene'. *Nucl. Acids Res.*, **16**, 10927.

Haider, M. Z. and Ellar, D. J. (1989). 'Functional mapping of an entomocidal δ-endotoxin. Single amino acid changes produced by site-directed mutagenesis influence toxicity and specificity of the protein'. *J. Mol. Biol.*, **208**, 183–194.

Hallet, B., Rezsöhazy, R. and Delcour, J. (1991). 'IS*231*A from *Bacillus thuringiensis* is functional in *Escherichia coli*: transposition and insertion specificity'. *J. Bacteriol.*, **173**, 4526–4529.

Herrnstadt, C., Soares, G. G., Wilcox, E. R. and Edwards, D. L. (1986). 'A new strain of *Bacillus thuringiensis* with activity against coleopteran insects'. *Bio/Technology*, **4**, 305–308.

Herrnstadt, C., Gilroy, T. E., Sobieski, D. A., Bennett, B. D. and Gaertner, F. H. (1987). 'Nucleotide sequence and deduced amino acid sequence of a coleopteran-active delta-endotoxin gene from *Bacillus thuringiensis* subsp. *san diego*'. *Gene*, **57**, 37–46.

Hess, G. F. and Graham, R. S. (1990). 'Efficiency of transcriptional terminators in *Bacillus subtilis*'. *Gene*, **95**, 137–141.

Hofmann, C., Vanderbruggen, H., Höfte, H., Van Rie, J., Jansens, S. and Van Mellaert, H. (1988). 'Specificity of *Bacillus thuringiensis* δ-endotoxins is correlated with the presence of high-affinity binding sites in the brush border membrane of target insect midguts'. *Proc. Natl Acad. Sci. USA*, **85**, 7844–7848.

Höfte, H. and Whiteley, H. R. (1989). 'Insecticidal crystal proteins of Bacillus thuringiensis'. Microbiol. Rev., 53, 242–255.

Höfte, H., de Greve, H., Seurinck, J., Jansens, S., Mahillon, J., Ampe, C., Vandekerckhove, J., Vanderbruggen, H., Van Montagu, M., Zabeau, M. and Vaeck, M. (1986). 'Structural and functional analysis of a cloned delta endotoxin of Bacillus thuringiensis berliner 1715'. Eur. J. Biochem., 161, 273–280.

Höfte, H., Seurinck, J., Van Houtven, A. and Vaeck, M. (1987). 'Nucleotide sequence of a gene encoding an insecticidal protein of Bacillus thuringiensis var. tenebrionis toxic against coleoptera'. Nucl. Acids Res., 15, 7189.

Höfte, H., Van Rie, J., Jansens, S., Van Houtven, A., Vanderbruggen, H. and Vaeck, M. (1988). 'Monoclonal antibody analysis and insecticidal spectrum of three types of lepidopteran-specific insecticidal crystal proteins of Bacillus thuringiensis'. Appl. Env. Microbiol., 54, 2010–2017.

Höfte, H., Soetaert, P., Jansens, S. and Peferoen, M. (1990). 'Nucleotide sequence and deduced amino acid sequence of a new lepidoptera-specific crystal protein gene from Bacillus thuringiensis'. Nucl. Acids Res., 18, 5545.

Honée, G., van der Salm, T. and Visser, B. (1988). 'Nucleotide sequence of crystal protein gene isolated from B. thuringiensis subspecies entomocidus 60.5 coding for a toxin highly active against Spodoptera species'. Nucl. Acids Res., 16, 6240.

Ibarra, J. E. and Federici, B. A. (1986). 'Isolation of a relatively nontoxic 65-kilodalton protein inclusion from the parasporal body of Bacillus thuringiensis subsp. israelensis'. J. Bacteriol., 165, 527–533.

Jahn, N., Schnetter, W. and Geider, K. (1987). 'Cloning of an insecticidal toxin gene of Bacillus thuringiensis subsp. tenebrionis and its expression in Escherichia coli cells'. FEMS Microbiol. Lett., 48, 311–315.

Jannière, L., Bruand, C. and Ehrlich, S. D. (1990). 'Structurally stable Bacillus subtilis cloning vectors'. Gene, 87, 53–61.

Jarrett, P. and Stephenson, M. (1990). 'Plasmid transfer between strains of Bacillus thuringiensis infecting Galleria mellonella and Spodoptera littoralis'. Appl. Env. Microbiol., 56, 1608–1614.

Josson, K., Soetaert, P., Michiels, F., Joos, H. and Mahillon, J. (1990). 'Lactobacillus hilgardii plasmid pLAB1000 consists of two functional cassettes commonly found in other Gram-positive organisms'. J. Bacteriol., 172, 3089–3099.

Klier, A., Fargette, F., Ribier, J. and Rapoport, G. (1982). 'Cloning and expression of the crystal protein genes from Bacillus thuringiensis strain berliner 1715'. EMBO J., 1, 791–799.

Klier, A., Bourgouin, C. and Rapoport, G. (1983). 'Mating between Bacillus subtilis and Bacillus thuringiensis and transfer of cloned crystal genes'. Mol. Gen. Genet., 191, 257–262.

Krieg, A., Huger, A. M., Langenbruch, G. A. and Schnetter, W. (1983). 'Bacillus thuringiensis var. tenebrionis: ein neuer gegenuber larven von Coleopteren wirksamer athotyp'. Z. Ang. Entomol., 96, 500–508.

Krieg, A., Schnetter, W., Huger, A. M. and Langenbruch, G. A. (1987). 'Bacillus thuringiensis subsp. tenebrionis, strain BI 256–82: a third pathotype within the H-serotype 8a8b'. System. Appl. Microbiol., 9, 138–141.

Kronstad, J. W. and Whiteley, H. R. (1984). 'Inverted repeat sequences flank a Bacillus thuringiensis crystal protein gene'. J. Bacteriol., 160, 95–102.

Lecadet, M.-M. (1970). 'Bacillus thuringiensis toxins. The proteinaceous crystal'. In Microbial Toxins (Eds. T. C. Montie, S. Kadis and S. J. Aijl), Vol. 3, pp. 437–471, Academic Press, New York.

Lecadet, M.-M. and de Barjac, H. (1981). 'Bacillus thuringiensis Beta-exotoxin'. In

Pathogenesis of Invertebrate Microbial Diseases (Ed. E. Davidson), pp. 293–316, Allanheld, Osmun and Co., New Jersey.

Lecadet, M.-M., Chaufaux, J., Ribier, J. and Lereclus, D. (1992). 'Construction of novel *Bacillus thuringiensis* strains with different insecticidal specificities by transduction and by transformation'. *Appl. Env. Microbiol.*, **58**, 840–849.

Lereclus, D. and Arantes, O. (1992). '*spbA* locus ensures the segregational stability of pHT1030, a novel type of Gram-positive replicon'. *Mol. Microbiol.*, **6**, 35–46.

Lereclus, D., Lecadet, M.-M., Ribier, J. and Dedonder, R. (1982). 'Molecular relationships among plasmids of *Bacillus thuringiensis*: conserved sequences through 11 crystalliferous strains'. *Mol. Gen. Genet.*, **186**, 391–398.

Lereclus, D., Menou, G. and Lecadet, M.-M. (1983). 'Isolation of a DNA sequence related to several plasmids from *Bacillus thuringiensis* after a mating involving the *Streptococcus faecalis* plasmid pAMβ1'. *Mol. Gen. Genet.*, **191**, 307–313.

Lereclus, D., Ribier, J., Klier, A., Menou, G. and Lecadet, M.-M. (1984). 'A transposon-like structure related to the δ-endotoxin gene of *Bacillus thuringiensis*'. *EMBO J.*, **3**, 2561–2567.

Lereclus, D., Mahillon, J., Menou, G. and Lecadet, M.-M. (1986). 'Identification of Tn*4430*, a transposon of *Bacillus thuringiensis* functional in *Escherichia coli*'. *Mol. Gen. Genet.*, **204**, 52–57.

Lereclus, D., Guo, S., Sanchis, V. and Lecadet, M.-M. (1988). 'Characterization of two *Bacillus thuringiensis* plasmids whose replication is thermosensitive in *B. subtilis*'. *FEMS Microbiol. Lett.*, **49**, 417–422.

Lereclus, D., Arantes, O., Chaufaux, J. and Lecadet, M.-M. (1989a). 'Transformation and expression of a cloned δ-endotoxin gene in *Bacillus thuringiensis*'. *FEMS Microbiol. Lett.*, **60**, 211–218.

Lereclus, D., Bourgouin, C., Lecadet, M.-M., Klier, A., and Rapoport, G. (1989b). 'Role, structure and molecular organization of the genes coding for the parasporal δ-endotoxins of *Bacillus thuringiensis*'. In *Regulation of Procaryotic Development* (Eds. I. Smith, R. A. Slepecky and P. Setlow), pp. 255–276, American Society for Microbiology, Washington, D.C.

Lereclus, D., Vallade, M., Chaufaux, J., Arantes, O. and Rambaud, S. (1992). 'Expansion of insecticidal host range of *Bacillus thuringiensis* by *in vivo* genetic recombination'. *Bio/Technology*, **10**, 418–421.

Levinson, B. L., Kasyan, K. J., Chiu, S. S., Currier, T. S. and González J. M., Jr (1990). 'Identification of β-exotoxin production, plasmids encoding β-exotoxin and a new exotoxin in *Bacillus thuringiensis* by using high-performance liquid chromatography'. *J. Bacteriol.*, **172**, 3172–3179.

Losick, R. and Kroos, L. (1989). 'Dependence pathways for the expression of genes involved in endospore formation in *Bacillus subtilis*'. In *Regulation of Procaryotic Development* (Eds. I. Smith, R. A. Slepecky and P. Setlow), pp. 223–241, American Society for Microbiology, Washington DC.

Mahillon, J. and Lereclus, D. (1988). 'Structural and functional analysis of Tn*4430*: identification of an integrase-like protein involved in the co-integrate-resolution process'. *EMBO J.*, **7**, 1515–1526.

Mahillon, J. and Seurinck, J. (1988). 'Complete nucleotide sequence of pG12, a *Bacillus thuringiensis* plasmid containing Tn*4430*'. *Nucl. Acids Res.*, **16**, 11827–11828.

Mahillon, J., Seurinck, J., Van Rompuy, L., Delcour, J. and Zabeau, M. (1985). 'Nucleotide sequence and structural organization of an insertion sequence element (IS*231*) from *Bacillus thuringiensis* strain Berliner 1715'. *EMBO J.*, **4**, 3895–3899.

Mahillon, J., Seurinck, J., Delcour, J. and Zabeau, M. (1987). 'Cloning and nucleotide sequence of different iso-IS*231* elements and their structural association with the Tn*4430* transposon in *Bacillus thuringiensis*'. *Gene*, **51**, 187–196.

Mahillon, J., Hespel, F., Pierssens, A. M. and Delcour, J. (1988). 'Cloning and partial characterization of three small cryptic plasmids from *Bacillus thuringiensis*'. *Plasmid*, **19**, 169–173.

Martin, P. A. W. and Travers, R. S. (1989). 'Worldwide abundance and distribution of *Bacillus thuringiensis* isolates'. *Appl. Env. Microbiol.*, **55**, 2437–2442.

Masson, L., Marcotte, P., Préfontaine, G. and Brousseau, R. (1989). 'Nucleotide sequence of a gene cloned from *Bacillus thuringiensis* subspecies *entomocidus* coding for an insecticidal protein toxic for *Bombyx mori*'. *Nucl. Acids Res.*, **17**, 446.

Masson, L., Préfontaine, G., Péloquin, L., Lau, P. C. K. and Brousseau, R. (1990). 'Comparative analysis of the individual protoxin components in P1 crystals of *Bacillus thuringiensis* subsp. *kurstaki* isolates NRD-12 and HD-1'. *Biochem. J.*, **269**, 507–512.

McDowell, D. G. and Mann, N. H. (1991). 'Characterization and sequence analysis of a small plasmid from *Bacillus thuringiensis* var. *kurstaki* strain HD1-DIPEL'. *Plasmid*, **25**, 113–120.

McLean, K. M. and Whiteley, H. R. (1987). 'Expression in *Escherichia coli* of a cloned crystal protein gene of *Bacillus thuringiensis* subsp. *israelensis*'. *J. Bacteriol.*, **169**, 1017–1023.

McPherson, S. A., Perlak, F. J., Fuchs, R. L., Marrone, P. G., Lavrik, P. B. and Fischhoff, D. A. (1988). 'Characterization of the coleopteran-specific protein gene of *Bacillus thuringiensis* var. *tenebrionis*'. *Bio/Technology*, **6**, 61–66.

Menou, G., Mahillon, J., Lecadet, M.-M. and Lereclus, D. (1990). 'Structural and genetic organization of IS*232*, a new insertion sequence of *Bacillus thuringiensis*'. *J. Bacteriol.*, **172**, 6689–6696.

Minnich, S. A. and Aronson, A. I. (1984). 'Regulation of protoxin synthesis in *Bacillus thuringiensis*'. *J. Bacteriol.*, **158**, 447–454.

Moar, W. J., Masson, L., Brousseau, R. and Trumble, J. T. (1990). 'Toxicity to *Spodoptera exigua* and *Trichoplusia ni* of individual P1 protoxins and sporulated cultures of *Bacillus thuringiensis* subsp. *kurstaki* HD-1 and NRD-12'. *Appl. Env. Microbiol.*, **56**, 2480–2483.

Moran, C. P. J. (1989). 'Sigma factors and the regulation of transcription'. In *Regulation of Procaryotic Development* (Eds. I. Smith, R. A. Slepecky and P. Setlow), pp. 167–184, American Society for Microbiology, Washington, D.C.

Nagamatsu, Y., Itai, Y., Hatanaka, C., Funatsu, G. and Hayashi, K. (1984). 'A toxic fragment from the entomocidal crystal protein of *Bacillus thuringiensis*'. *Agric. Biol. Chem.*, **48**, 611–619.

Pao-intara, M., Angsuthanasombat, C. and Panyim, S. (1988). 'The mosquito larvicidal activity of 130 kDa delta-endotoxin of *Bacillus thuringiensis* var. *israelensis* resides in the 72 kDa amino-terminal fragment'. *Biochem. Biophys. Res. Commun.*, **153**, 294–300.

Pfannenstiel, M. A., Couche, G. A., Ross, E. J. and Nickerson, K. W. (1986). 'Immunological relationships among proteins making up the *Bacillus thuringiensis* subsp. *israelensis* crystalline toxin'. *Appl. Env. Microbiol.*, **52**, 644–649.

Préfontaine, G., Fast, P., Lau, P. C. K., Hefford, M. A., Hanna, Z. and Brousseau, R. (1987). 'Use of oligonucleotide probes to study the relatedness of delta-endotoxin genes among *Bacillus thuringiensis* subspecies and strains'. *Appl. Env. Microbiol.*, **53**, 2808–2814.

Reddy, A., Battisti, L. and Thorne, C. B. (1987). 'Identification of self-transmissible plasmids in four *Bacillus thuringiensis* subspecies'. *J. Bacteriol.*, **169**, 5263–5270.

Ribier, J. and Lecadet, M.-M. (1973). 'Etude ultrastructurale et cinétique de la sporu-lation de *Bacillus thuringiensis* var. *Berliner* 1715. Remarques sur la formation de l'inclusion parasporale'. *Ann. Microbiol. (Inst. Pasteur)*, **124A**, 311–344.

Ribier, J. and Lecadet, M.-M. (1981). '*Bacillus thuringiensis* var. Berliner 1715. Isole-ment et caractérisation de mutants de sporulation'. *C. R. Acad. Sc. Paris (Série III)*, **292**, 803–808.

Rupar, M. J., Donovan, W. P., Groat, R. G., Slaney, A. C., Mattison, J. W., Johnson, T. B., Charles, J.-F., Cosmao-Dumanoir, V. and de Barjac, H. (1991). 'Two novel coleopteran toxic strains of *Bacillus thuringiensis*'. *Appl. Env. Microbiol.*, **57**, 3337–3344.

Sanchis, V., Lereclus, D., Menou, G., Chaufaux, J. and Lecadet, M.-M. (1988). 'Mul-tiplicity of δ-endotoxin genes with different specificities in *Bacillus thuringiensis aizawai* 7.29'. *Mol. Microbiol.*, **2**, 393–404.

Sanchis, V., Lereclus, D., Menou, G., Chaufaux, J., Guo, S. and Lecadet, M.-M. (1989). 'Nucleotide sequence and analysis of the N-terminal coding region of the *Spodoptera*-active δ-endotoxin gene of *Bacillus thuringiensis aizawai* 7.29'. *Mol. Microbiol.*, **3**, 229–238.

Schnepf, H. E. and Whiteley, H. R. (1985). 'Delineation of a toxin-encoding segment of a *Bacillus thuringiensis* crystal protein gene'. *J. Biol. Chem.*, **260**, 6273–6280.

Schnepf, H. E., Wong, H. C. and Whiteley, H. R. (1985). 'The amino acid sequence of a crystal protein from *Bacillus thuringiensis* deduced from the DNA base sequence'. *J. Biol. Chem.*, **260**, 6264–6272.

Schwab, G. E. and Halloran, T. P. (1989). 'Use of a gene product of *Bacillus thurin-giensis* var. *israelensis*'. *European patent application*, EP 0 352 052 A2.

Sebesta, K. and Horska, K. (1970). 'Mechanisms of inhibition of DNA dependent RNA polymerase by exotoxin of *Bacillus thuringiensis*'. *Biochim. Biophys. Acta*, **209**, 357–376.

Sekar, V. and Carlton, B. C. (1985). 'Molecular cloning of the delta-endotoxin gene of *Bacillus thuringiensis* var. *israelensis*'. *Gene*, **33**, 151–158.

Sekar, V., Thompson, D. V., Maroney, M. J., Bookland, R. G. and Adang, M. J. (1987). 'Molecular cloning and characterization of the insecticidal crystal protein gene of *Bacillus thuringiensis* var. *tenebrionis*'. *Proc. Natl Acad. Sci. USA*, **84**, 7036–7040.

Sen, K., Honda, G., Koyama, N., Nishida, M., Neki, A., Sakai, H., Himeno, M. and Komano, T. (1988). 'Cloning and nucleotide sequences of the two 130 kDa insec-ticidal protein genes of *Bacillus thuringiensis* var. *israelensis*'. *Agric. Biol. Chem.*, **52**, 873–878.

Sherratt, D. (1989). 'Tn3 and related transposable elements: site-specific recom-bination and transposition'. In *Mobile DNA* (Eds. D. E. Berg and M. M. Howe), pp. 163–184, American Society for Microbiology, Washington, DC.

Shivakumar, A. G., Vanags, R. I., Wilcox, D. R., Katz, L., Vary, P. S. and Fox, J. L. (1989). 'Gene dosage effect on the expression of the delta-endotoxin genes of *Bacillus thuringiensis* subsp. *kurstaki* in *Bacillus subtilis* and *Bacillus megaterium*'. *Gene*, **79**, 21–31.

Sick, A., Gaertner, F. and Wong, A. (1990). 'Nucleotide sequence of a coleopteran-active toxin gene from a new isolate of *Bacillus thuringiensis* subsp. *tolworthi*'. *Nucl. Acids Res.*, **18**, 1305.

Smith, R. A. and Couche, G. A. (1991). 'The phyllophane as a source of *Bacillus thuringiensis* variants'. *Appl. Env. Microbiol.*, **57**, 311–315.

Stahly, D. P., Dingman, D. W., Bulla, L. A., Jr and Aronson, A. I. (1978). 'Possible origin and function of the parasporal crystals in *Bacillus thuringiensis*'. *Biochem. Biophys. Res. Commun.*, **84**, 581–588.

Stragier, P. and Losick, R. (1990). 'Cascades of sigma factors revisited'. *Mol. Microbiol.*, **4**, 1801–1806.

Stragier, P., Kunkel, B., Kroos, L. and Losick, R. (1989). 'Chromosomal rearrangement generating a composite gene for a developmental transcription factor'. *Science*, **243**, 507–512.

Tailor, R., Tippett, J., Gibb, G., Pells, S., Pike, D., Jordan, L. and Ely, S. (1992). 'Identification and characterisation of a novel *Bacillus thuringiensis* δ-endotoxin entomocidal to coleopteran and lepidopteran larvae'. *Mol. Microbiol.*, **6**, No. 9, 1211–1217.

Thomas, W. E. and Ellar, D. J. (1983). '*Bacillus thuringiensis* var. *israelensis* crystal δ-endotoxin: effects on insect and mammalian cells *in vitro* and *in vivo*'. *J. Cell Sci.*, **60**, 181–197.

Thorne, L., Garduno, F., Thompson, T., Decker, D., Zounes, M., Wild, M., Walfield, A. M. and Pollock, T. J. (1986). 'Structural similarity between the Lepidoptera- and Diptera-specific insecticidal endotoxin genes of *Bacillus thuringiensis* subsp. "*kurstaki*"- and "*israelensis*"'. *J. Bacteriol.*, **166**, 801–811.

Trieu-Cuot, P., Carlier, C., Martin, P. and Courvalin, P. (1987). 'Plasmid transfer by conjugation from *Escherichia coli* to Gram positive bacteria'. *FEMS Microbiol. Lett.*, **48**, 289–294.

Tungpradubkul, S., Settasatien, C. and Panyim, S. (1988). 'The complete nucleotide sequence of a 130 kDa mosquito-larvicidal delta-endotoxin gene of *Bacillus thuringiensis* var. *israelensis*'. *Nucl. Acids Res.*, **16**, 1637–1638.

van Frankenhuyzen, K., Gringorten, J. L., Milne, R. E., Gauthier, D., Pusztai, M., Brousseau, R. and Masson, L. (1991). 'Specificity of activated CryIA proteins from *Bacillus thuringiensis* subsp. *kurstaki* HD-1 for defoliating forest lepidoptera'. *Appl. Env. Microbiol.*, **57**, 1650–1655.

Van Rie, J., Jansens, S., Höfte, H., Degheele, D. and Van Mellaert, H. (1990). 'Receptors on the brush border membrane of the insect midgut as determinants of the specificity of *Bacillus thuringiensis* delta-endotoxins'. *Appl. Env. Microbiol.*, **56**, 1378–1385.

Visick, J. E. and Whiteley, H. R. (1991). 'Effect of a 20-kilodalton protein from *Bacillus thuringiensis* subsp. *israelensis* on production of the CytA protein by *Escherichia coli*'. *J. Bacteriol.*, **173**, 1748–1756.

Visser, B., van der Salm, T., van den Brink, W. and Folkers, G. (1988). 'Genes from *Bacillus thuringiensis entomocidus* 60.5 coding for insect-specific crystal proteins'. *Mol. Gen. Genet.*, **212**, 219–224.

Visser, B., Munsterman, E., Stoker, A. and Dirkse, W. G. (1990). 'A novel *Bacillus thuringiensis* gene encoding a *Spodoptera exigua*-specific crystal protein'. *J. Bacteriol.*, **172**, 6783–6788.

Von Tersch, M. A., Loidl Robbins, H., Jany, C. S. and Johnson, T. B. (1991). 'Insecticidal toxins from *Bacillus thuringiensis* subsp. *kenyae*: gene cloning and characterization and comparison with *B. thuringiensis* subsp. *kurstaki* CryIA(c) toxins'. *Appl. Env. Microbiol.*, **57**, 349–358.

Waalwijk, C., Dullemans, A. M., Van Workum, M. E. S. and Visser, B. (1985). 'Molecular cloning and the nucleotide sequence of the Mr 28000 crystal protein gene of *Bacillus thuringiensis* subsp. *israelensis*'. *Nucl. Acids Res.*, **13**, 8207–8217.

Wabiko, H., Raymond, K. C. and Bulla, L. A., Jr (1986). '*Bacillus thuringiensis* entomocidal protoxin gene sequence and gene product analysis'. *DNA*, **5**, 305–314.

Ward, E. S. and Ellar, D. J. (1986). '*Bacillus thuringiensis* var. *israelensis* δ-endotoxin. Nucleotide sequence and characterization of the transcripts in *Bacillus thuringiensis* and *Escherichia coli*'. *J. Mol. Biol.*, **191**, 1–11.

Ward, E. S. and Ellar, D. J. (1987). 'Nucleotide sequence of a *Bacillus thuringiensis* var. *israelensis* gene encoding a 130 kDa delta-endotoxin'. *Nucl. Acids Res.*, **15**, 7195.

Ward, E. S. and Ellar, D. J. (1988). 'Cloning and expression of two homologous genes of *Bacillus thuringiensis* subsp. *israelensis* which encode 130-kilodalton mosquitocidal proteins'. *J. Bacteriol.*, **170**, 727–735.

Ward, E. S., Ellar, D. J. and Todd, J. A. (1984). 'Cloning and expression in *Escherichia coli* of the insecticidal δ-endotoxin gene of *Bacillus thuringiensis* var. *israelensis*'. *FEBS Lett.*, **175**, 377–382.

Ward, E. S., Ridley, A. R., Ellar, D. J. and Todd, J. A. (1986). '*Bacillus thuringiensis* var. *israelensis* δ-endotoxin. Cloning and expression of the toxin in sporogenic and asporogenic strains of *Bacillus subtilis*'. *J. Mol. Biol.*, **191**, 13–22.

Whiteley, H. R. and Schnepf, H. E. (1986). 'The molecular biology of parasporal crystal body formation in *Bacillus thuringiensis*. *Ann. Rev. Microbiol.*, **40**, 549–576.

Widner, W. R. and Whiteley, H. R. (1989). 'Two highly related insecticidal crystal proteins of *Bacillus thuringiensis* subsp. *kurstaki* possess different host range specificities'. *J. Bacteriol.*, **171**, 965–974.

Widner, W. R. and Whiteley, H. R. (1990). 'Location of the dipteran specificity region in a lepidopteran-dipteran crystal protein from *Bacillus thuringiensis*'. *J. Bacteriol.*, **172**, 2826–2832.

Wong, H. C. and Chang, S. (1986). 'Identification of a positive retroregulator that stabilizes mRNAs in bacteria'. *Proc. Natl Acad. Sci. USA.*, **83**, 3233–3237.

Wong, H. C., Schnepf, H. E. and Whiteley, H. R. (1983). 'Transcriptional and translational start sites for the *Bacillus thuringiensis* crystal protein gene'. *J. Biol. Chem.*, **258**, 1960–1967.

Wu, D., Cao, X. L., Bay, Y. Y. and Aronson, A. I. (1991). 'Sequence of an operon containing a novel δ-endotoxin gene from *Bacillus thuringiensis*'. *FEMS Microbiol. Lett.*, **81**, 31–36.

Yamamoto, T. and McLaughlin, R. E. (1981). 'Isolation of a protein from the paraporal crystal of *Bacillus thuringiensis* var. *kurstaki* toxic to the mosquito larvae *Aedes taeniorhynchus*'. *Biochem. Biophys. Res. Commun.*, **103**, 414–421.

Yamamoto, T., Ehmann, A., González, J. M., Jr and Carlton, B. C. (1988a). 'Expression of three genes coding for 135-kilodalton entomocidal proteins in *Bacillus thuringiensis kurstaki*'. *Curr. Microbiol.*, **17**, 5–12.

Yamamoto, T., Watkinson, I. A., Kim, L., Sage, M. V., Stratton, R., Akande, N., Li, Y., Ma, D. P. and Roe, B. A. (1988b). 'Nucleotide sequence of the gene coding for a 130 kDa mosquitocidal protein of *Bacillus thuringiensis israelensis*'. *Gene*, **66**, 107–120.

3 Domain–Function Studies of *Bacillus thuringiensis* Crystal Proteins: A Genetic Approach

[1]BERT VISSER, [1]DIRK BOSCH AND [2]GUY HONÉE
[1]*Centre for Plant Breeding and Reproduction Research CPRO, PO Box 16, 6700 AA Wageningen, The Netherlands*
[2]*Department of Phytopathology, Agricultural University, Binnenhaven 9, 6709 PD Wageningen, The Netherlands*

INTRODUCTION

For more than two decades the application of *Bacillus thuringiensis* (*B. t.*) for pest control has been accompanied by ongoing research into the basis of its entomocidal activity and its potential additional applications. However, the molecular genetics of *B. t.* and its insecticidal properties have only been fully developed in the last 10 years. In 1983 the first, partial, nucleotide sequence of a *B. t.* crystal protein gene was published (Wong *et al.*, 1983), and by 1986 the simultaneous presence of more than one crystal protein gene in the same bacterial isolate was first demonstrated (Kronstad and Whiteley, 1986). Since then, several thousands of isolates have been analysed (Martin and Travers, 1989) and the number of crystal protein genes described and sequenced has rapidly expanded. This led to a new nomenclature and classification based on the primary structure and the host range of the crystal proteins (Höfte and Whiteley, 1989). An updated version of this classification scheme now contains 20 different classes and subclasses (Visser *et al.*, 1990; Sick *et al.*, 1990; Rupar *et al.*, 1991; Smulevitch *et al.*, 1991; Brown and Whiteley, 1992). The four major classes are Lepidoptera-specific (CryI), Lepidoptera- and Diptera-specific (CryII), Coleoptera-specific (CryIII) and Diptera-specific (CryIV). A fifth class has recently been reported (Tailor *et al.*, 1992).

   A basic knowledge of the mode of action of the crystal proteins, briefly described in the next paragraph, resulted from physiological and biochemical research, carried out along with molecular genetic studies. In order to improve our understanding of the mode of action, and, in particular, its molecular basis, domain–function studies on the crystal proteins

Bacillus thuringiensis, An Environmental Biopesticide: Theory and Practice. Edited by P. F. Entwistle, J. S. Cory, M. J. Bailey and S. Higgs

(relating the primary protein structure to its biological properties) may prove helpful. A thorough analysis of the primary and, as far as possible, higher order structure of the protein forms a necessary prerequisite for such an approach to be successful. This includes a comparison of the primary sequence of the different crystal protein subclasses to identify conserved and variable regions. Chemical studies of the protein conformation by determination of protease sensitivity and of protein (un)folding have extended our insight. In addition, the X-ray crystallography data on the structure of CryIIIA, obtained by Li *et al.* (1991), form a major contribution to the structure–function analysis of crystal proteins.

Two types of domain–function studies can be distinguished, involving either amino acid substitution and deletion mutants, or chimeric proteins in which part of one 'parent' molecule has been replaced by the homologous part from another, related protein. Although the study of substitution mutants can directly relate an altered property to a defined amino acid residue, a drawback of this approach is that a detailed knowledge of the protein structure is needed to render the analysis of results sufficiently meaningful. On the other hand, studies of deletion mutants and chimeric proteins may identify amino acid regions, involved in specific protein functions, thus providing a schematic view on domain–function relations. Both deletion mutants and substitution mutants often exhibit loss of function, whereas chimeric proteins, depending on the genetic distance of their parents, may also show additional functions. In general, the level of homology between two parent molecules determines to a large extent the chances for obtaining functional chimeras.

The study of chimeric proteins has been shown to be very useful in several other systems; for instance for the identification of regions in the *Escherichia coli* outer membrane pore proteins determining pore specificity (Tommassen *et al.*, 1985), and for the study of nodulins present in *Rhizobium* species, involved in the specificity of plant–microbe symbiosis (Spaink *et al.*, 1989). A detailed knowledge of domain–function relationships can lead to the design of new chimeric proteins in which characteristics of two parental proteins are combined. This has been demonstrated with *Pseudomonas* exotoxin A (Pastan and FitzGerald, 1989). In some cases the domains of entirely unrelated proteins may be combined to form chimeras exhibiting a specific set of functions, for example as with the Barnase toxin, consisting of *Pseudomonas* exotoxin A and the extracellular ribonuclease of *Bacillus amyloliquefaciens* (Prior *et al.*, 1991).

A better understanding of domain–function relationships in *B. t.* crystal proteins might aid the selection of crystal protein genes for incorporation into transgenic organisms, in order to prevent, or at least postpone, the development of crystal protein resistant insect populations. Alternatively, it might facilitate the design of new chimeric or mutant crystal proteins showing an increased toxicity or expanded insecticidal spectrum.

A description and discussion of the results obtained from domain–function studies forms the major part of this review. Since the crystal proteins used in these studies all belong to the classes CryI and CryII, the review focuses only on these two classes.

MODE OF ACTION

Upon ingestion of a *B. t.* crystal containing either class CryI or CryII proteins, several steps, which ultimately lead to the death of sensitive lepidopteran larvae, can be distinguished. Firstly, the crystal dissolves in the alkaline midgut of the insect, generating 130–140 kDa CryI proteins or 71 kDa CryII proteins. The efficiency of this process, influencing specificity, is co-determined by the midgut environment and the composition of the crystals (Jacquet *et al.*, 1987; Aronson *et al.*, 1991). For example, the presence of CryIA(b) amongst other proteins was shown to increase total crystal solubility in several Lepidoptera.

Secondly, the solubilized crystal protein or protoxin is proteolytically processed to produce the actual toxic fragment of 60–65 kDa. For the CryIA(c) protoxin, conversion *in vitro* by trypsin into the toxic fragment has been reported to occur via seven specific cleavages, each removing fragments of approximately 10 kDa which are rapidly proteolysed to small peptides (Choma *et al.*, 1990). The final toxic fragment probably comprises the sequence ranging from the isoleucine residue 29 to the first trypsin-sensitive site downstream of the alanine residue 608 (Höfte and Whiteley, 1989; Visser *et al.*, unpublished). It is evident that the protease composition of the insect midgut together with the crystal protein structure determines the efficiency of this step.

Thirdly, the toxic fragment binds, at least in the case of the CryI class, to specific receptors present on the membranes of epithelial midgut cells. Several studies, making use of isolated brush border membrane vesicles, convincingly demonstrate that the interaction of the toxic fragment with high-affinity binding sites on the midgut epithelium is important for determining the insecticidal spectrum of *B. t.* crystal proteins (Hofmann *et al.*, 1988; Van Rie *et al.*, 1990).

Finally, the membrane-bound toxic fragment induces the formation of pores, either aspecific or K^+-specific, in the midgut epithelial cell membrane (Sacchi *et al.*, 1986; Hendrickx *et al.*, 1990). The X-ray crystallography data on CryIIIA suggest that the toxic fragment itself is able to induce pore formation via insertion into the membrane. It is likely that, after receptor binding of the toxic fragment, a conformational change is needed for pore formation to occur. Whether the membrane receptor plays a role in these processes, or whether its role is restricted to the binding of the toxic fragment to the membrane, is unclear. As a result of pore formation the cells die, eventually leading to death of the larvae.

From the mode of action, described above, it can be inferred that at least four parameters are involved in crystal protein function: effectiveness of solubilization, efficiency of protoxin–toxin conversion, specific membrane receptor binding and pore formation. Together, and in interaction with the midgut environment, these parameters determine the specificity—or the insecticidal spectrum—of a crystal protein.

STRUCTURAL FEATURES OF CRYSTAL PROTEINS

Although in principle domain–function studies can be performed without extensive prior knowledge of the protein structure, a more direct approach is heavily dependent on structural data. Several groups have investigated the crystal protein structure both at the primary structure and at the higher order level.

Comparative analysis of the amino acid composition of the CryI crystal proteins shows a high degree of homology (>90%) for the C-terminal part of the molecule ranging approximately from residues 600 to 1150. The N-terminal part of the crystal protein, encompassing the toxic fragment, is much less conserved, with homology varying from 40 to 90%. These two regions of the CryI proteins appear to represent two distinct structural domains, the N-terminal half of the protoxin being predominantly hydrophobic and the C-terminal part more hydrophilic.

Whereas 16 cysteine residues are present in the intact CryIA(c) protoxin, the trypsin-resistant CryIA(c) toxic fragment, generated by removal of the first 28 amino acids and the C-terminal half of the protoxin, contains no such residues, a biased distribution characteristic for all CryI class molecules. Moreover, 12 out of these 16 residues are positionally conserved. Indications have been obtained for a dimeric organization (Huber *et al.*, 1981; Visser *et al.*, 1986), formed by intermolecular disulphide bridges (Bietlot *et al.*, 1990). It has therefore been suggested that this highly conserved part of the protoxin might play a role in crystal formation and maintenance and contribute to the unusual solubility properties of the crystals (Lüthy and Ebersold, 1981). In addition, only three of the 34 lysine residues remain in the trypsin-generated CryIA(c) toxic fragment. The fact that the toxic fragment is resistant to further trypsin degradation indicates that these three remaining lysine residues, as well as the many arginine residues, are probably inaccessible. Modification of approximately 12 of the estimated 24 tyrosine residues of a mixture of toxic fragments of unknown composition resulted in a decrease in cytolytic activity, whereas an attempted, but unproven, modification of lysine and cysteine residues had no effect on toxicity (Yan and McCarthy, 1991). This indicates an external position of the tyrosine residues in the toxin molecule.

Few studies into the protoxin and toxin conformation have been carried out (Convents *et al.*, 1990, 1991; Choma and Kaplan, 1990). Characterization of monoclonal antibodies to the protoxins solubilized from a *B. t.* subspecies *kurstaki* strain containing CryIA proteins showed that eight out of ten antibodies bound to both protoxin and toxin, suggesting that the toxic moiety is at least partially exposed within the protoxin (Huber-Lukač *et al.*, 1986). Unfolding of the CryIA(c) protoxin in the presence of urea or guanidinium chloride was measured by fluorescence emission and sensitivity to proteolysis. Results suggest that *in vitro* folding and unfolding of the N-terminal toxic fragment proceeds independently of the C-terminal protoxin part and that the protoxin–toxin conversion does not lead to a major conformational change in the toxic segment. However, a minor conformational change is suggested by differential scanning calorimetry of the protoxin and toxic fragment of CryIA(c) (Choma *et al.*, 1991). Like the protoxin that is built up from two different segments, the trypsin-resistant toxin, located in the N-terminal half of the protoxin molecule, consists of distinct domains. Experiments on the toxic segments of CryIA(c), CryIA(b) and CryIC, similar to those described for the protoxins, led to the identification of two or three independent structural domains, possibly linked by a more or less protease-sensitive region. Proteolysis of the toxic fragments using different proteases under mild denaturing conditions generated two major fragments. Sequence analysis of the N-termini showed that one fragment originated from the N-terminus of the toxin molecule whereas the other major isolated fragment appeared to start in the region of residues 371–373 (CryIA(b); Convents *et al.*, 1991) and at residue 327 respectively (CryIA(c); Choma *et al.*, 1990). The X-ray crystallography data on CryIIIA do not confirm the hypothesis that either of these sites is located in a region linking two protein domains. The reason for this discrepancy remains unclear. An initial cleavage of the toxin molecule at the boundary between the first and second fragment followed by a limited digestion of the N-terminus of the C-terminal fragment might explain the identification of these downstream located residues. Another explanation might be that sequences linking two protein domains are not a priori more sensitive to proteolysis and that a correlation between these two parameters will not always occur.

In accordance with the experimental data outlined above, a model for the secondary structure of the toxic fragments was proposed based on the analysis of the primary structure of seven different crystal proteins belonging to all four major classes (Convents *et al.*, 1990). Despite their divergence, a consensus structure could be produced which predicts that the N-terminal halves of these toxic fragments are relatively rich in alpha-helix structure, as experimentally proven by circular dichroism studies,

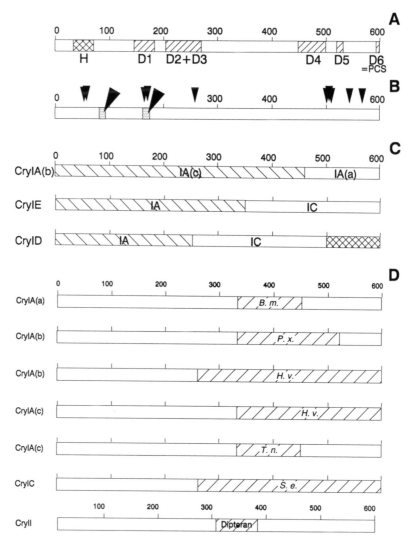

Figure 3.1. An alignment of structural and functional domains of class CryI and CryII toxins. A: Conserved sequence blocks based on the comparison of crystal proteins after Höfte and Whiteley (1989). B: Introduced mutations described in the text and leading to changes in toxicity have been indicated by arrows. The shaded areas represent short sequences in which multiple mutations have been monitored. C: A depiction of the indicated toxins as 'natural' hybrids of two homologous parents, based on the sequence analysis programme PCGene (Intelligenetics). The numbers in the bars indicate to which Cry protein the sequence is most homologous; ID: A depiction of identified insect specificity domains as identified in the text. *B. m., Bombyx mori; P. x., Plutella xylostella; H. v., Heliothis virescens; T. n., Trichoplusia ni; S. e., Spodoptera exigua.* PCS = Proteolytic cleavage site.

and the C-terminal parts contain alternating beta-strand and coil structures. In addition, the N-terminal parts show very conserved hydrophobicity patterns, whereas the C-terminal parts show less hydrophobicity of a more variable pattern. These predictions are largely in accordance with the protein structure of CryIIIA derived from the crystallography studies (Li *et al.*, 1991), which contains an N-terminal domain consisting of seven amphipathic and hydrophobic alpha-helices and two additional domains mainly consisting of beta-strands. Assuming a similar three-dimensional structure for the CryI toxins, the first domain would run from residue 29 to 260, the second domain from residue 260 to 460 and the third domain from residue 460 to 620 (approximate figures).

Sequence alignment of all Cry proteins reveals the presence of five conserved blocks separated by highly variable sequences of various lengths (Figure 3.1A). The CryII proteins only show homology to the first block. The structure of CryIIIA would imply that the first highly conserved block serves to maintain the hydrophobicity of the central helix whereas the others are involved in the interaction between the three domains of the toxic fragment. On the basis of sequence comparisons a potential transmembrane sequence has been postulated to occur near the N-terminus of all CryI and CryII toxic fragments, although in this region only the hydrophobic character, and not the actual sequence, is conserved (Höfte and Whiteley, 1989). The CryIIIA structure does only confirm this hypothesis to the extent that this region forms part of a domain containing several potential membrane spanning segments.

CRYSTAL PROTEIN MUTANTS

Some crystal proteins may be considered as 'natural' mutants only differing by one or a few residues from the holotype proteins as defined by the classification of Höfte and Whiteley (1989). However, with two exceptions it has not been reported whether these mutations (assuming they are not the result of sequencing errors) affect toxic properties. The first exception concerns a report on CryIA(c) proteins exhibiting seven amino acid differences compared to the holotype and showing slightly altered toxicity levels against some insects (Von Tersch *et al.*, 1991). The second exception is the CryIA(b) protein present in the *B. t.* subspecies *aizawai* isolate IC1, which differs from the holotype in only three amino acid residues but shows dual specificity against lepidopteran and dipteran insects unlike the holotype (Haider and Ellar, 1989).

The first mutants to be constructed were deletion mutants of the protoxins which were used to identify the minimal sequence necessary for full toxicity to larvae sensitive to the natural crystal protein. It was shown that the toxicity was still retained in a peptide produced from a truncated cryIA(a) gene ending with the codon 645, whereas a similar truncated gene ending with codon 603 appeared to be non-toxic (Schnepf and Whiteley, 1985). The latter observation was also made for a cryIC gene truncated behind the codon homologous to residue 603 of the cryIA(a) gene (Sanchis et al., 1989). Similar truncated versions of the cryIA(b) gene showed that a gene ending with the alanine codon 607 (homologous to cryIA(a) codon 606) encoded an active toxin, whereas a mutant gene ending with codon 604 (homologous to cryIA(a) codon 603) produced a dysfunctional protein which was not properly dissolved (Höfte et al., 1986; Visser et al., unpublished).

A 'second generation' of mutants was isolated for the identification of functional domains and/or essential amino acids (Figure 3.1B). These mutants all constitute substitution or small deletion mutants, except for some large deletion mutants of the CryIA(b) protein encoded by isolate B. t. aizawai IC1 (Haider and Ellar, 1989). In most cases one or a few residues in the conserved blocks were substituted or deleted in attempts to iden-tify amino acid residues essential for toxin function. A double mutation, in which two adjacent residues (phenylalanine and valine at positions 50 and 51) located in the structurally conserved N-terminal hydrophobic domain were replaced by aspartate, showed reduced toxicity (Ahmad and Ellar, 1990). Although data were not shown, it was suggested that mem-brane binding properties of the mutant proteins were unchanged indi-cating an impaired pore formation capability of these mutants. Similar observations were made for the protein CryIA(c) in which mutations either between positions 84 and 93, located in the third putative alpha-helix, or between positions 163 and 171, located in the fifth, central alpha-helix were generated (Wu and Aronson, 1990). Out of 25 mutations in the third helix tested, mutations of only two residues, alanine at position 92 or arginine at position 93, resulted in loss of toxicity. Loss of toxicity was prevented when a positive charge at these positions was retained. Surpris-ingly, the mutant, in which alanine had been replaced by aspartate at posi-tion 92, showed impaired binding to tobacco hornworm, Manduca sexta, vesicles. A substitution of either of two amino acid residues in the fifth alpha-helix at positions 165 (alanine replaced by methionine) and 167 (leu-cine replaced by proline) of the CryIA(a) protein resulted in mutant CryIA(a) proteins showing reduced toxicity (Ahmad and Ellar, 1990). Four-teen out of 31 mutant CryIA(c) proteins containing single mutations between residues 162 and 171 in the fifth alpha-helix showed a severe loss of toxicity (Wu and Aronson, 1990). Several mutant toxins tested success-fully competed with the wild-type toxin for binding to M. sexta vesicles, whereas all were deficient in the inhibition of K^+-dependent transport into midgut vesicles, confirming the hypothesis that this protein domain is primarily involved in pore formation. A mutant of a CryIA(b)/CryIC chimera containing a single mutation at position 259 (threonine replaced by proline) located at the boundary between the first and the second domain showed complete loss of toxicity due to instability of the toxic fragment (Honée *et al.*, 1991).

In the C-terminal part of the toxic fragment, it was shown that four out of 10 proteins containing a single mutation at serine residues 503 or 504, located in a variable region of the third domain, showed 5- to 10-fold lower activity towards the tobacco hornworm, *M. sexta*, and the tobacco budworm, *Heliothis virescens* (Wu and Aronson, 1990). Furthermore, a CryIA(c) mutant containing three adjacent substitutions in the second domain at positions 439–441 (glycine to alanine, phenylalanine to alanine, serine to glycine) showed a 110-fold reduction in toxicity towards *M. sexta*, and only a ten-fold decrease towards *H. virescens* (Schnepf *et al.*, 1990). Several more substitution and deletion mutants showed relatively small changes in toxicity patterns. Unfortunately, nearly all substitutions involved non-conserved changes. Therefore, it is impossible to conclude from this type of mutation analysis whether or when the loss of (or change in) function results from an overall change in conformation of the toxic fragment leading to impaired processing or decreased toxin stability *in vivo*, or that the mutation involved directly influences receptor binding.

Site-directed mutagenesis of the gene encoding the IC1 crystal protein exhibiting dual specificity showed that substitution of isoleucine residue 545 by the corresponding residue present in the holotype CryIA(b) (proline) resulted in loss of toxicity towards the mosquito (Diptera) *Aedes aegypti* but not against the cabbage white butterfly (Lepidoptera), *Pieris brassicae*, whereas the reverse toxicity pattern was observed after substitution of isoleucine residue 568 by the CryIA(b) residue threonine. Both residues follow arginine residues known to be sensitive to trypsin-like enzymes. Moreover, lepidopteran gut proteases convert the protoxin to a 55 kDa toxic fragment, whereas dipteran gut juice converts the same protoxin to a 53 kDa dipteran toxic fragment. The holotype CryIA(b) toxic fragment shows a molecular weight of 60 kDa, resulting from cleavage of the protoxin downstream residue 607. Assuming that the three-dimensional structure of CryIIIA approximates the CryIA(b) structure and considering the data obtained with the truncated *cryI* genes, it is difficult to imagine how the IC1 toxic fragments can be generated by cleavage at a site located within the third domain and why they appear to be biologically active.

In conclusion, several amino acid residues from different regions of the toxic fragment, either conserved or variable, were shown to affect toxic activity. Some mutations clearly resulted in impairment of a single toxin function, i.e. pore formation, whereas for other mutations no distinction could be made between a direct effect on receptor binding, or an indirect effect via a conformational change leading to protein instability or impaired processing.

CRYSTAL PROTEIN HYBRIDS

'Natural' hybrids

As discussed above, the amino acid sequences of the crystal proteins were compared (Höfte and Whiteley, 1989). Except for the high homology in the C-terminal half of the CryI and CryIV protoxins, a number of conserved blocks within the first stretch of 600 amino acid residues were identified. Upon closer inspection of the homology patterns of the CryI protoxins, for which class over 20 sequences have been obtained, some of these proteins may be depicted as hybrid molecules arising from a recombination between the genes encoding two other CryI proteins (Figure 3.1C). A mosaic pattern, in which each of the CryIA subclass proteins contained unique sequences, was recognized by Höfte *et al.* (1986). These data might also be interpreted as if the gene *cryIA(b)* is the result of a recombination between the genes *cryIA(c)* and *cryIA(a)*, followed by a deletion of 78 bp in the 3'-terminal half of the gene. The recombination site should be located at amino acid residue position 460. Similarly, the *cryIE* gene might be conceived as a recombinant of either the *cryIA(b)* or the *cryIA(c)* gene and the *cryIC* gene, the recombination site being located at amino acid residue position 350. The *cryID* gene could represent recombination between a *cryIA* gene and the *cryIC* gene, after which the part of the gene encoding the amino acid residues 500–600 has diverged. In this example the recombination site is located around amino acid residue position 257. Finally, the most divergent member of the *cryI* class, *cryIB*, shows most homology with any other members of the class, i.e. *cryIA(b)* or *cryIA(c)*, in the gene segment encoding the C-terminal part of the toxic fragment between residues 500 and 600. The deduced amino acid sequence of gene *cryIF* shows considerable homology to the domains 1 and 3 of CryIA(a) but not to the second domain (Chambers *et al.*, 1991). The amino acid sequence of *cryIG* shows a low level of homology for all parts of the toxic fragment (Smulevitch *et al.*, 1991).

It should be stressed that no evidence has been obtained to support our view that some of the crystal proteins might have arisen from evolutionary late recombination events in addition to the mutational divergence process leading from ancestor gene to the currently described gene family. Nevertheless, it is useful to compare the properties of some of these 'natural' hybrids that show additional mutations in divergence from the putative parent proteins with genetically engineered hybrid proteins, because this may add to our knowledge of domain–function relationships. Additionally, the locations of putative recombination sites might indicate in which regions of the toxic fragment recombination leads to functional hybrids. This aids a rational approach to engineering and hybrid construction. In this respect it is interesting to note that two of the proposed recombination sites, i.e. those located at positions 260 and 460 approximately, coincide with the transition from the first to the second, and from the second to the third domain respectively, based on the structure for the related CryIII protein.

Constructed hybrids

A limited number of studies involving the construction of hybrid genes and consequently of hybrid crystal proteins have been reported, all concerning CryI class proteins (Ge *et al.*, 1989, 1991; Raymond *et al.*, 1990; Nakamura *et al.*, 1990; Schnepf *et al.*, 1990; Caramori *et al.*, 1991; Honée *et al.*, 1991) and in one case CryII proteins (Widner and Whiteley, 1990). All of these studies were aimed at identifying the insect specificity of a given protein with a particular sequence (Figure 3.1D). All involved the construction of complete protoxin genes, and the expression of these genes in the heterologous host *E. coli*. Toxicity was determined using either cell suspensions, purified crystal inclusions, solubilized protoxins or purified toxic fragments. Caution is needed in the evaluation of the specificity of the hybrid crystals or protoxins since several factors, i.e. solubilization, processing and membrane receptor binding, are known to influence specificity. Furthermore, it is striking that no studies with hybrid proteins have been devoted to the toxin's capacity of pore formation.

In only one study were solubilized hybrid protoxins evaluated with respect to their insect specificity (Raymond *et al.*, 1990). Two reciprocal hybrid genes derived from a *cryIA(a)* gene and a *cryIA(b)* gene were constructed, the junction being positioned at amino acid residues 449 and 450 respectively, and expressed in *E. coli*. The solubilized protoxins were tested for toxicity against the larvae of *M. sexta*, an insect slightly more sensitive to the CryIA(a) protein than to the CryIA(b) protein. Unfortunately, due to the hybrid nature of CryIA(b) itself, only minimal differences distinguish the two reciprocal chimeras from the two parents, resulting in two amino acid changes directly downstream from the junction used and the exchange of the C-terminal part of the protoxin. Although differences in toxicity were minimal, the authors conclude that the N-terminal part of the protoxins determines specificity. This would imply that these differences in toxicity are not related to either the two amino acid substitutions or the protoxin C-terminal differences mentioned. However, solubilization in the midgut of three other Lepidoptera is believed to be more efficient for CryIA(b) than for CryIA(a) (Aronson *et al.*, 1991), thereby implying differences between the C-terminal protoxin parts to explain differences in toxicity between these two protoxin subclasses, as observed by Raymond *et al.* (1990). A second study did not determine the toxicity of solubilized protoxin molecules but measured the solubility of several crystals containing a single hybrid protein derived from the same parentals CryIA(a) and CryIA(b) (Nakamura et al., 1990). Although evaluation of the results appeared complex, a contribution of the C-terminal half of the protoxin to the solubilization properties was confirmed when solubilization was carried out in vitro in an alkaline buffer. This study also analysed solubility and proteolytic processing in the presence of tobacco cutworm, Spodoptera litura, gut juice. Only sequences located in the toxin part of the protoxin molecule were implicated in correct processing. A protein segment of CryIA(b) running from amino acid residue 332 to 523, placed in a CryIA(a) background, was suggested to enhance toxicity towards the diamondback moth, Plutella xylostella (Nakamura et al., 1990).

Three different groups generated hybrids between the protoxins of CryIA(a) and CryIA(c). Insect specificity was studied with insect species showing a large difference in sensitivity towards these two crystal proteins. A detailed analysis of 18 reciprocal hybrids (Ge et al., 1989) clearly showed that a CryIA(a) segment from amino acid position 332 to 450, differing in 52 amino acid positions from the corresponding segment of CryIA(c), and transferred to a CryIA(c) background, was responsible for silkworm, Bombyx mori, specificity, an insect showing a 400 times difference in sensitivity towards these two proteins. In follow-up studies, receptor binding studies with trypsin-activated toxic fragments of the parent and engineered types showed that the B. mori receptor binding domain of CryIA(a) was located between residues 283 and 450, coinciding with the specificity domain determined by toxicity assays (Dean, personal communication). Furthermore, additional bioassays with mutant toxin proteins revealed that amino acids 335 to 450 of CryIA(c) were associated with the activity against the cabbage looper, Trichoplusia ni, whereas amino acids 335–615 of the same toxic fragment were required for full H. virescens specificity. Finally, a hybrid between CryIA(a) and CryIA(c), reciprocal to the natural hybrid CryIA(b) and containing the toxin sequences derived from CryIA(c) located between residues 450 and 615, appeared to be 30 times as toxic against H. virescens as CryIA(c) itself, showing a major role for this toxin segment in H. virescens specificity (Ge et al., 1991).

A second study involving hybrids between CryIA(a) and CryIA(c) tested the toxicity of a series of non-reciprocal hybrids against the test insects M. sexta and H. virescens (Schnepf et al., 1990). Whereas M. sexta is equally sensitive towards both parental crystal proteins, H. virescens is 50 times more sensitive towards CryIA(c). Unfortunately, these studies monitored loss of toxicity, which renders the positive identification of a domain with major responsibility for insect specificity rather difficult. However, segments spanning the entire C-terminal toxin domain from residue 332 towards 722 can be assumed to contribute to H. virescens specificity, in agreement with the findings of Ge et al. (1991), reported above. The authors state that preliminary results identify the CryIA(c) fragment between residues 332 and 447 as the *T. ni* specificity domain. Finally, hybrids generated by *in vivo* recombination (Caramori *et al.*, 1991) showed that recombination in the protein fragment between residues 333 and 498 influenced specificity against the Mediterranean flour moth, *Anagasta (Ephestia) kuehniella*, *T. ni*, and an unidentified *Heliothis* sp., but failed to determine a specificity domain.

Of 11 recombinant crystal proteins produced between CryIA(b) and the more distantly related CryIC, most did not show toxicity due to decreased stability of the protein upon solubilization or tryptic processing (Honée *et al.*, 1991). However, one purified recombinant toxic fragment, consisting of the first 258 residues of CryIA(b) and the entire C-terminal toxic fragment region from CryIC (amino acid residues 261 to 654), showed the specificity pattern of CryIC. Again, this illustrates the major involvement of the C-terminal part of the toxic fragment in specificity, in this case towards *Spodoptera littoralis* and *H. virescens*. Although delineation of the specificity domain is less accurate due to stability problems probably related to conformational changes in the toxic fragment, membrane binding studies on this hybrid clearly show that it is the C-terminal part of the toxic fragment that is involved in receptor binding and thus largely determines insect specificity. A reciprocal hybrid containing N-terminal sequences derived from CryIC and C-terminal toxin sequences from CryIA(b) showed similar results on insect specificity but appeared less stable during incubation with membrane vesicles, necessary for *in vitro* binding studies. A decrease in the toxicity of these two hybrids against sensitive species is at least partially caused by a lower binding affinity, and maybe additionally by a more general conformational change affecting some further step in the mode of action, e.g. pore formation.

An elegant study on the specificity differences between CryIIA, toxic to both Lepidoptera and Diptera, and CryIIB, only toxic for Lepidoptera, established that the CryIIA region corresponding to residues 307 to 382 was sufficient for altering specificity (Widner and Whiteley, 1990). The region conferred toxicity against mosquitoes and contained only 18 amino acid differences compared with CryIIB, again showing that only a few changes can have a substantial effect on the toxicity spectra.

Conclusions on hybrid studies

The C-terminal protoxin sequences, removed during processing of the protein, influence specificity to some extent. In some insects a difference in the efficiency of the solubilization process can be demonstrated, especially for CryIA(b).

From mutation analysis of the atypical CryIA(b) protein present in the *B. t. aizawai* strain IC1 it may be inferred that proteolytic processing is a determinant in insect specificity and that proteolysis is dependent on the sequences constituting the processing site. From an analysis of deletion mutants of the CryIA(b) IC1 protein it appeared that sequences located near the C-terminus of the toxic fragment, between residues 524 and 544 are sufficient for specificity towards the mosquito *Aedes aegypti*, although additional sequences might be required for maximal toxicity levels. Also, deletion of the amino acid sequence between residues 545 and 568 renders the molecule non-toxic to *P. brassicae*, implicating this sequence in lepidopteran specificity. The extent of the sequences determining the efficiency of processing is not exactly known but appears to include more than simply two amino acid residues at the processing site.

Receptor binding probably forms the major determinant in insect specificity. A specificity domain located between residues 332 and 450 in the CryI class proteins determines toxicity against *P. xylostella*, *B mori* and *T. ni*. This genetic delineation of the specificity domain coincides very well with the putative second domain of the CryI toxins which based on its structure can be supposed to be responsible for receptor binding. Sequences spanning the toxin segment between residues 450 and 615 seem to play a substantial additional role in *H. virescens* specificity. The CryIA(c) specific sequences of this third domain may provide especially effective protection against proteolysis in the *H. virescens* midgut.

The host range of CryIA(b), the putative hybrid of CryIA(a) and CryIA(c), might be explained as being in agreement with the model that the C-terminal part of the toxic fragment is responsible for insect specificity: the *H. virescens* specificity domain of CryIA(c) is partly present in CryIA(b), which shows a toxicity level towards *H. virescens* intermediate between that of CryIA(c) and of the other relatively ineffective parent CryIA(a) (Höfte and Whiteley, 1989). Similarly, the C-terminal part of CryIE from amino acid residue 350 onwards shows strong homology with the comparable part of the CryIC protein with which it shares its toxicity towards *Spodoptera* species (Visser *et al.*, 1990). On the other hand, the toxicity pattern of CryID which shares greatest homology with CryIC for most of the C-terminal part of the toxic fragment does not resemble that of CryIC. Of course, it may be hypothesized that additional divergence of the putative hybrid, especially in the region between residues 500 and 600, is responsible for this lack in toxicity pattern homology.

Many of the recombination sites of the viable genetically constructed hybrids appear to coincide with the putative recombination sites of the natural hybrids, e.g. the sites around residues 257, 350 and 460, and with the putative toxin domains, possibly indicating the constraints upon hybrid formation related to toxin conformation in general and protease resistance in particular.

GENERAL CONCLUSIONS AND PERSPECTIVES

A first series of experiments aimed at the elucidation of domain–function relationships has provided some initial insight. In CryI proteins, a specificity domain located between residues 332 and 450 is essential for toxicity towards several lepidopteran insects. However, for full toxicity against *H. virescens* additional sequences are needed. Furthermore, sequences determining the efficiency of the solubilization located in the C-terminal part of the protoxin and sequences involved in proteolytic processing appeared to influence insect specificity. Previously, little attention has been paid to the pore formation capability in relation to specific sequence domains. Studies on the properties of hybrid crystal proteins should include this question. For example, it is not known why the insect most sensitive to CryIC cabbage moth, *Mamestra brassicae*, requires four to seventy times more toxin than three species sensitive to CryIA(c). Is this simply a function of receptor concentration and affinity or does it also reflect a difference in pore formation efficiency of these toxic fragments?

A considerable proportion of the generated hybrids appeared to be non-toxic to any of the insects tested due to instability of the toxic fragment. Understandably, this shows the importance of the exact toxin conformation and stresses the fact that although different domains may be distinguished within the toxin moiety, these domains clearly show structural interaction. In general, the function of essential stretches of the toxic fragment is unknown. Whether these stretches simply serve a structural role, or whether some of these sequences exert a more specific function in toxic action, for instance by enlarging the host range or improving toxicity of the molecule, as seems to be the case for *H. virescens* specificity, remains unclear at the moment. The construction of hybrid crystal proteins has shown that this genetic approach is fruitful in determining the domain–function relationships of *B. t.* crystal proteins. Knowledge of these relationships is a necessary condition for a directed approach towards manipulating toxin function.

REFERENCES

Ahmad, W. and Ellar, D. J. (1990). 'Directed mutagenesis of selected regions of a *Bacillus thuringiensis* entomocidal protein'. *FEMS Microbiol. Lett.*, **68**, 97–104.

Aronson, A. I., Han, E.-S., McGaughey, W. and Johnson, D. (1991). 'The solubility of inclusion proteins from *Bacillus thuringiensis* is dependent upon protoxin composition and is a factor in toxicity to insects'. *Appl. Env. Microbiol.*, **57**, 981–986.

Bietlot, H. P. L., Vishnubhatla, I., Carey, P. R., Pozsgay, M. and Kaplan, H. (1990). 'Characterization of the cysteine residues and disulphide linkages in the protein crystal of *Bacillus thuringiensis*'. *Biochem. J.*, **267** 309–315.

Brown, K. L. and Whiteley, H. R. (1992). 'Molecular characterization of two novel crystal protein genes from Bacillus thuringiensis subsp. thompsoni'. J. Bacteriol, 174, 549–557.

Caramori, T., Albertini, A. M. and Galizzi, A. (1991). 'In vivo generation of hybrids between two Bacillus thuringiensis insect-toxin-encoding genes'. Gene, 98, 37–44.

Chambers, J. A., Jelen, A., Gilbert, M. P., Jany, C. S., Johnson, T. B., and Gawron-Burke, C. (1991). 'Isolation and characterization of a novel insecticidal crystal protein gene from Bacillus thuringiensis subsp. aizawai'. J. Bacteriol., 173, 3966–3976.

Choma, C.T. and Kaplan, H. (1990). 'Folding and unfolding of the protoxin from Bacillus thuringiensis: Evidence that the toxic moiety is present in an active conformation'. Biochemistry, 29, 10971–10977.

Choma, C. T., Surewicz, W. K., Carey, P. R., Pozsgay, M., Raynor, T. and Kaplan, H. (1990). 'Unusual proteolysis of the protoxin and toxin from Bacillus thuringiensis. Structural implications'. Eur. J. Biochem., 189, 523–527.

Choma, C. T., Surewicz, W. K. and Kaplan H. (1991) 'The toxic moiety of the Bacillus thuringiensis protoxin undergoes a conformational change upon activation'. Biochem. Biophys. Res. Commun., 179, 933–938.

Convents, D., Houssier, C., Lasters, I. and Lauwereys, M. (1990). 'The Bacillus thuringiensis δ-endotoxin. Evidence for a two domain structure of the minimal toxic fragment'. J. Biol. Chem., 254, 1369–1375.

Convents, D., Cherlet, M., Van Damme, J., Lasters, I. and Lauwereys, M. (1991). 'Two structural domains as a general fold of the toxic fragment of the Bacillus thuringiensis δ-endotoxins'. Eur. J. Biochem., 195, 631–635.

Garfield, J. L. and Stout, C. D. (1988) 'Crystallization and preliminary X-ray diffraction studies of a toxic crystal protein from a subspecies of Bacillus thuringiensis'. J. Biol. Chem., 263, 11800–11801.

Ge, A. Z., Shivarova, N. I. and Dean, D. H. (1989). 'Location of the Bombyx mori specificity domain on a Bacillus thuringiensis δ-endotoxin protein'. Proc. Natl Acad. Sci. USA, 86, 4037–4941.

Ge, A. Z., Rivers, D., Milne, R. and Dean D. (1991). 'Functional domains of Bacillus thuringiensis insecticidal proteins'. J. Biol. Chem., 266, 17954–17958.

Haider, M. Z. and Ellar, D. J. (1989). 'Functional mapping of an entomocidal δ-endotoxin. Single amino acid changes produced by site-directed mutagenesis influence toxicity and specificity of the protein'. J. Mol. Biol., 208, 183–194.

Hendrickx, K., De Loof, A. and Van Mellaert, H. (1990). 'Effects of Bacillus thuringiensis delta-endotoxin on the permeability of brush border membrane vesicles from tobacco hornworm (Manduca sexta) midgut'. Comp. Biochem. Physiol., 95C, 241–245.

Hofmann, C., Vanderbruggen, H., Höfte, H., Van Rie, J., Jansens, S. and Van Mellaert, H. (1988). 'Specificity of Bacillus thuringiensis δ-endotoxins is correlated with the presence of high-affinity binding sites in the brush border membrane of target insect midguts'. Proc. Natl Acad. Sci. USA, 85, 7844–7848.

Höfte, H. and Whiteley, H. R. (1989). 'Insecticidal crystal proteins of Bacillus thuringiensis'. Microbiol. Rev., 53, 242–255.

Höfte, H., De Greve, H., Seurinck, J., Jansens, S., Mahillon, J., Ampe, C., Vandekerckhove, J., Vanderbruggen, H., Van Montagu, M., Zabeau, M. and Vaeck, M. (1986). 'Structural and functional analysis of a cloned delta endotoxin of Bacillus thuringiensis berliner 1715'. Eur. J. Biochem., 161, 273–280.

Honée, G., Convents, D., Van Rie, J., Jansens, S., Peferoen, M. and Visser, B. (1991).

'The C-terminal domain of the toxic fragment of a *Bacillus thuringiensis* crystal protein determines receptor binding'. *Mol. Microbiol.*, **5**, 2799–2806.

Huber, H. E., Lüthy, P. Ebersold, H.-R. and Cordier, J.-L. (1981). 'The subunits of the parasporal crystal of *Bacillus thuringiensis*: size, linkage and toxicity'. *Arch. Microbiol.*, **129**, 14–18.

Huber-Lukač, M., Jacquet, F., Lüthy, P., Hütter, R. and Braun, D. G. (1986). 'Characterization of monoclonal antibodies to a crystal protein of *Bacillus thuringiensis* subsp. *kurstaki*'. *Infect. Immun.*, **54** (1), 228–232.

Jacquet, F., Hütter, R. and Lüthy, P. (1987). 'Specificity of *Bacillus thuringiensis* delta-endotoxin'. *Appl. Env. Microbiol.*, **53**, 500–504.

Kronstad, J. W. and Whiteley, H. R. (1986). 'Three classes of homologous *Bacillus thuringiensis* crystal protein genes'. *Gene*, **43**, 29–40.

Li, J., Carroll, J. and Ellar, D. J. (1991). 'Crystal structure of insecticidal δ-endotoxin from *Bacillus thuringiensis* at 2–5Å resolution'. *Nature (London)*, **353**, 815–821.

Lüthy, P. and Ebersold, H. R. (1981). '*Bacillus thuringiensis* delta-endotoxin: histopathology and molecular mode of action'. In *Pathogenesis of Invertebrate Microbial Diseases* (Ed. E. W. Davidson), pp. 235–267.

Martin, P. A. W. and Travers, R. S. (1989). 'Worldwide abundance and distribution of *Bacillus thuringiensis* isolates'. *Appl. Env. Microbiol.*, **55**, 2437–2442.

Nakamura, K., Oshie, K., Shimizu, M., Takada, Y., Oeda, K. and Ohkawa, H. (1990). 'Construction of chimeric insecticidal proteins between the 130-kDa and 135-kDa proteins of *Bacillus thuringiensis* subsp. *aizawai* for analysis of structure–function relationship'. *Agric. Biol. Chem.*, **54**, 715–724.

Pastan, I. and FitzGerald, D. (1989). '*Pseudomonas* exotoxin: Chimeric toxins'. *J. Biol. Chem.*, **264**, 15157–15160.

Prior, I. P., FitzGerald, D. J. and Pastan, I. (1991). 'Barnase toxin: A new chimeric toxin composed of *Pseudomonas* exotoxin A and Barnase'. *Cell*, **64**, 1017–1023.

Raymond, K. C., John, T. R. and Bulla, L. A., Jr (1990). 'Larvicidal activity of chimeric *Bacillus thuringiensis* protoxins'. *Mol. Microbiol.*, **4**, 1967–1973.

Rupar, M. J., Donovan, W. P., Groat, R. G., Slaney A. C., Mattison, J. W., Johnson, T. B., Charles J.-F., Cosmao Dumanoir, V. and de Barjac, H. (1991). 'Two novel strains of *Bacillus thuringiensis* toxic to coleopterans'. *Appl. Env. Microbiol.*, **57**, 3337–3344.

Sacchi, V. F., Parenti, P., Hanozet, G. M., Biordana, B., Lüthy, P. and Wolfersberger, M. G. (1986). '*Bacillus thuringiensis* toxin inhibits K^+-gradient-dependent amino acid transport across the brush border membrane of *Pieris brassicae* midgut cells'. *FEBS Lett.*, **204**, 213–218.

Sanchis, V., Lereclus, D., Menou, G., Chaufaux, J., Guo S. and Lecadet, M.-M. (1989). 'Nucleotide sequence and analysis of the N-terminal coding region of the *Spodoptera*-active delta-endotoxin gene of *Bacillus thuringiensis aizawai* 7.29'. *Mol. Microbiol.*, **3**, 229–238.

Schnepf, H. E. and Whiteley, H. R. (1985). 'Delineation of a toxin-encoding segment of a *Bacillus thuringiensis* crystal protein gene'. *J. Biol. Chem.*, **260**, 6273–6280.

Schnepf, H. E., Tomczak, K., Ortega, J. P. and Whiteley, H. R. (1990). 'Specificity-determining regions of a lepidopteran-specific insecticidal protein produced by *Bacillus thuringiensis*'. *J. Biol. Chem.*, **265**, 20923–20930.

Sick, A., Gaertner, F. and Wong, A. (1990). 'Nucleotide sequence of a coleopteran-active toxin gene from a new isolate of *Bacillus thuringiensis* subsp. *tolworthi*'. *Nucl. Acids Res.*, **18**, 1305.

Spaink, H. P., Wijffelman, C. A., Okker, R. J. H. and Lugtenberg, B. E. J. (1989). 'Localization of functional regions of the *Rhizobium* nodD product using hybrid *nodD* genes'. *Plant Mol. Biol.*, **12**, 59–73.

Smulevitch, S. V., Osterman, A. L., Shevelev, A. B., Kaluger, S. V., Karimin, A. I., Kadyrov, R. M., Zagnitko, O. F. and Chestukhiha, G. G. St. (1991). 'Nucleotide sequence of a novel delta-endotoxin gene *cryIG* of *Bacillus thuringiensis* spp. *galleriae'. FEBS Lett.*, **293** (1–2), 25–28.

Tailor, R., Tippett, J., Gibb, G., Pells, S., Pike, D., Jordan, L. and Ely, S. (1992). 'Identification and characterisation of a novel *Bacillus thuringiensis* δ-endotoxin entomocidal to coleopteran and lepidopteran larvae'. *Mol. Microbiol.*, **6** (9) (in press).

Tommassen, J., Van Der Ley, P., Van Zeijl, M. and Agterberg, M. (1985). 'Localization of functional domains in *E. coli* K-12 outer membrane porins'. *EMBO J.*, **4**, 1583–1587.

Van Rie, J., Jansens, S., Höfte, H., Degheele, D. and Van Mellaert, H. (1990). 'Receptors on the brush border membrane of the insect midgut as determinants of the specificity of *Bacillus thuringiensis* delta-endotoxins'. *Appl. Env. Microbiol.*, **56**, 1378–1385.

Visser, B., Van Workum, B., Dullemans, M. and Waalwijk, C. (1986). 'The mosquitocidal activity of *Bacillus thuringiensis* var. *israelensis* is associated with Mr 230,000 and 130,000 crystal proteins', *FEMS Microbiol. Lett.*, **30**, 211–214.

Visser, B., Munsterman, E., Stoker, A. and Dirkse, W. G. (1990). 'A novel *Bacillus thuringiensis* gene encoding a *Spodoptera exigua*-specific crystal protein'. *J. Bacteriol.*, **172**, 6783–6788.

Von Tersch, M. A., Robbins, H. L., Jany, C. S. and Johnson, T. B. (1991). 'Insecticidal toxins from *Bacillus thuringiensis* subsp. *kenyae*: Gene cloning and characterization and comparison with *B. thuringiensis* subsp. *kurstaki* CryIA(c) toxins'. *Appl. Env. Microbiol.*, **57**, 349–358.

Widner, W. R. and Whiteley, H. R. (1990). 'Location of the dipteran specificity region in a lepidopteran dipteran crystal protein from *Bacillus thuringiensis'. J. Bacteriol.*, **172**, 2826–2832.

Wong, H. C., Schnepf, H. E. and Whiteley, H. R. (1983). 'Transcriptional and translational start sites for the *Bacillus thuringiensis* crystal protein gene'. *J. Biol. Chem.*, **258**, 1960–1967.

Wu, D. and Aronson, A. I. (1990). 'Use of mutagenic oligonucleotides for defining regions of a *Bacillus thuringiensis* δ-endotoxin involved in toxicity'. In *Proceedings of the Vth International Colloquium on Invertebrate Pathology and Microbial Control* (Ed. D. E. Pinnock), pp. 273–277, Adelaide, Australia.

Yan, S. and McCarthy, J. (1991). 'Chemical modification of *Bacillus thuringiensis* (HD-524) trypsin-activated endotoxin: Implication of tyrosine residues in lepidopteran cell lysis'. *J. Invert. Pathol.*, **57**, 101–108.

4 Transgenic Bacteria, Viruses, Algae and Other Microorganisms as *Bacillus thuringiensis* Toxin Delivery Systems

WENDY GELERNTER AND GEORGE E. SCHWAB

Mycogen Corporation, 5451 Oberlin Drive, San Diego, CA 92121, USA

INTRODUCTION

It has now been 10 years since Schnepf and Whiteley (1981) first cloned and expressed a *Bacillus thuringiensis* (*B. t.*) delta (δ) endotoxin gene in *Escherichia coli*, demonstrating for the first time the potential of genetic engineering technology for biological control. Since that time, *B. t.* δ-endotoxin genes (Table 4.1) have been cloned and expressed in a wide variety of microorganisms—from baculoviruses to cyanobacteria—in attempts to improve their delivery and efficacy to insect pests. Progress has been rapid, and in 1991 the US Environmental Protection Agency's registration of two genetically engineered products based on *B. t.* toxin-expressing *Pseudomonas fluorescens*, in which the cells are killed before use, was an important first step in the process of commercializing recombinant or transgenic microbial pesticides. Because we are on the verge of a new era in the development of biological pest control agents, introspection and debate on the positive and negative implications of this new technology should be encouraged. In this chapter, progress on the development of transgenic microorganisms as *B. t.* toxin delivery systems will be reviewed, and issues and hurdles to their development will be outlined, in the hope of providing a basis for continuing progress and discussion on this topic.

CONVENTIONAL *B. THURINGIENSIS* PRODUCTS: DO THEY NEED TO BE IMPROVED?

Although *B. t.* was first commercialized more than 40 years ago (Rowe

Bacillus thuringiensis, An Environmental Biopesticide: Theory and Practice. Edited by P. F. Entwistle, J. S. Cory, M. J. Bailey and S. Higgs
© 1993 John Wiley & Sons Ltd

Table 4.1. *Bacillus thuringiensis* genes utilized in the development of transgenic microorganisms

Gene	Molecular mass (kDa)[a]	Subspecies[b]	Reference
Lepidopteran active			
crylA(a)	133.1	kurstaki	Schnepf et al., 1985
crylA(b)	130.6	kurstaki	Wabiko et al., 1986
crylA(c)	133.3	kurstaki	Adang et al., 1985
crylC	135.0	entomocidus	Honée and Visser, 1988
Dipteran active			
crylVA	134.4	israelensis	Ward and Ellar, 1987
crylVB	127.8	israelensis	Sekar and Carlton, 1985
crylVC	77.8	israelensis	Thorne et al., 1986
crylVD	72.4	israelensis	Donavan et al., 1988
cytA	27.4	israelensis	Waalwijk et al., 1985
Coleopteran active			
crylllA	73.1	san diego	Herrnstadt et al., 1987

[a] Refers to the calculated molecular weight of the deduced gene product.
[b] Subspecies from which the gene was first isolated.

and Margaritis, 1987), adoption of its use in mainstream agriculture has expanded significantly only in the past 10 years. This is primarily due to increased public demand during the last decade for pesticides that are less toxic for consumers, workers and the environment and to grower demand for products that effectively control insects that have developed resistance to synthetic chemical insecticides.

Simultaneously with increased acceptance for conventional (naturally occurring, non-engineered) *B. t.* based insecticides, research efforts have intensified, yielding improved production methods, formulations, and new strains capable of targeting a diverse array of pests—from beetles to nematodes (Edwards et al., 1990). For these reasons, the increased usage of *B. t.* products is projected to continue, with sales expected to reach close to $300 million by the year 2000 (McKemy, 1990).

Despite this unprecedented success, sales of *B. t.* based products are currently restricted almost exclusively to high value or environmentally sensitive niche markets such as fresh market vegetables, disease vector control and forestry. This lack of widespread adoption is due partly to several inherent biological characteristics of *B. t.* which include:

—host range specificity,
—inability to target pests that feed internally or on roots,
—lack of residual activity on foliage due to degradation by sunlight and other environmental factors,

—lack of residual activity in water due to rapid settling of spores and crystals and adsorption to organic particles,
—lack of residual activity in soil due to degradation by soil microorganisms.

These features (which are described more fully below) contribute to the now frequently cited disadvantages of using B. t., i.e. the requirement for increased frequency and number of applications, the need for improved application equipment and for intensive scouting to improve timing of applications; in other words, increased time and financial investment.

Thus, B. t. is still very much a boutique product that is rarely, if ever, utilized by growers of the world's largest acreage crops: cotton, grain, oilseed and other row crops (Gelernter, 1992). To penetrate these markets, where the environmentally friendly features of B. t. are perhaps less important than the cost and difficulty of implementing its use, certain features of B. t. must be changed through improved formulations, novel application methods, or perhaps most reliably by altering the biology of B. t. itself through the use of recombinant DNA (rDNA) techniques and development of transgenic microorganisms that express B. t. δ-endotoxins.

THE USE OF TRANSGENIC MICROORGANISMS TO IMPROVE INSECT HOST RANGE

Host range of conventional B. thuringiensis products

Despite continued advances in the discovery of new B. t. isolates with novel host range activities, each B. t. isolate is generally characterized by a very specific host spectrum:

Species specificity In general, each B. t. isolate is characterized by a distinct and specific host range. For example, the widely commercialized H-14 isolate of B. t. subspecies israelensis is highly active against dipteran larvae such as Aedes and Culex spp. mosquitoes, but has limited activity against other mosquito larvae. Similarly, coleopteran active isolates of B. t. subspecies tenebrionis and san diego are active against chrysomelid beetles such as the Colorado potato beetle, Leptinotarsa decemlineata, and the Elm leaf beetle, Xanthogaleruca luteola, but have almost no activity on other chrysomelid pests such as Diabrotica spp. (Herrnstadt et al., 1986).

Age specificity Most B. t. isolates are more effective against newly hatched and small larvae than they are against older life stages. For example, while early instar larvae of the Colorado potato beetle are sensitive to *B. t. san diego* and *tenebrionis* (see Keller and Langenbruch, Chapter 8, on synonymy), older larvae and adults are not significantly affected (Ferro and Lyon, 1991; Zehnder and Gelernter, 1989).

The outcome of this level of specificity is that growers must scout intensively to determine whether the appropriate life stages and species complexes are present to warrant a *B. t.* application. In contrast, this level of attention is not required to trigger the use of conventional chemical insecticides which have broad pest age and species spectra.

Use of transgenic microorganisms to improve insect host range

Efforts to improve insecticidal activity (summarized in Table 4.2) have been based on two different approaches: (a) transfer of *B. t.* genes into non-homologous isolates of *B. t.* as a means of combining δ-endotoxins to produce either additive or synergistic effects and thus expand the host range; and (b) transfer of *B. t.* genes into insect-specific baculoviruses as a means of overcoming species and age specificity, as well as for increasing baculovirus speed of kill.

Crickmore *et al.* (1990) have taken the first approach in their successful attempts to introduce cloned *B. t.* genes into native isolates of *B. t.*, thus increasing their host spectrum. Using electroporation, a plasmid bearing the beetle active *crylllA* gene was transferred to *B. t. israelensis*. The resulting organism demonstrated activity against mosquitoes and beetles as expected, but also showed unexpected activity against lepidopteran pests such as the cabbage white butterfly, *Pieris brassicae*. Similarly, when a

Table 4.2. Improving insect host range of δ-endotoxins: a summary of research projects

Endotoxin gene	Vector	Advantages	Author
crylllA (tenebrionis)	B. t. israelensis	Expand host range; δ-endotoxin synergy	Crickmore et al., 1990
crylA(b) (aizawai)	B. t. tenebrionis	Expand host range; δ-endotoxin synergy	Crickmore et al., 1990
crylllA (tenebrionis)	B. t. kurstaki	Corn borer and potato beetle activity; regulatory flexibility	Carlton et al., 1990
crylA(b) (aizawai)	Autographa californica NPV	Expand host range; activity on large larvae	Martens et al., 1990
crylA(c) (kurstaki)	Autographa californica NPV	Expand host range; activity on large larvae	Merryweather et al., 1990

plasmid bearing the lepidopteran active *crylA(b)* gene was introduced into *B. t. tenebrionis*, activity on Lepidoptera, Coleoptera and, unexpectedly, on mosquitoes, was observed. These experiments support the exciting concept that δ-endotoxin proteins may interact synergistically, expanding activity to groups of insects that are not affected by either of the toxin proteins when expressed alone.

To avoid the regulatory restrictions associated with the testing and use of organisms such as those described above which are produced using rDNA technology, Carlton *et al.* (1990) have utilized an intermediate strategy. A plasmid bearing the beetle active *crylllA* gene was transferred to lepidopteran active *B. t.* subspecies *kurstaki* cells, utilizing bacterial conjugation. The resulting transconjugant, the product Foil®, possesses both coleopteran and lepidopteran activity, and has demonstrated good activity in the field against the larvae of both the Colorado potato beetle and the European corn borer (ECB), *Ostrinia nubilalis*. Because the organism is considered by regulatory agencies to be genetically manipulated rather than genetically engineered, it was approved for use by the US Environmental Protection Agency in 1990.

Two different laboratories have taken an approach similar to that described above, using the lepidopteran active *Autographa californica* nuclear polyhedrosis virus (AcNPV) as an expression system for *B. t.* δ-endotoxins. Because AcNPV has a broad lepidopteran host range, and is also capable of killing both early and late instar lepidopteran larvae, its combination with *B. t.* toxins could result in expanded host and age specificity. Additionally, the typically slow activity of baculoviruses in their insect hosts could be accelerated through association with faster acting *B. t.* δ-endotoxins. Finally, AcNPV is a good vehicle for expressing *B. t.* and other insect specific proteins because it produces two highly expressed viral proteins—polyhedrin and p10. Since neither of these proteins is essential for viral replication, they can be replaced with a *B. t.* δ-endotoxin gene, which is inserted under the control of the highly active polyhedrin or P10 promoter. Alternatively, the promoters can be copied, resulting in a polyhedrin positive virus.

Merryweather *et al.* (1990) have achieved expression of the lepidopteran active *crylA(c)* gene in both polyhedrin plus and polyhedrin minus AcNPV recombinants. Expression of the 130 kDa δ-endotoxin protein was confirmed by immunological assays, and the presence of 62 kDa and 44 kDa breakdown products was also observed. Martens *et al.* (1990) also successfully expressed the lepidopteran active gene *crylA(b)* in a polyhedrin minus AcNPV recombinant. When insect tissue culture cells of fall armyworm, *Spodoptera frugiperda*, were infected with the recombinant, virus particles and protein crystals containing the immunologically reactive CrylA(b) 130 kDa protein were present in the cell cytoplasm. Relatively high levels of expression were observed, with the CrylA(b) protein accounting for approximately 5% of the total cell protein. Although these laboratories achieved relatively high levels of δ-endotoxin production with the transgenic baculoviruses, more rapid activity against lepidopteran larvae was not observed by either group. Further studies to understand the basis of this observed lack of additivity are ongoing. To be effective, the δ-endotoxin must come into contact with the microvillar membrane of the insect midgut. However, for the transgenic viruses described above, the δ-endotoxin is produced *inside* the virus infected cell, where it has little or no toxic effect. Efforts to develop a transgenic virus capable of transporting the δ-endotoxin outside infected midgut cells to the microvillar membrane should therefore also result in improved activity levels.

IMPROVING FOLIAR ACTIVITY AND DELIVERY

Limitations associated with conventional *B. thuringiensis* products

B. t. has been most successfully used when delivered as foliar sprays against lepidopteran pests in forestry and on vegetable crops. However, two important limitations, lack of delivery to cryptic or internal feeders and short residual activity, inhibit its more widespread use.

Delivery   *B. t.* products have been most successfully used to control insects such as the diamondback moth, *Plutella xylostella*, and the Colorado potato beetle which feed in exposed positions where they are easy to target with foliar applications. However, insects such as the cotton boll weevil, *Athonomus grandis*, which are susceptible to *B. t.* endotoxins in laboratory tests (Herrnstadt *et al.*, 1986), are still difficult to control with *B. t.* due to their habit of feeding internally within the plant. Even noncryptic insects can be difficult to control if they spend a portion of their life cycle feeding on the undersides of crop foliage as does the cabbage looper, *Trichoplusia ni*.

Residual activity   When applied outdoors to crop plant foliage, the activity of *B. t.* spore/crystal preparations and purified crystal preparations is significantly degraded in 24 to 48 hours (Pusztai *et al.*, 1991; Gelernter, 1990). Because of their proteinaceous composition, and because they are applied in an unprotected state, the δ-endotoxin crystals are a prime target for degradation by sunlight, primarily in the 300–380 nm region of the solar spectrum (Pusztai *et al.*, 1991), and heat (Dulmage, 1982). In addition, antagonistic interactions with other leaf colonizing bacteria probably play an important role in degradation of activity, as West (1984) has shown for *B. t.* and soil microorganisms. Finally, the δ-endotoxin dosage can be significantly diluted as the result of precipitation (dew, rain or irrigation), or as a result of plant growth and foliar expansion. Growers must therefore apply *B. t.* several times during the season to compensate for the product's brief residual activity. However, we must not lose sight of the fact that there is also often a need for several applications of chemical insecticides.

Use of transgenic microorganisms to improve foliar activity and delivery

To improve delivery and residual activity, researchers have expressed *B. t.* toxins in plants (see Ely, Chapter 5) and in plant colonizing microorganisms (Table 4.3) as summarized below.

The InCide® system

To target internally feeding insects, such as larvae of the European corn borer (ECB), more effectively, the US based Crop Genetics International Corporation has selected the endophytic (vascular system colonizing) bacterium *Clavibacter xyli* subspecies *cynodontis* as a δ-endotoxin delivery vehicle. In this system, a single copy of the *B. t. kurstaki* δ-endotoxin gene *crylA(c)* is inserted in the chromosome. Toxin expression levels of 0.025% of total plant cell protein were reported (Beach, 1990). When corn seed is inoculated with *C. xyli* recombinants that express the *crylA(c)* gene, the bacteria replicate and produce δ-endotoxin inside the growing plant. Colonization routinely occurs within 90% of the corn plants. ECB larvae, which may be difficult to target with foliar sprays of *B. t.* due to their habit of feeding inside the corn ear or stalk, are thus effectively targeted with an endotoxin to which they are susceptible.

Beginning in 1988, small plot tests have been conducted in several US locations with moderately good control of ECB larvae reported. It is likely that more effective control will be achieved once toxin expression can be increased above the current level (Beach, 1990).

Table 4.3. Improving foliar activity and delivery of δ-endotoxins: a summary of research projects

Endotoxin gene	Vector	Advantages	Author
crylA(c) (*kurstaki*)	Vascular system colonizer *Clavibacter xyli*	Delivery to internal feeders (European corn borer)	Beach, 1990
crylA(c) (*kurstaki*) and *crylIIA* (*san diego*)	Leaf colonizer *Pseudomonas fluorescens* (killed)	Delivery, foliar persistence, regulatory flexibility	Barnes and Cummings, 1987

Crop Genetics's product, known as InCide®, is one of the few reviewed in this chapter which has been evaluated in field tests. To receive regulatory approval to conduct field trials, Crop Genetics was required by the US Environmental Protection Agency to present detailed ecological information on the transgenic bacterium including: its ability to colonize other plants; pathogenicity to plants; dissemination to other plants; transmission in soil; insect transmission potential; survival in soil, water and plants; and horizontal transfer to other microorganisms. To date, no adverse environmental effects have been reported (Beach, 1990).

The CellCap® system

To increase the residual activity of *B. t.* δ-endotoxins, Mycogen Corporation has developed the CellCap® encapsulation system which increases the foliar persistence of insecticidal proteins by two- to three-fold (Gelernter, 1990). In this system a δ-endotoxin bearing plasmid is transferred to a non-pathogenic, leaf-colonizing isolate of *Pseudomonas fluorescens*. When the transgenic *P. fluorescens* cells are grown in fermenters, the endotoxin is produced as a crystal, and is expressed at levels of 10–20% total cell protein. The fermentation end-product is intact *P. fluorescens* cells, each bearing one or more δ-endotoxin crystals (Barnes and Cummings, 1987). This is in contrast to the production of *B. t.*, where the bacterial cells lyse at the completion of the replication cycle, releasing spores and unprotected endotoxin crystals into the fermentation medium.

Although originally intended as a living delivery system for *B. t.* endotoxins, regulatory restrictions on the release of living recombinants caused Barnes and Cummings (1987) to seek an intermediate solution. Thus, the final step in the CellCap® encapsulation process is a fixation step which kills the transgenic *P. fluorescens* cells while they are still in the fermenter. The result is a *P. fluorescens* cell made thicker and more rigid through cross-linking of cell wall proteins, which now serves as an effective protective capsule for the enclosed δ-endotoxin crystal. Additionally, because these transgenic microorganisms are dead, environmental concerns regarding the release and spread of living recombinant did not affect regulatory review for products based on this system. Thus, in 1991, two products based on the CellCap® system—MVP® bioinsecticide, which contains the lepidopteran active CryIA(c) protein, and M- Trak® bioinsecticide, which contains the coleopteran active CryIIIA protein— became the first genetically engineered biopesticides to be registered for use by the United States Environmental Protection Agency.

IMPROVING AQUATIC ACTIVITY AND DELIVERY

Limitations associated with conventional *B. thuringiensis* products

Products based on *B. t. israelensis* have been successfully used throughout the world for control of the aquatic larvae of disease-vectoring mosquitoes and blackflies (see Becker and Margalit, Chapter 7). However, two important limitations have restricted further adoption of this biological insecticide for control of aquatic pests.

Short residual activity Longevity of insect control is limited to 24 hours or less due to the fact that crystals and spores of *B. t. israelensis* rapidly settle, away from the air/water interface, where many mosquito and blackfly larvae feed (Ohana *et al.*, 1987).

Adsorption to organic particles *B. t. israelensis* crystals tend to adsorb rapidly to particles of mud or other organic materials. In this form, larvae do not ingest the crystals because they are now either too large or too unattractive as a larval food source (Ohana *et al.*, 1987).

Use of transgenic microorganisms to improve aquatic activity and delivery

To prolong the effectiveness of *B. t. israelensis* toxins, various groups have investigated the expression of selected *B. t. israelensis* toxin genes in microorganisms that inhabit the same aquatic environment as targeted mosquitoes or blackflies (see Table 4.4). In this way, dipteran active toxins would persist, and possibly increase in number at the appropriate water

Table 4.4. Improving aquatic activity and delivery of δ-endotoxins: a summary of research projects

Endotoxin gene	Vector	Advantages	Author
crylVB (*israelensis*)	Aquatic cyanobacterium *Agmenellum quadruplicatum*	Delivery to mosquito larvae; persistence	Angsuthansombat and Panyim, 1989
crylVB, cytA (*israelensis*)	Aquatic cyanobacterium *Synechocystis* PCC6803	Delivery to mosquito larvae; persistence	Chungjatupornchai, 1990
Gene mixture (*israelensis*)	Aquatic protozoan *Tetrahymena pyriformis*	Delivery to mosquito larvae, persistence; regulatory flexibility	Ben-Dov (personal communication)

depth thus increasing delivery of *B. t.* toxins to targeted aquatic insect species.

In their development of an aquatic transgenic microbe, Angsuthansombat and Panyim (1989) selected the cyanobacterium, *Agmenellum quadruplicatum*, which inhabits the same air/water interface as many mosquito species. When the mosquito active *crylVB* gene was introduced into the cyanobacterium using an *E. coli/A. quadruplicatum* bifunctional plasmid, immunologically reactive CrylVB protein was produced in low quantities and some toxicity to *Aedes aegypti* mosquito larvae was observed. Protein inclusions do not appear to have formed. Low toxin expression levels were probably due to a low plasmid copy number (of one per cell), as well as due to degradation of the toxin to a 42 kDa fragment within the cells.

In a similar vein, Chungjatupornchai (1990) integrated either the *crylVB* gene or the *cytA* gene into the chromosome of the cyanobacterium, *Synechocystis* PCC6803. In each instance, the expression of the toxin genes was under the control of the psbA promoter derived from the chloroplast of tobacco genomic DNA. Immunological analyses revealed that the gene expression product of *crylVB* accumulated to a level of 0.1–0.2% of total soluble protein of the cyanobacterium. Toxicity to *A. aegypti* larvae was observed, but once again toxin expression levels were too low to result in formation of a protein inclusion. Low expression levels are believed to be due to low activity of the psbA promoter, or to low gene dosage levels (only one copy of the gene was integrated into the *Synechocystis* chromosome). Similar analyses of the CytA gene expression product revealed barely detectable levels of immunologically reactive protein, and no activity for *A. aegypti* larvae. Low expression levels for the CytA toxin were attributed to the approximately five-fold decrease in the steady-state levels of mRNA for CytA, as compared to those levels for CrylVB.

To overcome the hurdles of low gene expression, regulatory restrictions for recombinant organisms and the apparent need for expression of multiple *B. t. israelensis* toxins (Wu and Chang, 1985), Ben-Dov (personal communication) has proposed an intermediate solution. By 'encapsulating' *B. t. israelensis* spore/crystal formulations in the aquatic protozoan, *Tetrahymena pyriformis*, these authors have demonstrated increased persistence of *B. t. israelensis* toxins under simulated field conditions. This concept is based on the observation that when conventional formulations of *B. t. israelensis* are applied to an aquatic system, *T. pyriformis* living at the air/water interface will ingest *B. t. israelensis* toxin crystals and store them in food vacuoles. When mosquito larvae, which normally prey on single-celled microbes, eat the *T. pyriformis* cells, they will also ingest *B. t. israelensis* toxins. The authors have shown that addition of *T. pyriformis* to a simulated aquatic system resulted in an increase in *B. t. israelensis* persistence from 24 hours to 71 hours. Thus, without the use of genetic engineering, a system has been developed that potentially can achieve the goal of increased *B. t. israelensis* persistence, without encountering regulatory and development hurdles.

IMPROVING SOIL ACTIVITY AND DELIVERY

Limitations associated with conventional *B. thuringiensis* products

B. t. based products are rarely used for control of soil dwelling insects for the following reasons:

1. Degradation of δ-endotoxin proteins is believed to occur rapidly in soils due to antagonistic interactions with soil microbes and the proteases they secrete (West, 1984).
2. The target insect, and therefore the target site, cannot be seen, so it is difficult to manipulate application conditions to achieve effective delivery.

Use of transgenic microorganisms to improve soil activity and delivery

To overcome these problems, several research groups have identified microorganisms which normally occur in close association with the roots of crop plants. By engineering these bacteria to produce the appropriate *B. t.* toxin(s), the issues of persistence and delivery described above could potentially be resolved (see Table 4.5).

Obukowicz *et al.* (1987) were among the first to successfully develop a δ-endotoxin producing transgenic microorganism. In this case, a corn root colonizing isolate of *P. fluorescens* was engineered to produce the lepidopteran active CryIA(b) toxin. To ensure stability of the gene, and to

Table 4.5. Improving soil activity and delivery of δ-endotoxins: a summary of research projects

Endotoxin gene	Vector	Advantages	Author
cryIA(b) (*kurstaki*)	Root colonizing *Pseudomonas fluorescens*	Delivery to root feeding insects; persistence	Obukowicz *et al.*, 1987
cryIVB (*morrisoni PG-14*)	Root colonizing *Pseudomonas fluorescens*	Delivery to crane fly larvae; persistence	Waalwijk *et al.*, 1991
cryIVD (*israelensis*)	Nodulating *Bradyrhizobium*	Delivery to *Rivellia* larvae in root nodules; persistence	Nambiar *et al.*, 1990

avoid horizontal transfer of the gene to other bacteria, the toxin gene was integrated, via homologous recombination, into the *Pseudomonas* chromosome. Production of the δ-endotoxin was confirmed via Western blot analysis and by toxicity against larvae of the tobacco hornworm, *Manduca sexta*. Relatively high levels of the 134 kDa protein were produced at concentrations of 0.5–1.0% of total soluble protein.

Unfortunately, this project was discontinued, and the recombinants were not further developed or evaluated. However, Waalwijk *et al.* (1991) have expanded the concept of improved soil delivery of *B. t.* endotoxins, again using a root colonizing strain of *P. fluorescens*, but this time as a delivery system for the dipteran active *crylVB* gene. The target insects in this case are soil dwelling larvae of the crane fly, *Tipula oleracea* (otherwise known as leatherjackets) which cause economic damage by feeding on the roots and leaves of pasture grasses. Successful expression of the 128 kDa toxin in the transformed *P. fluorescens* cells was confirmed through immunologic analysis, with breakdown products of 90 000 and 65 000 kDa also observed. Bioassays conducted on crane fly larvae and on mosquito larvae further confirmed toxin expression, although relatively low insect mortality levels of only 10–30%, were observed, most likely due to low concentrations of δ-endotoxin. In addition to successfully achieving expression of the dipteran active toxin, Waalwijk *et al.* (1991) also sought to decrease the mobility of the inserted gene through integration into the *P. fluorescens* chromosome via homologous recombination. The development and use of suicide vectors in the recombination process theoretically decreased the chances of gene mobilization via bacterial transposases, although stability of the transformed microbes has not yet been evaluated.

In an innovative approach towards improving delivery to soil insects, Nambiar *et al.* (1990) exploited the symbiotic relationship between pigeon pea *Cajanus indica*, a leguminous crop, and the nitrogen-fixing bacterium, *Bradyrhizobium*. Pigeon peas are an important source of protein in the tropics and, due to nodulation by *Bradyrhizobium* spp., grow well on poor soils where nitrogen is limited. However, larvae of the dipteran pest, *Rivellia angulata*, can significantly reduce yields by feeding on the root nodule contents of pigeon peas. These authors were able to reduce *Rivellia* infestations on pigeon peas through development of transgenic *Bradyrhizobium* cells that expressed the dipteran active *crylVD* gene. Transfer of the *crylVD* plasmid was achieved through conjugation of an *E. coli* strain containing the *crylVD* gene and *Bradyrhizobium* IC3554. Toxin expression was confirmed via reaction with a polyclonal antibody prepared against the parasporal inclusion of *B. t. israelensis*, although toxin expression levels were not reported. The immunoreactive species expressed in the transformed *Bradyrhizobium* had an apparent molecular mass of 45 kDa, versus an expected mass of 72.4 kDa. Despite this discrepancy, results of greenhouse tests revealed that pigeon pea plants grown in soil inoculated with the *cryIVD* transgenic *Bradyrhizobium* showed a 40% reduction in *R. angulata* infestation levels as compared to plants grown with non-engineered *Bradyrhizobium* spp. This is one of few studies where the authors were able to carry their experiment beyond the laboratory and into the greenhouse, where insecticidal activity and competitive performance of the transgenic microorganism could be assessed under more realistic conditions.

ISSUES IN THE DEVELOPMENT OF TRANSGENIC MICROORGANISMS

The research projects described in this review, and others like them, should provide the foundation for new and useful bioinsecticide products that will be introduced within the next decade. Yet there are several issues which deserve attention as these projects move from the laboratory to the field and, ultimately, to the marketplace.

Regulatory issues

The European Economic Community, the USA and Japan are currently in the process of developing data requirements for registration of genetically engineered biopesticides. Of key concern is the development of monitoring techniques that will track the ecological interactions of living engineered organisms (Tiedje *et al.*, 1989; Kluepfel *et al.*, 1991). These include:

—dispersal,
—survival and reproduction,
—interactions with other organisms (non-target effects),
—genetic stability.

At the time of writing, no official guidelines have been published and no pesticide products based on living recombinants have been registered. In fact, regulatory permission to conduct small plot outdoor field tests has been granted for only a few recombinant organisms to date.

Technical issues

At first glance it may appear that lack of experience in monitoring the behaviour of engineered organisms and lack of a regulatory structure for approving their use presents significant hurdles to the commercialization of transgenic microbes. However, it is probably technical problems which are the greatest impediments to product commercialization. Of the transgenic microbes described above, few have given economic δ-endotoxin yields and the proteins produced often degrade intracellularly; field efficacy has been demonstrated only for Crop Genetic's InCide® and Mycogen's CellCap® systems. For living recombinants, the question of fitness remains unresolved. Should a transgenic microbe be designed to outcompete indigenous microorganisms as a means of prolonging their efficacy? Or should the recombinant be purposely 'weak' so that it cannot effectively compete and upset the ecological balance of the environment in which it is placed? Another significant concern, that insect resistance could develop to δ-endotoxin producing transgenic microbes, and that management strategies are needed to avoid this, also remains unresolved (see Marrone and MacIntosh, Chapter 10).

Setting realistic expectations

While these technical hurdles are not insurmountable, when they are combined with the absence of regulatory guidelines for product registration, it seems likely that products based on living transgenic microbes will not be introduced before 1997. As we work together to overcome these hurdles, a study of several of the 'intermediate' strategies described in this chapter may help accelerate our learning curve. These intermediate systems, which utilize genetically altered organisms that are not necessarily living recombinants, have already met with some success. The transconjugant system described by Carlton *et al.* (1990), and the killed recombinant cell system known as CellCap® (Barnes and Cummings, 1987), have already been utilized to produce several registered biopesticides. These intermediate systems have also been useful in jumpstarting the development of regulatory guidelines for genetically altered and recombinant microbes. And through study of the 'intermediate' system proposed by Ben-Dov (personal communication), where naturally occurring protozoans are used to deliver *B. t.* toxins to mosquitoes, we can perhaps begin to gain an understanding of the behaviour of transgenic microbes in the environment.

REFERENCES

Adang, M. J., Staver, M. J., Rocheleau, T. A., Leighton, J., Barker, R. F. and Thompson, D. V. (1985). 'Characterized full length and truncated plasmid clones of the crystal protein of *Bacillus thuringiensis* subsp. *kurstaki* HD-73 and their toxicity to *Manduca sexta*'. *Gene*, **36**, 289–300.

Angsuthansombat, C. and Panyim, S. (1989). 'Biosynthesis of 130-kilodalton mosquito larvicide in the cyanobacterium *Agmenellum quadruplicatum* PR-6'. *Appl. Env. Microbiol.*, **55**, 2428–2430.

Barnes, A. C. and Cummings, S. E. (1987). US Patent No. 4695455.

Beach, R. M. (1990). Application for an experimental use permit to ship and use a pesticide for experimental purposes only. Permit number 58788-EUP-4 for InCide™ 586. Crop Genetics International, Hanover, Maryland.

Carlton, B. C., Gawron-Burke, C. and Johnson, T. B. (1990). 'Exploiting the genetic diversity of Bacillus thuringiensis for the creation of new bioinsecticides'. In Proceedings, Fifth International Colloquium on Invertebrate Pathology and Microbial Control, pp. 18–22, August 20–24, 1990, Society for Invertebrate Pathology, Adelaide, Australia.

Chungjatupornchai, W. (1990). 'Expression of the mosquitocidal protein genes of Bacillus thuringiensis subsp. israelensis and the herbicide resistance gene bar in Synechocystis PCC6803'. Curr. Microbiol., 21, 283–288.

Crickmore, N., Nicholls, C., Earp, D. J., Hodgman, T. C. and Ellar, D. J. (1990). 'The construction of Bacillus thuringiensis strains expressing novel entomocidal delta endotoxin combinations'. Biochem. J., 270, 133–136.

Donavan, W. P., Dankocsik, C. C. and Gilbert, M. P. (1988). 'Molecular characterization of a gene encoding a 72 kilodalton mosquito toxic crystal protein from Bacillus thuringiensis subsp. israelensis'. J. Bacteriol., 170, 4732–4738.

Dulmage, H. T. (1982). 'Distribution of Bacillus thuringiensis in nature'. In E. Kurstak, ed., Microbial and Viral Pesticides (Ed. E. Kurstak), pp. 209–237, Marcel Dekker, New York.

Edwards, D. L., Payne, J. and Soares, G. G. (1990). US Patent no. 4,948,734.

Ferro, D. N. and Lyon, S. M. (1991). 'Colorado potato beetle (Coleoptera: Chrysomelidae) larval mortality: operative effects of Bacillus thuringiensis subsp. san diego'. J. Econ. Entomol., 84, 806–809.

Gelernter, W. D. (1990). 'Targeting insecticide-resistant markets: new developments in microbial based products'. Managing Resistance to Agrochemicals: From Fundamental Research to Practical Strategies. (Eds. M. B. Green, W. K. Moberg and H. LeBaron). Series 4221, pp. 105–117, American Chemical Society, Washington, D.C.

Gelernter, W. D. (1992). Application of biotechnology for improvement of Bacillus thuringiensis based products and their use for control of Lepidopteran pests in the Caribbean'. Florida Entomologist (in press).

Herrnstadt, C., Soares, G. G., Wilcox, E. R. and Edwards, D. L. (1986). 'A new strain of Bacillus thuringiensis with activity against coleopteran insects'. Bio/Technology, 4, 305–308.

Herrnstadt, C., Gilroy, T. E., Sobieski, D. A., Bennet, B. D. and Gaertner, F. H. (1987). 'Nucloetide sequence and deduced amino acid sequence of a coleopteran active delta endotoxin gene from Bacillus thuringiensis subsp. san diego'. Gene, 57, 37–46.

Honée, G. T. and Visser, B. (1988). 'Nucleotide sequence of crystal protein gene isolated from B. thuringiensis subspecies entomocidus 60.5 coding for a toxin highly active against Spodoptera species'. Nucl. Acids Res., 16, 6240.

Kluepfel, D. A., Kline, E. L., Skipper, H. D., Hughes, T. A., Gooden, D. T., Drahos, D. J., Barry, G. F., Hemming, B. C. and Brandt, E. J. (1991). 'The release and tracking of genetically engineered bacteria in the environment'. Phytopathology, 81, 348–352.

Martens, J. W. M., Honée, G., Zuidema, D., van Lent, J. W. M., Visser, B. and Vlak, J. (1990). 'Insecticidal activity of a bacterial crystal protein expressed by a recombinant baculovirus in insect cells'. Appl. Env. Microbiol., 56, 2764–2770.

McKemy, C. (1990). 'Bt: keeps bugs from bouncing back'. The Grower, August, pp. 22–24.

Merryweather, A. T., Weyer, U., Harris, M. P. G., Hirst, M., Booth, T. and Possee, R. D. (1990). 'Construction of genetically engineered baculovirus insecticides containing the *Bacillus thuringiensis* subsp. *kurstaki* HD-73 delta endotoxin'. *J. Gen. Virol.*, **71**, 1535–1544.

Nambiar, P. T. C., Ma, S. W. and Iyer, V. N. (1990). 'Limiting an insect infestation of nitrogen-fixing root nodules of the pigeon pea (*Cajanus cajan*) by engineering the expression of an entomocidal gene in its root nodules'. *Appl. Env. Microbiol.*, **56**, 2866–2869.

Obukowicz, M. G., Perlak, F. J., Bolten, S. L., Kusano-Kretzmer, K., Mayer, E. J. and Watrud, L. S. (1987). 'IS50L as a non-self-transposable vector used to integrate the *Bacillus thuringiensis* delta-endotoxin gene into the chromosome of root-colonizing pseudomonads'. *Gene*, **51**, 91–96.

Ohana, B., Margalit, J. and Barak, Z. (1987). 'Fate of *Bacillus thuringiensis* subsp. *israelensis* under simulated field conditions'. *Appl. Env Microbiol.*, **53**, 828–831.

Pusztai, M., Fast, P., Gringorten, L., Kaplan, H., Lessard, T. and Carey, R. (1991). 'The mechanism of sunlight mediated inactivation of *Bacillus thuringiensis* crystals'. *Biochem. J.*, **273**, 43–47.

Rowe, G. E. and Margaritis, A. (1987). 'Bioprocess developments in the production of bioinsecticides by *Bacillus thuringiensis*'. *CRC Critical Reviews in Biotechnology*, **6**, 87–127.

Schnepf, H. E. and Whiteley, H. R. (1981). 'Cloning and expression of the *Bacillus thuringiensis* protein gene in *Escherichia coli*'. *Proc. Natl Acad. Sci. USA.*, **78**, 2893–2897.

Schnepf, H. E., Wong, C. and Whiteley, H. R. (1985). 'The amino acid sequence of a crystal protein from *Bacillus thuringiensis* deduced from the DNA base sequence'. *J. Biol. Chem.*, **260**, 6264–6272.

Sekar, V. and Carlton, B. C. (1985). 'Molecular cloning of the delta endotoxin gene of *Bacillus thuringiensis* var. *israelensis*'. *Gene*, **33**, 151–158.

Thorne, L., Garduno, F., Thompson, T., Decker, D., Zounes, M., Wild, M., Walfield, A. M. and Pollock, T. (1986). 'Structural similarity between the lepidoptera and diptera specific insecticidal endotoxin genes of *Bacillus thuringiensis* subsp. *kurstaki* and *israelensis*'. *J. Bacteriol.*, **166**, 801–811.

Tiedje, J. M., Colwell, R. K., Grossman, Y. L., Hodson, R. E., Lenski, R. E., Mack, R. N. and Regal, P. J. (1989). 'The planned introduction of genetically engineered organisms: ecological considerations and recommendations'. *Ecology*, **70**, 298–315.

Waalwijk, C., Dullemans, A., van Workum, M. E. S. and Visser, B. (1985). 'Molecular cloning and the nucleotide sequence of the Mr 28,000 crystal protein gene of *Bacillus thuringiensis* subsp. *israelensis*'. *Nucl. Acids Res.*, **13**, 8206–8217.

Waalwijk, C., Dullemans, A. and Maat, C. (1991). 'Construction of a bioinsecticidal rhizosphere isolate of *Pseudomonas fluorescens*'. *FEMS Microbiol. Lett.*, **77**, 257–264.

Wabiko, H., Raymond, K. C. and Bulla, L. A. (1986). '*Bacillus thuringiensis* entomocidal protoxin gene sequence and gene product analysis'. *DNA*, **5**, 305–314.

Ward, S. and Ellar, D. (1987). 'Nucleotide sequence of a *Bacillus thuringiensis* var. *israelensis* gene encoding a 130 kDa delta endotoxin'. *Nucl. Acids Res.*, **15**, 7195.

West, A. W. (1984). 'Fate of the insecticidal proteinaceous parasporal crystal of *Bacillus thuringiensis* in soil'. *Soil Biol. Biochem.*, **16**, 357–360.

Wu, D. and Chang, F. N. (1985). 'Synergism in mosquitocidal activity of 26 and 65 kDa proteins from *Bacillus thuringiensis* subsp. *israelenis* crystal'. *FEBS Lett.*, **190**, 232–236.

Zehnder, G. W. and Gelernter, W. D. (1989). 'Activity of the M-One formulation of a new strain of *Bacillus thuringiensis* against the Colorado potato beetle: relationship between susceptibility and insect life stage'. *J. Econ. Entomol.*, **82**, 756–761.

5 The Engineering of Plants to Express *Bacillus thuringiensis* δ-Endotoxins

SUSAN ELY

*ICI Seeds, Jealott's Hill Research Station, Bracknell, Berks RG12 6EY, UK
and Plant Science Center, Cornell University, Ithaca, NY, USA*

INTRODUCTION

Bacillus thuringiensis (*B. t.*) has been safely used for almost three decades as a microbial insecticide. Recent environmental concerns about the extensive use of chemical pesticides has stimulated an enhanced interest in biological control agents (BCAs). Although conventionally formulated *B. t.* insecticides are increasingly likely to comprise a component of integrated pest management systems, there are certain limitations to this form of plant protection. A major constraint is the inability of conventional BCA applications to protect inaccessible plant tissue. A number of economically significant insect pests feed on internal tissue and are thus sequestered from surface pesticide applications. On maize, for example, European corn borer, *Ostrinia nubilalis*, larvae migrate to the relative protection of the whorls and then bore into the stalk. As shown in Figure 5.1, larvae can completely destroy the structural integrity of the maize stalk; this will eventually lead to plant lodging and yield loss. Many agronomically important crop plants are attacked by pests which feed internally; other examples include stalk borers such as *Chilo* spp. and *Busseola fusca* which attack maize, rice and sorghum (Hill, 1975, 1987), and cabbage sand weevil, *Ceutorhynchus assimilis* which feeds on developing ovules within the pods of oilseed rape (Gatehouse *et al.*, 1991).

Recent developments in plant transformation technology have allowed the stable introduction of foreign genes into many important crop species, including monocotyledons (Gasser and Fraley, 1989). To date, stable transformation has been reported for over 50 plant species (Ellis, in press). Table 5.1 lists those plants which have been stably transformed and

Bacillus thuringiensis, An Environmental Biopesticide: Theory and Practice. Edited by P. F. Entwistle, J. S. Cory, M. J. Bailey and S. Higgs
© 1993 John Wiley & Sons Ltd

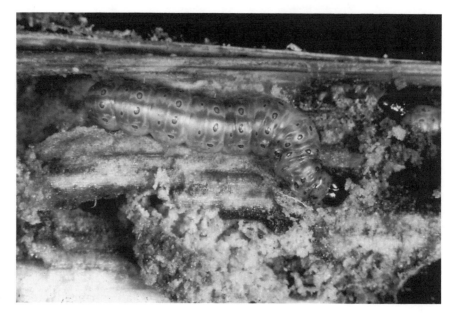

Figure 5.1. Cross-section showing extensive maize stalk damage by a late instar *Ostrinia nubilalis* (European corn borer) larva

which are attacked by pests with known susceptibility to control by *B. t.* A proportion of insect species listed in Table 5.1 have developed resistance to one or more widely used chemical pesticides (Hill, 1987; Rigby, 1991). There is clearly considerable potential for the genetic engineering of plants to express introduced *B. t.* δ-endotoxin genes. The more obvious advantages of engineered insect resistance in plants include (a) the protection of target tissues even when internal, (b) weather-independent protection, and (c) limited environmental distribution since the BCA is confined to plant tissue. Other advantages may include greater stability of δ-endotoxin proteins, and lack of dependence on application timing for treatment success. Finally, protection is possible for the entire season and only those insects which feed on the engineered crop will be exposed. Although often cited as advantages (e.g. Hilder *et al.*, 1990), these latter factors could also contribute to the development of pest resistance to *B. t.* δ-endotoxins, especially in the case of monophagous insects. Responsible and informed deployment of transgenic plants expressing δ-endotoxins should, however, result in the extended use of this natural insecticide in integrated pest management systems, providing that the principles of pest resistance management are also employed (see Roush and Tabashnik, 1990; Marrone and MacIntosh, Chapter 10).

Table 5.1. Plant species which have been stably transformed and corresponding insect pests susceptible to control by *B. t.*

DICOTYLEDONS	
Crop groups	Susceptible insect pests

Roots and tubers

Potato	*Solanum tuberosum*	Loopers (L) (*Autographa*
Sugarbeet	*Beta vulgaris*	*californica, Trichoplusia ni*);
Carrot	*Daucus carota*	omnivorous leafroller (L)
		(*Platynota stultana*); horn-
		worms (L) (*Manduca* spp.)

Crucifiers

Kale and cauliflower	*Brassica juncea*	Cabbageworms (L) (*Pieris*
Celery	*Apium graveolens*	spp.); diamondback moths (L)
Lettuce	*Lactuca sativa*	(*Plutella xylostella*); green
		cloverworm (L) (*Plathypena*
		scabra); webworms (L)
		(*Loxostege sticticalis*);
		saltmarsh caterpillar (L)
		(*Estigmene acrea*); armyworms
		(L) (*Spodoptera* spp.);
		cutworms (L) (*Agrotis* spp.);
		cabbage moths (L) (*Plusia*
		spp.); Colorado potato beetle
		(C) (*Leptinotarsa decemlineata*)

Fruiting vegetables

Tomato	*Lycopersicon esculentum*	Loopers (L); armyworms (L);
Eggplant	*Solanum melongena*	saltmarsh caterpillar (L);
Pepper	*Capsicum annuum*	hornworms (L); fruitworms (L)
		(*Heliothis* spp.); pinworms (L)
		(*Keiferia lycopersicella*);
		Colorado potato beetle (C)

Cucurbits

Melon	*Cucumis melo*	Loopers (L); armyworms (L);
Cucumber	*C. sativus*	melonworm (L) (*Diaphania*
		hyalinata); cucumber beetles
		(C)[a] (*Diabrotica*
		undecimpunctata
		undecimpunctata,
		D. balteata);

Continued

Table 5.1. (*Continued*)

DICOTYLEDONS		
Crop groups		Susceptible insect pests
		Black vine weevil (C) (*Otiorhynchus sulcatus*); melon beetle (C) (*Epilachna varivestis*)
Legume vegetables		
Soybean	*Glycine canescens*	Loopers (L); green cloverworm
Pea	*Pisum sativum*	(L); velvetbean caterpillar (L)
Moth bean	*Vigna aconitifolia*	(*Anticarsia gemmatalis*); armyworms (L); bollworms (L) (*Heliothis* spp.)
Stone and pome (top) fruits		
Apple	*Malus pumila*	Redhumped caterpillar (L)
Plum	*Prunus domestica*	(*Schizura concinna*); tent caterpillar (L) (*Hylesia nigricans*); omnivorous leafroller (L)
Tree nuts and citrus		
Walnut	*Juglans regia*	Fall webworm (L) (*Hyphantria*
Orange	*Citrus sinensis*	*cunea*); cankerworms (L)
	C. jambhiri	(*Alsophila pometaria, Paleacrita vernata*); gypsy moth (L) (*Lymantria dispar*)
Miscellaneous		
Poplar	*Populus* spp.	Fruit tree leafrollers (L) (*Ascia monuste, Archips argyrospilus*); orangedog (L) (*Papilio cresphontes*); apple leaf skeletonizer (L) (*Simaethis pariana*); walnut caterpillar (L) (*Datana integerrima*)
Small (soft) fruits		
Currant	*Ribes nigrum*	Loopers (L); saltmarsh
Strawberry	*Fragaria ananassa*	caterpillar (L); omnivorous
Raspberry and black raspberry	*Rubus* spp.	leafroller (L); grape leafroller (L) (*Desmia funeralis*); vine
Cranberry	*Vaccinium macrocarpon*	moth (grape berry moth) (L)
Grapevine	*Vitis vinifera*	(*Clysiana ambiguella*)

Table 5.1. (*Continued*)

DICOTYLEDONS		
Crop groups		Susceptible insect pests

Tropical fruit

Papaya	*Carica papaya*	Hornworms (L); leafrollers (L); loopers (L)

Ornamentals

Carnation	*Dianthus* spp.	Loopers (L); tobacco budworm
Morning glory	*Convolvulus arvensis*	(L) (*H. virescens*);
Periwinkle	*Catharanthus roseus*	diamondback moth (L);
Petunia	*Petunia hybrida*	armyworms (L); hornworm (L);
Tobacco	*Nicotiana* spp.	omnivorous leafroller (L); Io Moth (L) (*Automeris io*); oleander moth (L)

Other crops

Cotton	*Gossypium hirsutum*	Budworms and bollworms (L) (*Heliothis* spp., *Pectinophora gossypiella*); loopers (L); saltmarsh caterpillars (L); armyworms (L); cotton worm (L) (*Anomis texana*); boll weevil (C) (*Anthonomus grandis*)
Tobacco	*Nicotiana tabacum*	Bollworms and budworms (L) (*Heliothis* spp.); tobacco hornworm (L) (*Manduca sexta*); semiloopers (L) (*Plusia* spp.); cutworms and leafworms (L) (*Spodoptera* spp.)
Sunflower	*Helianthus anuus*	Old World bollworm (L) (*Heliothis armigera*); cutworms and leafworms (L) (*Spodoptera* spp.)
Alfalfa	*Medicago sativa*	Armyworms (L) (*Spodoptera* spp.)
Rape (canola)	*Brassica napus*	Cutworms (L) (*Agrotis* spp.); armyworms and leafworms (L) (*Spodoptera* spp.); bollworms (L) (*Heliothis* spp. (L));
Mint	*Mentha citrata*	Armyworms (L)

Continued

Table 5.1. (*Continued*)

MONOCOTYLEDONS		
Crop		Susceptible insect pests
Maize	*Zea mays*	Borers (L) (*Ostrinia nubilalis, Busseola fusca, Chilo partellus*); armyworm (L) (*Spodoptera frugiperda*); corn earworm (L) (*Helicoverpa zea*); Cutworms (L) (*Agrotis* spp.) rootworms (C)[a] (*Diabrotica vergifera vergifera, Diabrotica undecimpunctata howardi*)
Rice	*Oryza sativa*	Stalkborer (L) (*Chilo* spp.); armyworms (L) (*Spodoptera* spp.)
Asparagus	*Asparagus officinalis*	Armyworms (L); asparagus beetle (C) (*Crioceris asparagi*)

[a] = Weak susceptibility (Herrnstadt *et al.*, 1986); (L) = Lepidoptera, (C) = Coleoptera. The susceptible insect pest list includes major and minor pests from North America, Latin America, Europe, South Africa and the Far East. Latin names are usually only listed once. Neither list is comprehensive and not all insects attack all crops listed in a given group. References: transformed crops (Ellis, in press; Gasser and Fraley, 1989; McCown *et al.*, 1991); susceptible insects (Abbott Laboratories Product Label Guide, 1990 and Dipel™ Technical Manual, 1978; CIMMYT, 1989; Delannay *et al.*, 1989; Herrnstadt *et al.*, 1986; Hill, 1975, 1987; Huger *et al.*, 1986; Krysan, 1986; Perlak *et al.*, 1990).

EXPRESSION OF δ-ENDOTOXINS IN DICOTYLEDONS

δ-Endotoxins in tobacco

Tobacco (*Nicotiana* spp.) is very amenable to *Agrobacterium*-mediated transformation, and was the first plant species to be stably transformed with a δ-endotoxin gene (Vaeck *et al.*, 1987; Barton *et al.*, 1987).

Vaeck *et al.* (1987) used the lepidopteran-active *bt2* δ-endotoxin (Höfte *et al.*, 1986), also designated *crylA(b)* (Höfte and Whiteley, 1989). In these experiments, the *crylA(b)* gene was introduced under the control of the constitutive *Agrobacterium* 2' mannopine synthase promoter, either as a full-length protoxin gene, as a truncated gene, or as a truncated gene in 5' translational fusion with the selectable marker gene *neo*, also designated *nptII*, encoding neomycin phosphotransferase. The expression of the selectable marker gene either in *cis*, or in translational fusion, conferred resistance to the antibiotics neomycin and kanamycin. Kanamycin-resistant plants from the initial transformed population (T_1) were evaluated directly at the 6–8 leaf stage for insecticidal phenotype using a tobacco hornworm, *Manduca sexta*, bioassay. Results indicated that although there was a large variation in insecticidal efficacy, 45–66% of T_1 plants resistant to ≥ 500 μg kanamycin/ml and carrying either of the truncated *crylA(b)* gene constructs caused bioassay mortality levels of $\geq 50\%$. There was, however, no significant larvicidal activity in the kanamycin-resistant T_1 plant population carrying the full-length *crylA(b)* gene.

Barton *et al.* (1987) reported tobacco transformation experiments in which a truncated *crylA(a)* gene (Schnepf *et al.*, 1985) under the control of the constitutive cauliflower mosaic virus (CaMV) 35S promoter was shown to confer a wide range of insect resistance levels to T_1 plants. As with the work described by Vaeck *et al.* (1987), insect-resistant plants could not be identified in the T_1 population transformed with a full-length *crylA(a)* gene.

In a similar experiment, a truncated *crylA(c)* gene encoding the N-terminal 716 amino acids of the δ-endotoxin was cloned between the CaMV 35S promoter (Odell *et al.*, 1985) and the *Agrobacterium* nopaline synthase gene transcriptional terminator (Depicker *et al.*, 1982) in the binary vector Bin19 (Bevan, 1984) (Figure 5.2). This construct was then introduced into *Agrobacterium tumefaciens* LBA4404 in a triparental mating with the mobilizing plasmid pRK2013. Tobacco transformation by leaf disc infection (Horsch *et al.*, 1985) resulted in a T_1 population of 48 plants which rooted successfully on kanamycin-containing medium. These plants were evaluated at the 7–9 leaf stage for resistance to damage by tobacco budworm, *Heliothis virescens*, assessing leaf damage 7 days after infestation by late first instar larvae. Damage to T_1 plants is shown

Figure 5.2. Tobacco transformation construct: The CaMV 35S promoter is in transcriptional fusion with a truncated *crylA(c)* gene from *B. t.* strain HD-73 encoding the N-terminal 716 amino acids of the δ-endotoxin. The polyadenylation signal for *in planta* transcriptional termination is provided by the *Agrobacterium* nopaline synthase gene (NOS) terminator (shaded areas). Direction of transcription is shown by the arrows. The plant-selectable marker gene is neomycin phosphotransferase (*npt*II) derived from *Escherichia coli* transposon Tn5. The *Agrobacterium* left and right border sequences are designated LB and RB respectively and are positioned as indicated

in Figure 5.3; mean damage was 38.7% leaf area eaten, with a standard deviation of 18.2%. Mean damage of untransformed control plants was 54.1% leaf area eaten, with a standard deviation of 14.3%. Assuming the damage observed in control plants is normally distributed, there is only a 5% chance of observing a control plant with less than 30% leaf area eaten. In the T_1 population, however, there are 22 out of 48 plants (46% of the T_1 plants) with $\leqslant 30\%$ leaf area eaten, indicating that some of these plants are likely to be transformed and to express δ-endotoxin at levels which afford some plant protection.

The experiment described above and those reported by Vaeck *et al.* (1987), and Barton *et al.* (1987), indicate that insect bioassays can be used to assess δ-endotoxin expression in T_1 populations containing some proportion of hemizygous transgenic plants. Even in the hemizygous state, some of these initial transformants exhibit an insect-resistant (or tolerant) phenotype, although levels of insect resistance vary greatly from plant to plant. Not all apparently kanamycin-resistant plants in a T_1 population will be transformed, especially if explants are only required to root once on kanamycin (as in the experiment described above). The histogram in Figure 5.3 suggests that there are probably two populations represented among the kanamycin-selected T_1 plants, only one of which is in fact transformed. Once insect-resistant plants are identified,

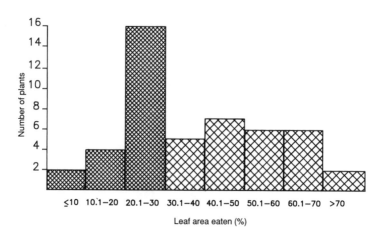

Figure 5.3. Assessment of leaf damage by *Heliothis virescens*, tobacco budworm on T_1 *Nicotiana tabacum* plants. The construct shown in Figure 5.2 was introduced into *Nicotiana tabacum* var *samsun* by leaf disc infection (Horsch *et al.*, 1985). Plants rooting in kanamycin-containing medium were clonally propagated and tested for resistance to attack by *H. virescens*. Bioassays were conducted on T_1 plants at the 7–9 leaf stage, at 25°C under continuous lighting. Plants were infested with eight late first instar larvae and assessed after 7 days. Control plant values for percentage leaf area eaten were: 72, 34, 62, 53, 47, 70, 44, 53, 71, 35

molecular characterization of transformants is undertaken to confirm the integrity of the expression cassette, to estimate introduced gene copy number and to determine δ-endotoxin expression levels as described below.

δ-Endotoxins in tomato, potato and cotton

Agrobacterium-mediated transformation of tomato was reported by Fischoff *et al.* (1987) and indicated that T$_1$ populations transformed with the *B. t.* strain HD-1, *crylA(b)* gene under the control of the CaMV 35S promoter included plants resistant to attack by *M. sexta*, with a subset of plants which showed some resistance to *H. virescens* and *Helicoverpa (Heliothis) zea* (variously known as tomato fruitworm, cotton bollworm, or corn earworm). Selected homozygous T$_2$ and T$_3$ progeny plant lines were then field tested in two subsequent seasons (Delannay *et al.*, 1989) as were homozygous T$_2$ plants from another tomato line transformed with the *crylA(b)* gene under the control of the mannopine synthase promoter. These trials indicated that transgenic tomato plants expressing the *crylA(b)* gene are resistant to damage by lepidopteran pests under field conditions. Complete control was observed for the sensitive pest, *M. sexta*, with some control of the less sensitive but economically significant pests *H. zea* and tomato pinworm, *Keiferia lycopersicella*.

The truncated *crylA(b)* gene constructs used by Vaeck *et al.* (1987) to produce transgenic tobacco, were also used to transform tomato and potato (Vaeck *et al.*, 1989); these experiments resulted in insect-resistant plants with measurable levels of δ-endotoxin expression.

Finally, Perlak *et al.* (1990) have reported the *Agrobacterium*-mediated transformation of cotton with truncated forms of the *crylA(b)* and *crylA(c)* genes. In contrast to the results obtained with tomato (Fischoff *et al.*, 1987; Delannay *et al.*, 1989), expression of the truncated, but otherwise wild-type *crylA(b)* was insufficient to control even the most sensitive pests in transgenic cotton. Changes to both the promoter and the δ-endotoxin structural genes were used to obtain expression levels sufficient for insect control. Structural gene modifications, which did not change the encoded amino acid sequence, were extensive and produced an altered gene retaining only 20% DNA sequence homology to the wild-type gene. These changes, which were intended to alter codon preference and eliminate possible regulatory sequences, resulted in a marked effect on δ-endotoxin expression. As judged by Western blot analysis, wild-type expression was ≤ 0.002% δ-endotoxin per total soluble protein, whereas the modified gene was expressed at 0.05–0.1% in the best plants. In leaf bioassays, transgenic cotton carrying either the *crylA(c)* or *crylA(b)* genes under the control of an enhanced (Kay *et al.*, 1987) CaMV 35S promoter was completely resistant to larvae of the relatively sensitive pest cabbage looper,

Trichoplusia ni; these plants also showed some resistance to the less sensitive pest, beet armyworm, *Spodoptera exigua*. In whole plant bioassays using artificial infestation of *H. zea* under conditions of high insect pressure, the best results obtained with both truncated gene types was 70–87% protection of squares and bolls.

EXPRESSION OF δ-ENDOTOXINS IN MONOCOTYLEDONS

Polyethylene glycol-mediated uptake of DNA containing one of two truncated, presumably *cryI*-like, δ-endotoxin genes under the control of the CaMV 35S promoter and fused to the marker gene β-glucuronidase (GUS), was used to transform rice cell suspension cultures (Hong *et al.*, 1989). Transformed calli were identified on the basis of high GUS activity, and regenerated plants were shown by DNA hybridization analysis to contain the introduced δ-endotoxin gene. To date, this constitutes the sole published report on transgenic monocotyledons containing δ-endotoxin genes. As shown in Table 5.1, however, a number of monocotyledonous plant species have been stably transformed. Transformation technology in other major crops such as maize is currently the focus of intensive research efforts at numerous institutes worldwide; in view of this activity, full reports of δ-endotoxin gene expression in monocotyledons are likely to be imminent.

QUANTIFICATION OF δ-ENDOTOXIN EXPRESSION *IN PLANTA*

Fortunately, the initial experiments with transgenic dicotyledons engineered for insect resistance afforded some degree of plant protection even though unmodified, truncated δ-endotoxin genes were being evaluated for expression in hemizygous plants. As indicated in Table 5.2, insecticidal efficacy was generally evident even at extremely low δ-endotoxin expression levels, and insect control was often observed in the absence of a detectable δ-endotoxin signal in immunological assays (Perlak *et al.*, 1991). Purified, cloned δ-endotoxin gene products produce mortality in diet incorporation bioassays at levels of 0.004–0.1% w/v (S. Ely, unpublished results; MacIntosh *et al.*, 1990). Expression of δ-endotoxin in transgenic dicotyledons has largely been within this range (Table 5.2) even in the absence of engineering to enhance expression, or the production of plants homozygous for the introduced gene. Generally, however, effective field levels of plant protection will require modification of the δ-endotoxin gene and associated control sequences.

Table 5.2. Summary of δ-endotoxin expression levels in transgenic dicotyledons

Reference and plant	Gene—structure	Detection method	% δ-endotoxin per total soluble protein	Insecticidal efficacy
(1) TB	*crylA(b)* full-length	E	$0.0001–0.0005^a$	−
	crylA(b) truncated	E	$0.003–0.012^a$	+
	crylA(b) :: *neo* fusion 1	E	$0.003–0.013^a$	+
	crylA(b) :: *neo* fusion 2	E	$0.004–0.019^a$	+
(2) TB	*crylA(a)* full + Kozak^c	W	—	−
	crylA(a) trunc. + Kozak^c	W	0.00125^b	+
(3) TB	*crylA(c)* truncated	W	[< 0.014]	+
(4) TO	*crylA(b)* truncated$_{(4B)}$	NS	$0.0001_{(4A)}$	+
(5) CO	*crylA(b)* trunc. WT	W	< 0.002	−
	crylA(b) trunc. PM	W	0.05–0.1	+
(6) TO, TB	*crylA(b)* trunc. WT	W/E	0.002	+
	crylA(b) trunc. PM	W/E	0.002–0.2	+
	crylA(b) trunc. FM	W/E	up to 0.3	+
	crylA(c) trunc. FM	W/E	up to 0.3	+

^{a} Higher values are from field-grown plants.
^{b} Reported as an estimate.
^{c} Partial Kozak (1983, 1987) mammalian consensus sequence.
[ ] Assay detection limit, NS = not specified, W = Western blot, E = ELISA, TB = tobacco, TO = tomato, CO = cotton, WT = wild-type, PM = partially modified, FM = fully modified.
References: (1) Vaeck *et al.*, 1987; (2) Barton *et al.*, 1987; (3) this chapter; (4A) Delannay *et al.*, 1989, (4B) Fischhoff *et al.*, 1987; (5) Perlak *et al.*, 1990; (6) Perlak *et al.*, 1991.

ENHANCEMENT OF δ-ENDOTOXIN EXPRESSION *IN PLANTA*

Numerous factors will affect the expression of heterologous genes in transgenic plants. These include major features such as the codon usage bias of the host species, molecular context at the translational start, promoter strength, mRNA sequence and structure, and the presence of introns in monocotyledons. Many other factors may also be relevant including patterns of methylation, the presence of extraneous start codons, the topology of the transforming DNA, and positional effects specific to the site of stable integration. The current state of plant transformation technology does not permit the manipulation of some of these factors (e.g. positional effects). Similarly, heterologous gene copy number and number of integration sites cannot be manipulated. Apparently, however, in the majority of transgenic plants, introduced δ-endotoxin genes segregate as a single dominant Mendelian trait (e.g. Perlak *et al.*,

1990, 1991) and this single site of heterologous gene integration often contains only one, or very few (≤3), gene copies (e.g. Barton *et al.*, 1987). Routine methods in molecular biology can, however, be used to considerable advantage in the alteration of certain other features, some of which are described below.

Codon preference

The redundancy of the universal genetic code results in 18 of the 20 amino acids being encoded by two to six synonymous codons. The non-random nature of codon preference was hypothesized by Grantham (1980) to be both genome-specific and related to taxonomic order. The primary structure determination of a large number of plant genes has allowed the examination of codon usage bias in plants. A study of 207 plant genes (53 monocotyledon and 154 dicotyledon genes) led Murray *et al.* (1989) to conclude that monocotyledons and dicotyledons differ in the relative use of synonymous codons. In higher plants, it is possible to identify a set of 38 'preferred' monocotyledon codons, the majority of which have G or C in the third position. In contrast, there is an identifiable set of 44 'preferred' dicotyledon codons, with a marginal preference for those ending in A or T (Campbell and Gowri, 1990). Furthermore, within a single species, codon preference is usually related to the expression level of a given gene (Murray *et al.*, 1989). These factors will clearly affect the heterologous *in planta* expression of introduced genes derived from prokaryotic organisms.

B. t. δ-endotoxin genes have an *average* A + T content of 64% as compared to that of about 44% and 55% in monocotyledon and dicotyledon exons respectively. These differences in base composition may lead to aberrant processing of transcripts; in addition, the preferred amino acid codon for an average δ-endotoxin is often the least preferred codon in plants, particularly in an extremely GC-rich monocotyledon such as maize (Wada *et al.*, 1990). Not surprisingly then, codon usage modification of B. t. δ-endotoxins genes should result in enhanced heterologous expression in monocotyledons and dicotyledons.

As discussed above, codon alterations were required for the expression of the *crylA(b)* endotoxin gene in cotton (Perlak *et al.*, 1990). More recently this has been elegantly demonstrated for the *crylA(b)* endotoxin gene in tomato and tobacco (Perlak *et al.*, 1991). In these experiments, a truncated partially modified (PM) *crylA(b)* gene and a truncated fully-modified (FM) *crylA(b)* gene were compared to the truncated wild-type gene. The PM gene had been altered to remove sequences possibly involved in processing (e.g. polyadenylation signals) and codons which were extremely rare in plants (9.5% codons altered). The FM gene had 60% codons altered. A fully modified *crylA(c)* gene was also described.

Expression of the PM *crylA(b)* gene in dicotyledons was up to tenfold greater than that of wild-type *crylA(b)*, and expression of the FM gene was up to 100-fold greater than wild-type. The enhanced expression levels of the PM and FM δ-endotoxin genes (see Table 5.2) resulted in corresponding increases in insect resistance in *M. sexta* bioassays. In these experiments, increases in δ-endotoxin expression from the PM and FM genes were proportionally greater than increases in corresponding mRNA levels, indicating that translational efficiency was the more crucial improvement. In other reports, however, mRNA instability was considered the salient contribution to low expression levels of unmodified, truncated δ-endotoxin genes in transgenic tobacco (Barton *et al.*, 1987; Murray *et al.*, 1991).

Molecular context at the translational start

A eukaryotic translational initiation consensus sequence ((GCC)GCC A/G CC*AUG*G) has been identified and further characterized by Kozak (e.g. Kozak, 1983, 1987). Barton *et al.* (1987) modified the translational start region of the truncated *crylA(a)* gene to incorporate part of this sequence (CC*AUG*G) prior to introduction into tobacco. Subsequently, however, it has become clear that there are significant differences in the consensus sequences within plant and animal translational start regions (Luetcke *et al.*, 1987). The initial plant consensus sequence (AACA*AUG*GC) was based on 61 plant genes and, excluding 100% conservation of the AUG, is most highly conserved at the +4 (G) and +5 (C) positions. An extended sequence (UAAACA*AUG*GCU) based on 79 plant genes confirmed the high degree of conservation at positions +4 and +5 (Pautot *et al.*, 1989). Although it should be recognized that consensus sequences are (a) biased by the presence of related gene families in the sequence databases, and (b) not necessarily optimal, it is tempting to incorporate some consensus sequence features while engineering synthetic δ-endotoxin genes into plant transformation vectors. Use of the +4 and +5 consensus sequence GC, however, will alter the secondary structure of the δ-endotoxin, making the second residue alanine; this may be undesirable from an environmental and/or regulatory perspective.

Transcriptional features

The AT-rich base composition of δ-endotoxin genes results in the presence of fortuitous plant polyadenylation signals (AAUAAA and AAUAAU) (Dean *et al.*, 1986) within wild-type coding regions; these could lead to aberrant processing when present in plant cells. Clearly, in the construction of a fully or partially synthetic δ-endotoxin gene destined for introduction into plants, the removal of as many of these sites as possible would be an obvious and prudent measure. Similarly, fortuitous intron/exon splice junctions should also be eliminated when 'correctly' positioned in δ-endotoxin structural genes. These are more difficult to delineate than are polyadenylation signals, but a comparison of 177 plant introns (Pautot *et al.*, 1989) has defined the 5' exon : intron splice junction as C/A AG:*GU*AAU and the 3' intron : exon splice junction as UUUUUUUUUUUGC*A*C : G where the colon represents the splice point, and the bases in italics are 100%, 99%, 100% and 100% conserved respectively. Since most plant introns are between 70 and 1000 nucleotides in length (Goodall and Filipowicz, 1990) *cry* genes will require close examination to identify fortuitously 'correct' positioning of splice sites. In addition, plant introns are relatively AU-rich in comparison to exon sequences (Goodall and Filipowicz, 1989), although it may be that the exon : intron differential is more important than absolute base composition. Finally, it may also be generally useful to remove AUUUA sequences, as in the work reported by Perlak *et al.* (1990, 1991), since this motif has been shown to cause mRNA destabilization in other systems (Shaw and Kamen, 1986).

Intron enhancement of heterologous gene expression in monocotyledons

A comprehensive study of intron enhancement of the maize alcohol dehydrogenase *adh*1 gene has been accomplished through the use of transient expression assays in protoplasts of black Mexican sweet (BMS) corn (Callis *et al.*, 1987). Intron 1 of the *adh*1 gene is located at the extreme N-terminus of the gene; when reintroduced into an intronless *adh*1 cDNA, intron 1 restored expression levels to that of the intact gene with all nine introns present. This effect was also seen when the CaMV 35S promoter was substituted for the homologous *adh*1 promoter. Intron 1 placement, however, was shown to be correlated with the degree of expression enhancement. In general, intron-mediated enhancement of heterologous gene expression in maize cells is likely to depend on the particular promoter, the specific intron, intron placement and the nature of the heterologous gene.

The ability to test specific components of heterologous gene expression is fundamental to the successful engineering of plant transformation constructs for the expression of introduced genes, including δ-endotoxins. Ideally, this is accomplished by generating sets of rigorously parallel constructs in which the element to be tested is the only variable. The ability of the maize *adh*1 gene intron 1 to enhance expression of the marker gene GUS is demonstrated in Figure 5.4 using purpose-built vectors. In these experiments, intron 1 was removed from construct pIC254 to generate the analogous intronless construct, pIC255 (Figure 5.4A). These constructs were then separately introduced into maize L056 protoplasts, and transient GUS expression measured 48 hours after DNA uptake. As seen in

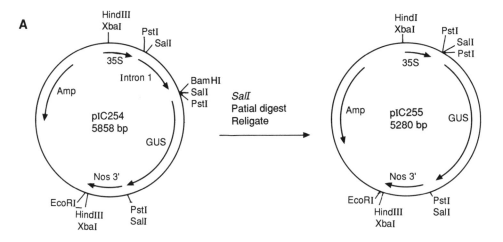

Figure 5.4. Assessment of *adh*1 intron 1 enhancement of GUS expression in maize cells. A. The intronless construct, pIC255, was derived from construct pIC254 after partial digestion with restriction endonuclease *Sal*I, isolation of the appropriate-sized DNA fragment, and religation. B. pIC254 and pIC255 were separately introduced into maize L056 protoplasts by PEG-mediated DNA uptake. Transient expression of GUS activity was measured 48 hours after uptake by quantification of the fluorescent product 4-methylumbelliferone (4-MU)

Figure 5.4B, the presence of intron 1 resulted in a 26-fold enhancement in GUS expression. Similar results were observed when the maize *adh*1 intron 1 was compared in parallel constructs for the ability to enhance *npt*II gene expression in both transiently and stably transformed maize cells.

Summary

There is great potential for the enhancement of δ-endotoxin expression in transformed plants. The most effective documented means of altering expression is changing codon preference to conform closely to that of the host plant. The benefit of removing sequences which may result in the aberrant processing of transcripts has not been rigorously documented in the absence of other variables. The advantages of creating plant consensus sequences at the translational start are, likewise, unknown. However, since the first two types of modifications necessarily involve the production of fully or partially synthetic δ-endotoxin genes, it seems reasonable to include at least some elements of the plant consensus sequence around the translational start. This level of δ-endotoxin gene engineering is sufficiently major to encourage the development of early evaluation methods, circumventing the need for whole-plant bioassays on all transformants generated in order to evaluate step-wise changes. Although immunological methods can be used to detect the physical presence of heterologously expressed δ-endotoxin proteins in plant cells, δ-endotoxin function must be evaluated in insect bioassays. In maize, for example, modified δ-endotoxin genes which are constitutively expressed can be phenotypically evaluated at an early stage using transgenic calli feeding bioassays.

ENDOPHYTE EXPRESSION OF δ-ENDOTOXIN GENES IN PLANTS

Plants which are amenable to colonization by endophytic bacteria are theoretically able to become insect resistant by virtue of δ-endotoxin expression in a resident, recombinant endophyte. Although many important crop plants (e.g. maize, rice, soybeans, wheat and cotton) can be colonized by endophytes, the only documented example of δ-endotoxin expression in a genetically engineered endophyte is the use of recombinant *Clavibacter xyli* subsp. *cynodontis* to express *crylA* gene product in maize (Rigby, 1991).

CONCLUSIONS

Experiments to generate and characterize transgenic dicotyledons engineered to express δ-endotoxins have provided an encouraging beginning to the genetic improvement of crop plants. Both laboratory and field tests indicate that the *in planta* expression of δ-endotoxins confers protection from insect pest damage. The recent confirmation of Grantham's (1980) prediction that heterologous gene expression levels in dicotyledons would be significantly enhanced by primary structure modifications which do not alter secondary structure (Perlak *et al.*, 1991) has set the stage for the engineering of fully insect-resistant transgenic plants. The best transgenic plants produced to date are probably expressing δ-endotoxin at levels approaching those sufficient for commercial use in some plant species. Higher levels of *cry* gene expression are certainly attainable with further manipulation, although a yield penalty due to biosynthetic drain on the transgenic plant is not an acceptable consequence of achieving extremely high heterologous gene expression. It is imperative, however, that δ-endotoxin levels in transgenic plants are sufficient to produce complete, or very near complete insect control and that wild-type refuge plants are provided. These factors are critical, since suboptimal control levels and lack of refugia could result in an increased pressure on insect populations to develop resistance to δ-endotoxins.

The domestication of most crop plants has resulted in the virtual elimination of natural forms of plant-derived insect resistance (Feeny, 1976). The commercialization of transgenic crops which are endogenously insect-resistant due to δ-endotoxin expression will afford protection to plant tissues previously inaccessible to conventionally applied insecticide treatments. The expression of δ-endotoxins at adequate levels in appropriate crop plants should productively contribute to the wider and prolonged use of this biological control agent. It is essential, however, that both conventional and *in planta B. t.* δ-endotoxin applications are responsibly used as components of integrated pest management systems which also incorporate the principles of pest resistance management.

ACKNOWLEDGEMENTS

I am grateful to my colleagues at ICI Seeds, and Garst Research Center in Slater, Iowa, USA, for technical and intellectual contributions. In particular I wish to thank W. Schuch, J. Ray and J. Foundling for tobacco transformation and bioassay results; J. Thompson and C. Sparks for maize protoplast results; I. Evans for an excellent review of plant gene expression; R. Ellis for documenting stably transformed plant species to date; and R. Tailor, G. Gibb, S. Baird, K. Brebner, L. Jordan and J. Tippett for *B. t.* work.

REFERENCES

Barton, K. A., Whiteley, H. R. and Yang, N.-S. (1987). '*Bacillus thuringiensis* δ-endotoxin expressed in transgenic *Nicotiana tabacum* provides resistance to lepidopteran insects'. *Plant Physiol.*, **85**, 1103–1109.

Bevan, M. (1984). 'Binary Agrobacterium vectors for plant transformation'. *Nucl. Acids Res.*, **12**, 8711–8721.

Callis, J., Fromm, M. and Walbot, V. (1987). 'Introns increase gene expression in cultured maize cells'. *Genes and Dev.*, **1**, 1183–1200.

Campbell, W. H. and Gowri, G. (1990). 'Codon usage in higher plants, green algae, and cyanobacteria'. *Plant Physiol.*, **92**, 1–11.

CIMMYT (1989). *Toward Insect Resistant Maize for the Third World: Proceedings of the International Symposium on Methodologies for Developing Host Plant Resistance to Maize Insects*. Mexico, D. F.: CIMMYT.

Dean, C., Tamaki, S., Dunsmuir, P., Favreau, M., Katayama, C., Dooner, H. and Bedbrook, J. (1986). 'Messenger RNA transcripts of several plant genes are poly-adenylated at multiple sites *in vivo*'. *Nucl. Acids Res.*, **14**, 2229–2240.

Delannay, X., LaVallee, B. J., Proksch, R. K., Fuchs, R. L., Sims, S. R., Greenplate, J. T., Marrone, P. G., Dodson, R. B., Augustine, J. J., Layton, J. G. and Fischhoff, D. A. (1989). 'Field performance of transgenic tomato plants expressing the *Bacillus thuringiensis* var. *kurstaki* insect control protein'. *BioTechnology*, **7**, 1265–1269.

Depicker, A., Stachel, S., Dhaese, P., Zambryski, P. and Goodman, H. M. (1982). 'Nopaline synthase: transcript mapping and DNA sequence'. *J. Mol. Appl. Genet.*, **1**, 561–573.

Ellis, J. R. (In Press). 'Tissue culture and transformation', In *The Plant Molecular Biology LabFax* (Ed. R. R. D. Croy) BIOS Scientific Publishers Ltd., Oxford, UK.

Feeny, P. P. (1976). 'Plant apparency and chemical defence'. *Recent Adv. Phytochem.*, **10**, 1–40.

Fischoff, D. A., Bowdish, K. S., Perlak, F. J., Marrone, P. G., McCormick, S. M., Niedermeyer, J. G., Dean, D. A., Kusano-Kretzmer, K., Mayer, E. J., Rochester, D. E., Rogers, S. G. and Fraley, R. T. (1987). 'Insect tolerant transgenic tomato plants'. *BioTechnology*, **5**, 807–813.

Gasser, C. S. and Fraley, R. T. (1989). 'Genetically engineering plants for crop improvement'. *Science*, **244**, 1293–1299.

Gatehouse, A. M. R., Boulter, D. and Hilder, V. A. (1991). 'Novel insect resistance using protease inhibitor genes'. In *Molecular Approaches to Crop Improvement* (Eds. E. S. Dennis and D. J. Llewellyn) pp. 63–77, Springer-Verlag, New York.

Goodall, G. J. and Filipowicz, W. (1989) 'The AU-rich sequences present in the introns of plant nuclear pre-RNAs are required for splicing'. *Cell*, **58**, 473–483.

Goodall, G. J. and Filipowicz, W. (1990). 'The minimum functional length of pre-mRNA introns in monocots and dicots', *Plant Mol. Biol.*, **14**, 727–733.

Grantham, R. (1980). 'Workings of the genetic code'. *Trends Biochem. Sci.*, **5**, 327–331.

Herrnstadt, C., Soares, G. G., Wilcox, E. R. and Edwards, D. L. (1986). 'A new strain of *Bacillus thuringiensis* with activity against coleopteran insects'. *BioTechnology*, **4**, 305–308.

Hilder, V. A., Gatehouse, A. M. R. and Boulter, D. (1990). 'Genetic engineering of crops for insect resistance using genes of plant origin'. In *Genetic Engineering of Crop Plants* (Eds. G. W. Lycett and D. Grierson), pp. 51–66, Butterworth, London.

Hill, D. S. (1975). *Agricultural Insect Pests of the Tropics and their Control*, 2nd edn, Cambridge University Press, Cambridge.

Hill, D. S. (1987). *Agricultural Insect Pests of Temperate Regions and their Control*, Cambridge University Press, Cambridge.

Höfte, H. and Whiteley, H. R. (1989). 'Insecticidal crystal proteins of *Bacillus thuringiensis*'. *Microbiol. Rev.*, **53**, 242–255.

Höfte, H., de Greve, H., Seurinck, J., Jansens, S., Mahillon, J., Ampe, C., Vanderkerckove, J., Vanderbruggen, M., Van Montagu, M., Zabeau, M. and Van Mellaert, H. (1986). 'Structural and functional analysis of a cloned δ-endotoxin of *Bacillus thuringiensis* Berliner 1715'. *Eur. J. Biochem.*, **161**, 273–280.

Hong, Y., Jiaxin, L., Sandui, G., Xuejun, C. and Yunliu, F. (1989). 'Transgenic rice plants produced by direct uptake of δ-endotoxin protein gene from *Bacillus thuringiensis* into rice protoplasts'. *Scientia Agricultura Sinica*, **22**, 1–5.

Horsch, R. B., Fry, J. E., Hoffman, N. L., Eichholz, D., Rogers, S.G. and Fraley, R. T. (1985). 'A simple and general method for transforming genes into plants'. *Science*, **227**, 1229–1231.

Huger, A. M., Kreig, A., Langenbruch, G. A. and Schnetter, W. (1986) 'Discovery of a new strain of *Bacillus thuringiensis* effective against Coleoptera'. In *Symposiums in memorium Dr. Ernst Berliner on the occasion of the 75th anniversery of a primary description of Bacillus thuringiensis*, pp. 83–96. Biologische Bundesanstaft für Land- und Forstwirtschaft, Berlin-Dahlem.

Kay, R., Chan, A., Daly, M. and McPherson, J. (1987) 'Duplication of CaMV 35S promoter sequences creates a strong enhancer for plant genes'. *Science*, **236**, 1299–1303.

Kozak, M. (1983). 'Comparison of initiation of protein synthesis in procaryotes, eucaryotes, and organelles'. *Microbiol. Rev.*, **47**, 1–45.

Kozak, M. (1987). 'At least six nucleotides preceding the AUG initiator codon enhance translation in mammalian cells'. *J. Mol. Biol.*, **196**, 947–950.

Krysan, J. L. (1986). 'Introduction: biology, distribution, and identification of pest *Diabrotica*'. In *Methods for the Study of Pest Diabrotica* (Eds. J. L. Krysan and T. A. Miller) p. 2, Springer-Verlag, New York.

Luetcke, H. A., Chow, K. C., Mickel, F. S., Moss, K. A., Kern, H. F. and Scheele, G. A. (1987). 'Selection of AUG initiation codons differs in plants and animals'. *EMBO J.*, **6**, 43–48.

MacIntosh, S. C., Stone, T. B., Sims, S. R., Hunst, P. L., Greenplate, J. T., Marrone, P. J., Perlak, F. J., Fischhoff, D. A. and Fuchs, R. L. (1990). 'Specificity and efficacy of purified *Bacillus thuringiensis* proteins against agronomically important insects'. *J. Invert. Pathol.*, **56**, 258–266.

McCown, B. H., McCabe, D. E., Russell, D. E., Robison, D. J., Barton, K. A. and Raffa, K. F. (1991). 'Stable transformation of *Populus* and incorporation of pest resistance by electric discharge particle acceleration'. *Plant Cell Rep.*, **9**, 590–594.

Murray, E. E., Lotzer, J. and Eberle, M. (1989). 'Codon usage in plant genes'. *Nucl. Acids Res.*, **17**, 477–498.

Murray, E. E., Rocheleau, T., Eberle, M., Stock, C., Sekar, V. and Adang, M. (1991). 'Analysis of unstable RNA transcripts of insecticidal crystal protein genes of *Bacillus thuringiensis* in transgenic plants and electroporated protoplasts'. *Plant Mol. Biol.*, **16**, 1035–1050.

Odell, J. T., Nagy, F. and Chua, N. H. (1985). 'Identification of DNA sequences required for activity of the CaMV 35S promoter'. *Nature*, **313**, 810–812.

Pautot, V., Brzezinski, R. and Tepfer, M. (1989). 'Expression of a mouse metallothionein gene in transgenic plant tissue'. *Gene*, **77**, 133–140.

Perlak, F. J., Deaton, R. W., Armstrong, T. A., Fuchs, R. L., Sims, S. R., Greenplate, J. T. and Fischhoff, D. A. (1990). 'Insect resistant cotton plants'. *BioTechnology*, **8**, 939–943.

Perlak, F. J., Fuchs, R. L., Dean, D. A., McPherson, S. L. and Fischhoff, D. A. (1991). 'Modification of the coding sequence enhances plant expression of insect control protein genes'. *Proc. Natl Acad. Sci. USA*, **88**, 3324–3328.

Rigby, S. (1991). *Bt in Crop Protection*, pp. 22–24, PJB Publications, Richmond, UK.

Roush, R. T. and Tabashnik, B. E. (Eds.) (1990). *Pesticide Resistance in Arthropods*, Chapman and Hall, New York.

Schnepf, H. E., Wong, H. C. and Whiteley, H. R. (1985). 'The amino acid sequence of a crystal protein from *Bacillus thuringiensis* deduced from the DNA base sequence'. *J. Biol. Chem.*, **260**, 6264–6272.

Shaw, G. and Kamen, R. (1986). 'A conserved AU sequence from the 3'-untranslated region of GM-CSF messenger RNA mediates selective messenger RNA degradation'. *Cell*, **46**, 659–667.

Vaeck, M., Reynaerts, A., Höfte, H., Jansens, S., DeBeuckeleer, M., Dean, C., Zabeau, M., Van Montagu, M. and Leemans, J. (1987). 'Transgenic plants protected from insect attack'. *Nature (London)*, **328**, 33–37.

Vaeck, M., Reynaerts, A. and Höfte, H. (1989). 'Protein engineering in plants: Expression of *Bacillus thuringiensis* insecticidal protein genes'. In *Cell Culture and Somatic Cell Genetics of Plants, Vol. 6* (Eds. J. Schell and I. K. Vasil) pp. 425–439, Academic Press, London.

Wada, K., Aota, S., Tsuchiya, R., Ishibashi, F., Gojobori, T. and Ikemura, T. (1990). 'Codon usage tabulated from the GenBank genetic sequence data'. *Nucl. Acids Res.*, **18**, *Supp.*, 2357–2411.

6 Control of Lepidopteran Pests with *Bacillus thuringiensis*

AMOS NAVON
Department of Entomology, Agricultural Research Organization, The
Volcani Center, Bet Dagan 50250, Israel

CONVENTIONAL AND GENETICALLY ENGINEERED *B. THURINGIENSIS* PRODUCTS

Most conventional *Bacillus thuringiensis* (*B. t.*) products are based on the subspecies *kurstaki* strain HD-1, which was introduced in 1971 by Abbott Laboratories and subsequently produced by other leading manufacturers of agricultural and chemical products. Currently, at least five major companies produce this strain today in the USA and Western Europe (Thompson, 1989). The decision to use HD-1 strain for over 20 years was made because this microbial product was active against more than 100 lepidopteran species. Other *B. t.* types exhibited narrower host ranges with a resultant lower economic value. Sandoz introduced the HRD12 strain of *B. t. kurstaki* to improve the control of *Spodoptera* species not sensitive to HD-1. A *B. t.* product based on a strain of *B. t.* subspecies *aizawai* is used specifically for the control of the wax moth, *Galleria mellonella* (Table 6.1). In Eastern Europe (Lipa, 1991) the commercial *B. t.* products differ from those used in Western Europe and the USA; Dendrobacillin based on *B. t.* subspecies *dendrolimus* was produced to ccontrol the pine defoliator Siberian silkmoth, *Dendrolimus sibiricus*. Several other products have been based on serotype 1 of *B. t.* subspecies *thuringiensis* containing β-exotoxin at levels ranging from 0.6% to 20%. In Japan, *B. t.* preparations are selected for low toxicity to the silkworm, *Bombyx mori*, to avoid possible damage to sericulture. Several *B. t.* products based on domestic strains of *B. t. kurstaki* (Takaki, 1985) have been used on vegetables and fruits. A. *B. t. kurstaki* strain in which the spores were inactivated, Toarow CT® (Toagosei Chemical Industry Co.), was registered and is being used in plant protection (K. Aizawa, personal communication); it is of low toxicity to the silkworm. Other *B. t. kustaki*-based products, such as Thuricide® and Dipel® are also used in Japan.

Bacillus thuringiensis, An Environmental Biopesticide: Theory and Practice. Edited by P. F. Entwistle,
J. S. Cory, M. J. Bailey and S. Higgs
© 1993 John Wiley & Sons Ltd

Table 6.1. Examples of conventional and genetically engineered *Bacillus thuringiensis* products and delivery systems for the control of lepidopteran pests (in Western Europe and the USA)[a]

No.	Commercial name/crop	Producer	*B. t.* subsp./strain	Target insects
Conventional products				
1.	Dipel®	Abbott	*kurstaki*,HD-1	> 100 species
2.	Thuricide®	Sandoz	*kurstaki*,HD-1	> 100 species
3.	Biobit®	Novo	*kurstaki*,HD-1	> 100 species
4.	Javelin®	Sandoz	*kurstaki*,NRD12	> 100 species + *Spodoptera* spp.
5.	Certan®	Sandoz	*aizawai*	*Galleria mellonella*
Genetically engineered products				
6.	Condor®	Ecogen	EG2348	Forest and vegetable pests
7.	Foil®	Ecogen	EG2424	*Ostrinia nubilalis*
8.	MVP™	Mycogen	*kurstaki*	Vegetable pests
Plant delivery system (experimental)				
9.	Tomato (transgenic)	Monsanto	*kurstaki*	*Manduca sexta, Helicoverpa zea Keiferia lycopersicella*
10.	Cotton (transgenic)	Monsanto	*kurstaki*	*Heliothis virescens, H. zea, Trichoplusia ni, Pectinophora, Gossypiella*
11.	Corn InCide™	Crop Genetic International	*kurstaki*	*Ostrinia nubilalis*

[a] Nos 1–7: Thompson, 1989; 8: Mycogen Technical Information; 9: Delannay *et al.*, 1989; 10: abstr. Beltwide Cotton Conference, San Antonio, 1991; 11: InCide™—Technical Bulletin.

The intensive research and development invested in *B. t.* genetics has resulted in several genetically engineered products of prokaryotic and plant systems (Table 6.1). To broaden the insect host range, Ecogen developed genetically modified products based on conjugation of two distinct *B. t.* strains. Condor® is active against forest pests such as the spruce budworm, *Choristoneura fumiferana*, and the gypsy moth, *Lymantria dispar* and Foil® is active against both lepidopteran and coleopteran pests. Mycogen encapsulated the microbial crystal in a non-viable cell of *Pseudomonas fluorescens*; persistence of this product against the diamondback moth, *Plutella xylostella*, was longer than that of conventional *B. t. kurstaki* products (Gelernter, 1990). The transfer of *B.t. kurstaki* HD-1 *cryIA(b)* and HD-73 *cryIA(c)* genes into cotton protected the plant against several important lepidopterous pests (Perlak *et al.*, 1990). Trials of *B. t.* transgenic cotton, carried out in five states in the cotton belt of the USA (Arizona, Texas, Mississippi, Louisiana and Missouri), demonstrated significant control of lepidopterous pests in these areas, according to reports submitted at the annual Beltwide Cotton Conference, San Antonio, 1991. Performance of transgenic tomato plants in the field showed significant protection against *Helicoverpa (Heliothis) zea*, cotton bollworm, and *Keiferia lycopersicella*, tomato pinworm (Delannay *et al.*, 1989). Crop Genetic International transformed the *B. t.* gene into endophytic bacteria from Bermuda grass. To express the *B. t.* δ-endotoxin in the vascular system, the endophyte/*B. t.* genetic product was introduced into corn by seed inoculation. Experiments with this delivery system in corn to control *Ostrinia nubilalis*, European corn borer, are in progress (see Gelernter and Schwab, Chapter 4).

MICROBIAL SPECIFICITY OF *B. THURINGIENSIS* IN LEPIDOPTERAN PESTS

B. thuringiensis classification

The flagellar antigen classification based on the reliable and specific character of the H-antigen has contributed to microbial identification for more than 25 years. The updated list comprises 34 serovars (de Barjac and Frachon, 1990) of which 22 are active against lepidopteran pests (Table 6.2).

Table 6.2. Serovar classification of *Bacillus thuringiensis* with activity against lepidopteran pests

H-antigen	Subspecies	H-antigen	Subspecies
1	*thuringiensis*[a]	8a,8c	*ostriniae*
2	*finitimus*	9	*tolworthi*
3a	*alesti*	10	*darmstadiensis*
3a3b	*kurstaki*	11a11c	*kyushuensis*[a]
4a4b	*sotto*[b]	12	*thompsoni*
4a4c	*kenyae*[a]	17	*tohokuensis*
5a5b	*galleriae*	20a20b	*yunnanensis*
5a5c	*canadensis*	20a20c	*pondicheriensis*
6	*entomocidus*[c]	21	*colmeri*
7	*aizawai*[a]	22	*shandongiensis*
8a8b	*morrisoni*[a]	—	*wuhanensis*

[a] Serovars active against Diptera.
[b] 4a4b, *sotto* and 4a4b, *dendrolimus* are identical serotypes.
[c] 6, *entomocidus* and 6, *subtoxicus* are identical serotypes.

The serovars 1, *thuringiensis*; 4a4c, *kenyae*; 7, *aizawai*; 8a8b, *morrisoni*; 9, *tolworthi*; and 11a11c, *kyushuensis* are also active against Diptera. Serovar 8a8b, *morrisoni*, also has some activity against Coleoptera. Serovar 3a3b, *kurstaki*, exhibits a wide insect host range whereas the other serovars are only active against a few lepidopteran species. With the establishment of international *B. t.* collections, the strains were coded and numbered. The HD listing belongs to the United States Department of Agriculture culture stock (Nakamura and Dulmage, 1988) and is widely used in *B. t.* screening programmes. Toxin gene sequencing led to the classification of the genes according to crystal protein structure and insecticidal spectra (Höfte and Whiteley, 1989). In most lepidopteran pests, the crystal protein is the major cause of insect mortality whereas the spore effect is marginal. However, the spore activity is essential for expressing the *B. t.* insecticidal activity in insects such as *G. mellonella* (Li *et al.*, 1987). This specific spore effect has been recorded, so far, in very few species.

Potency bioassays

Two potency bioassays have been used to test *B. t.* activity against lepidopteran pests. A French bioassay is based on the reference standard of *B. t. thuringiensis* (E61), with an assigned potency of $\times 1000$ international units (IU)/mg and *Anagasta (Ephestia) kuehniella*, Mediterranean flour moth, as the test insect, and the American bioassay is based on *B. t. kurstaki* serovar 3a3b strain HD-1 as the reference standard with potency of 16 000 IU/mg for the HD-1-S-80, and *Trichoplusia ni*, cabbage looper, as the test insect (Dulmage *et al.*, 1971). However, in view of the high insect specificity and the urgent need to select potent strains for control of lepidopteran pests, the official test insect had to be replaced by other pest species.

The general bioassay protocol is based on the following:

1. Insect colony supplying larvae for the bioassay.
2. Microbial sources: commercial *B. t.* products, powders of domestic and international strains, purified or solubilized crystals and activated toxins.
3. Bioassay diet supporting larval feeding and preserving the microbial activity.
4. Aqueous dilution series of the *B. t.* to obtain at least five on-curve mortality points with at least two above and two below the LC_{50}.
5. Larval age: up to 12h-old larvae for the neonate bioassay, third-instar larvae for the official bioassay.
6. Bioassay time: 3–4 days for the neonate bioassay, 6–8 days for the official bioassay both at 25–26 °C.
7. Preliminary screening using the LC_{50} of HD-1 as an activity reference for *B. t. kurstaki*-sensitive insects, and the LC_{50} of another *B. t.*

subspecies (*aizawai, kenyae, tolworthi*) against insects not sensitive
to *B. t. kurstaki.*

8. Potency bioassays with the most effective strain obtained from the
    preliminary bioassay.
9. Dose/probit analyses to determine mortalities with 95% fiducial limits
    not exceeding twice the LC_{50}.
10. Potency calculation in IU/mg by means of an international formula
    (Dulmage *et al.*, 1971).

Neonate larvae, the insect stage most sensitive to *B. t.*, can be used
instead of third instar larvae in order to reduce the bioassay time from 8
to 3 days and to raise the sensitivity of the bioassay (Navon *et al.*, 1990).
However, neonates die rapidly of starvation whereas third instar larvae
will survive for long periods suggesting that the pathogenicity for neonate
larvae is different from that for older larvae. Therefore, testing *B. t.* types
against third or fourth instars should not be overlooked in developing
microbial control programmes. Bioassays with mature larvae would better
reflect the effectiveness of *B. t.* preparations in field practice.

In some laboratories, the *B. t.* mixture is spread on the diet surface
instead of being introduced into the diet itself. Mixing commercial *B. t.*
products (Navon *et al.*, 1983; Navon, 1989) into the diet is essential for
masking undesired feeding effects of the fermentation and formulation
adjuvants. On the other hand surface-applied purified spores and/or
crystals, dissolved crystals, and activated toxin have been bioassayed with
satisfactory results. Recently, the bioassay diet for screening of *B. t.* strains
against lepidopteran pests in agriculture has been standardized (Navon *et
al.*, 1990). The standardized diets can be used to enable more reliable comparisons between potency to be recorded from different laboratories.

Selection of *B. thuringiensis* strains

Screening has two objectives: (a) selection of *B. t. kurstaki* strains more
potent than the HD-1 for the management of *kurstaki*-sensitive pests, and
(b) selection of strains for insects not sensitive to *B. t. kurstaki* (Table 6.3).
The *kurstaki* strain HD-263 showed superior activity to HD-1 against several major agricultural pests although the insecticidal spectrum of the
latter is broader. HD-73 is more active than HD-1 against *Heliothis* spp.
and also *Cydia pomonella*, codling moth (Andermatt *et al.*, 1988). HD-73
bioassayed against *T. ni* was as active as HD-1 in diet bioassays but less
so in field bioassays (Beegle *et al.*, 1982). *B. t. kenyae* was highly active
against tobacco budworm, *Heliothis virescens*, and the Bertha armyworm,
Mamestra configurata. The control of *Spodoptera* species and several
other pests of fibre, oil and vegetable crops is difficult because they
are not sensitive to commercial *B. t. kurstaki* strains. Therefore, a special

Table 6.3. Bacillus *thuringiensis* strains and subspecies with activity superior to HD-1

Target insect	B. t. subsp.	Strain	Microbial preparation	Comparative activity to HD-1	Ref.[a]
Boarmia selenaria	kurstaki	HD-263	Spore + cryst.	9.3	(7)
Earias insulana	kurstaki	HD-263	Spore + cryst.	2.2	(6)
Heliothis armigera	kurstaki	HD-263	Spore + cryst.	2.6	(6)
Heliothis virescens	kurstaki	HD-73	Spore + cryst.	11.3	(2)
Heliothis virescens	kenyae	—[b]	Crystals	14.3 (1.0)[c]	(2)
Heliothis virescens	kenyae	—	Protoxin	9.7 (3.8)	(2)
Heliothis virescens	kenyae	—	Toxin	12.5 (6.0)	(2)
Mamestra brassicae	aizawai	HD-283	Spore + cryst.	10.0	(1)
Mamestra configurata	aizawai	—	Spore + cryst.	2.4	(5)
Mamestra configurata	kenyae	—	Spore + cryst.	2.7	(5)
Pieris brassicae	thuringiensis	—	Crystals	1.9	(2)
Plodia interpunctella	aizawai	HD-133	Spore + cryst.	2.1[d]	(4)
Spodoptera littoralis	aizawai	—	Spore + cryst.	4.5	(6)
Spodoptera littoralis	aizawai	HD-137	Spore + cryst.	6.4	(1)
Spodoptera littoralis	entomocidus	—	Crystals	5.4	(2)
Spodoptera littoralis	entomocidus	HD-601	Crystals	5–10	(3)
Spodoptera littoralis	aizawai	7-29	Crystals	5 –10	(3)

[a] (1) Jarrett and Burges, 1986; (2) Jaquet et al., 1987; (3) Lecadet and Martouret, 1987; (4) McGaughey and Johnson, 1987; (5) Trottier et al., 1988; (6) Navon et al., 1990; (7) Wysoki and Scheepens, 1990.

[b] Domestic collection.

[c] Activity ratio for crystals, protoxin and toxin of strain HD-1.

[d] HD-133 was 57.9 times more active than HD-1 in P. interpunctella resistant to HD-1.

effort has been made to select strains of other *B. t.* subspecies. Several laboratories reported that strains of the subspecies *aizawai* and *entomocidus* were five to seven times more potent than HD-1 against the larvae of *Spodoptera littoralis*, Egyptian cotton leafworm (Table 6.3). An *entomocidus* strain was effective on cotton against young larvae of *S. littoralis* (Broza *et al.*, 1984), and experimental products of *B. t. aizawai* against this species were more potent than the HD-1 strain (Navon *et al.*, 1983). An *aizawai* strain was effective against both *B. t.* susceptible and resistant Indian mealmoth, *Plodia interpunctella* (McGaughey and Johnson, 1987). This strain could be used alternately with other control agents to delay the development of the pests' resistance to commercial *B. t. kurstaki* strains.

The screening of purified crystals has increased in recent years. Toxicity bioassays are conducted with either intact or dissolved crystals (= protoxin) or with an activated toxin. These *in vitro* activations of the toxin represent post-ingestion activities occurring in the larval midgut. Activity ratios of the crystal/protoxin/toxin determined for HD-1 (Table 6.3) showed that *H. virescens* was not a good activator, because the crystal toxicity in the insect was lower than that obtained with the *in vitro* activated toxin. It is probable that the larval midgut fluids only partially activate the toxin in this insect. In contrast to this, minor differences between the activities of the crystal, the protoxin and the toxin of *B. t. morrisoni* in *Pieris brassicae*, large cabbage white butterfly, suggests activation of the crystal by this species to be effective (Jaquet *et al.*, 1987). Using this type of screening, the insect's capacity to convert the crystal into an activated toxin can be determined. However, the need to evaluate spore effects should not be overlooked. The high insecticidal activities obtained with the spores in *G. mellonella* (Li *et al.*, 1987) could also be found in other target insects.

Another function of microbial screening involves selection of the single Cry proteins for expression in prokaryotic and plant delivery systems. The insecticidal activity of the CryIA(c) protein of strain HD-73 in several lepidopteran pests was higher than that of CryIA(b) protein isolated from strain HD-1 (MacIntosh *et al.*, 1990). The activity of CryIA(c) was two- to five-fold greater than that of CryIA(b) in *T. ni*, *H. virescens*, *H. zea* and *Agrotis ipsilon*, black cutworm. However, *O. nubilalis* was 10 times more sensitive to CryIA(b) than CryIA(c). The extremely high sensitivity of *H. virescens* to CryIA(c) protein was recorded with the growth inhibition bioassay (MacIntosh *et al.*, 1990). The use of this bioassay in addition to the potency bioassay improved the evaluation of the microbial bioactivity in the test. The 1000-fold difference in the control level between, tobacco hornworm, *Manduca sexta*, and *Spodoptera exigua*, beet armyworm, indicates the range of genes needing to be expressed to manage complexes of pest species. Screening of the Cry proteins' specificity will have further use in the selection of the protein genes for developing genetically modi-fied *B. t.* products. Combinations of the Cry proteins for expression in microbial and plant delivery products will have to be determined according to the composition of the fauna infesting the crop in its various geographical areas, and the rate at which resistance to the crystal protein may be developed.

ENTOMOLOGICAL ASPECTS OF LEPIDOPTERAN PEST CONTROL WITH *B. THURINGIENSIS*

General considerations

The potential of *B. t.* to control lepidopteran pests in their different habitats was evaluated by Burges (1981). The information available con-cerning this potential suggests that *B. t.* is generally more effective in forests than in agriculture. There are several factors responsible for the differences between forest and agricultural pest control (Table 6.4). For example, the infestation period of *O. nubilalis* and *Heliothis armigera*, Old World cotton bollworm, in the Mediterranean region occurs in the late spring and summer whereas the defoliation season of processionary

Table 6.4 Limitations of pest management with *Bacillus thuringiensis*: comparison between agriculture and forestry

Agriculture	Forestry
1. Major subtropical and tropical pests are polyphagous and multivoltine	1. Coniferous pest are oligophagous and monovoltine
2. Low *B. t.* persistence during the pest season (warm/hot spring–summer). Negative effects: sunlight, dew	2. *B. t.* persistence is longer than in agriculture. (pest season: short < warm summer or mild winter, shaded canopy). Negative effects: rain, high and low temperature
3. Limited control of bollworm and borer pests	3. Effective control of defoliators
4. Major pests are not susceptible to subsp. *kurstaki* Hd-1 products	4. Defoliators are controlled with commercial subsp. *kurstaki* products
5. Larval damage to vegetables and fruits is not tolerable	5. Limited defoliation caused by several pests is acceptable
6. Public pressure to replace chemical pesticides by biocontrol products in agricultural areas is limited	6. Public pressure to apply *B. t.* instead of chemical insecticides in residential areas is effective

moths, *Thaumatopoea* spp., in Mediterranean pine forests occurs in the winter and early spring. Because of these differences in pest phenology, the environmental problems related to *B. t.* persistence also vary. In the forest, rainfall (van Frankenhuyzen and Nystrom, 1989) and low temperatures (van Frankenhuyzen, 1990) cause pest management difficulties, such as delays in *B. t.* application, until the weather is suitable. Such delays can reduce the efficacy of pest control. Dew effects on *B. t.* in agricultural row crops have not been fully evaluated but a heavy dew has been found to limit *B. t.* efficacy on the plant (Navon, unpublished data). Sunlight inactivation of *B. t.* in the forest seems to be a minor factor in the decline of *B. t.* activity (van Frankenhuyzen and Nystrom, 1989). In agriculture, however, *B. t.* photoinactivation is probably a major cause for low *B. t.* persistence. Commercial *B. t.* products are not protected against potentially inactivating UV in sunlight at 290–400 nm wavelength. The addition of UV absorbers to the commercial product can increase biologically its persistence on plant surfaces (Morris, 1983). Encapsulation of *B. t.* and the use of a chromophore as a UV screener prolonged residual activity in plant activity (Dunkle and Shasha, 1989). Information on *B. t.* stability on row crops (half-life, < 1–2 days) was up to five-fold lower than in forest trees (Fuxa, 1989).

Sixty to seventy per cent of the *B. t.* products are used in forests. The market value of *B. t.* used in North American forests in 1984–5 was $8.1 million, as compared with $1.9 million in agriculture (Lethbridge, 1989). In 1985, about 870 000 hectares were sprayed with *B. t.* against spruce budworm, *Choristoneura* spp., in Canadian forests alone, replacing approximately half the amount of chemical insecticides used in previous years. The use of *B. t.* against gypsy moth, *L. dispar*, along the Pacific coast of the USA was effective even without chemical insecticides (Dreistadt and Dahlsten, 1989). Large amounts of *B. t.* were used for the control of *D. sibiricus* in the USSR and of *Lymantria* spp. in Bulgaria and Poland (Lipa, 1991). In several countries of Western Europe the control of *Thaumetopoea pityocampa*, pine processionary moth, with *B. t.* in coniferous forests is now well established.

Insect feeding behaviour and the use of *B. thuringiensis* for pest control

Forest defoliators and some vegetable pests are effectively controlled by commercial *B. t.* products. However, management of bollworm and borer pests with *B. t.* is either limited or totally unsuccessful, because the larvae rapidly invade the inner tissue of the plant and so do not acquire a lethal dose. Some of the bollworms and borers (*Heliothis* spp., *O. nubilalis*) feed on leaves and leaf terminals at the preflowering stage. Therefore, the control of such insects is more effective than that of the obligatory bollworms and borers.

Nevertheless, mature larvae of the wood-boring leopard moth, *Zeuzera pyrina*, can be adequately controlled in the galleries by commercial *B. t.* products (Navon, unpublished data). With the codling moth, even initial ingestion of *B. t.* was avoided because the larva feeds little on the apple skin (Andermatt *et al.*, 1988). Another example of unusual feeding behaviour is found with the cutworm *A. ipsilon* (Table 6.5). The larvae of this insect feed on hypogeal parts of the plant and on leaves close to the ground, where spray droplet penetration is poor and they are therefore little affected by the microbial spray.

Table 6.5. Larval feeding behaviour and the use of *Bacillus thuringiensis* in pest control

Plant/ commodity	Target insect	Infestation	Pest control[a]
Cotton	*Spodoptera* spp.	Leaf	+ +
	Heliothis/Helicoverpa spp.	Leaf terminals, boll	+
	Earias spp.	Boll, square	+/0
	Pectinophora gossypiella	Boll, square	0
Corn	*Helicoverpa* spp.	Leaf, fruit	+/0
	Ostrinia nubilalis	Leaf, stalk, fruit	+
	Sesamia spp.	Stalk, fruit	0
	Chilo spp.	Stalk	0
	Agrotis ipsilon	Hypogeal, leaf	+
Apple	*Cydia pomonella*	Fruit	0
	Zeuzera pyrina	Wooden parts	0 + +[b]
Cabbage	*Trichoplusia ni*	Leaf	+ + +
	Plutella xylostella	Leaf	+ + +
	Pieris rapae	Leaf	+ + +
	Agrotis ipsilon	Hypogeal, leaf	+
Avocado	*Boarmia selenaria*	Leaf, fruit	+ + +[c]
	Cryptoblabes gnidiella	Fruit peel	+ + +[c]
Stored products	*Plodia interpunctella*	Grain	+ +/+ + +[d]
	Ephestia cautella	Grain	+ +/+ + +[d]
	Sitotroga cerealella	Inside grain kernel	+[d]
	Galleria mellonella	Beeswax	+ +/+ + +
Forest	*Choristoneura fumiferana*	Leaf	+ + +
	Thaumatopoea spp.	Leaf	+ +/+ + +
	Lymantria dispar	Leaf	+ + +

[a] Ref.: Krieg and Langenbruch, 1981; Burges, 1986, personal survey.
[b] Control of mature larvae in galleries (Navon, unpublished).
[c] Wysoki, 1989.
[d] McGaughey, 1985.

H. virescens which were exposed to *B. t.* for 72 h showed remarkable recovery from gut paralysis after the *B. t.* source was removed (Dulmage *et al.*, 1978). This phenomenon was supported by the observation that resumption of feeding occurred while regeneration of the gut tissue was in progress (Spies and Spence, 1985). Improving formulations can increase larval mortality so that the extent of sublethal poisoning or recovery will be minimal. The management of coniferous defoliators was improved by increasing the *B. t.* concentration in the spray fluid and reducing the number of droplets per needle (van Frankenhuyzen, 1990). This improvement probably resulted from larval ingestion of a lethal *B. t.* concentration during a single leaf meal before feeding had been arrested. The addition of an adjuvant mixture to stimulate larval feeding on *B. t.* products has been a useful strategy in agriculture. The feeding stimulant Coax® increased the effectiveness of pest control in cotton (Bell and Romine, 1980). Moreover, the use of *B. t.* in concentrated forms (capsules, granules, concentrated drops) with feeding stimulants helps to ensure that larvae ingest a lethal quantity before feeding cessation. This approach fits in very well with the strategy of breeding host plants for insect resistance. Distasteful plants may induce the larva to prefer the *B. t.* bait product over the plant tissues.

Acquisition of sublethal amounts of *B. t.* is common where larvae are exposed to *B. t.* residues on the plant. Prolonged larval development was one of the most common sublethal effects (Morris, 1982). The continuous exposure of the insect to high sublethal *B. t.* concentrations greatly reduced pupal weight, adult emergence and fertility (Salama *et al.*, 1981). Prolongation of larval stages caused by sublethal intoxication enhanced the joint action of *B. t.* with entomophagous control agents, since the delay in larval development increases the availability of the pest to parasitoids and predators (Weseloh *et al.*, 1983).

INTEGRATION OF *B. THURINGIENSIS* IN LEPIDOPTERAN PEST CONTROL STRATEGIES

Interactions and joint action of *B. thuringiensis* with microbial or chemical control agents

Nuclear polyhedrosis viruses (NPVs) form a major subgroup of the insect pathogenic family Baculoviridae. NPVs are particularly important as pathogens of harmful Lepidoptera and have been much investigated as pest control agents (Entwistle and Evans, 1985). The effectiveness of the joint action of *B. t.* and NPV was either additive or antagonistic for several agricultural pests such as *T. ni* (Jaques and Morris, 1981), and *H. virescens* (Table 6.6). The source of NPV did not affect the interactions. Incompatibility between *B. t.* and the virus lay in feeding arrestment caused by lethal

Table 6.6. Interaction of commercial *Bacillus thuringiensis* subsp. *kurstaki* with other microbial agents

Target insect	Assay	Microbial insecticide	Interaction/ joint action	Ref.[a]
Spodoptera exigua	*B. t.* infected larvae	*Steinernema feltia*	Antagonistic	(5)
Spodoptera exigua	Diet/neonates	β-exotoxin	Potentiation	(3)
Spodoptera frugiperda	Diet/neonates, third and fourth instars	β-exotoxin	Additive/ synergistic	(4)
Heliothis virescens	Cotton field/ neonate larvae	NPV-*H. virescens* NPV-*A. californica*	< Additive	(1)
Heliothis virescens	Cotton field	NPV-*A. californica*	Additive	(2)

[a] (1) Luttrell *et al.*, 1982; (2) Bell and Romine, 1980; (3) Moar *et al.*, 1986; (4) Gardner, 1988; (5) Kaya and Burlando, 1989.

B. t. doses which prevented larvae from ingesting virus in sufficient amounts to enhance the efficiency of *B. t.*

It is not known whether midgut damage by *B. t.* affects the course of the virus disease. The additive interaction of the two microorganisms might be the result of NPV infection in larvae which have either recovered from *B. t.* poisoning or were initially affected by it. For example, where *B. t.* plus NPV were used to control *H. virescens*, crop yield was increased (Bell and Romine, 1980). Although the effect of *B. t.* alone was not tested, it was demonstrated that the joint action of *B. t.* and NPV (plus Coax® as a feeding adjuvant) increased the economic value of the crop more than the virus treatment alone. The joint action of *B. t.* and NPV in sublethal amounts provided better control of pests in brassicas than normal rates of either microbe alone (Tompkin *et al.*, 1986); ingestion of sublethal amounts of *B. t.* toxin probably did not reduce feeding. Larval resistance to NPV did not affect susceptibility to *B. t.* (Fuxa and Richter, 1990). In view of the increasing importance of insect resistance to *B. t.*, it would be useful to evaluate the sensitivity of *B. t.* resistant larvae to viruses and other microbial agents. Incompatibility between *B. t.* affected larvae and entomophilic nematodes has been observed (Kaya and Burlando, 1989). The nematodes did not produce progeny in the *B. t.* affected larvae.

The insecticidal activity of *B. t.* β-exotoxin has attracted increasing attention in recent years. Thuringiensis® (Abbott), a commercial β-exotoxin product, has been used in interaction experiments with the spore-crystal product: effects varied with larval instar, exposure time and concentration of toxins/crystals. In a 7-day assay with neonates, potentiation of *B. t.* activity was recorded (Moar *et al.*, 1986). In a 3-day assay with third instar larvae, the effect was additive but became synergistic after 7 days

(Gardner, 1988), probably reflecting the slow action of β-exotoxin which was expressed in moulting failure. Nevertheless, a shorter lethal time was obtained by the joint action of the two products. On plants, the β-exotoxin plus the spore-crystal was less effective than in dietary bioassays, but sublethal effects prolonged development in mature larvae. During the 1970s it was shown that a low chemical insecticide dosage combined with *B. t.* provided good crop protection (Jaques and Morris, 1981). Most of the interactions of chemical insecticides with *B. t.* have been additive or synergistic. The chemical insecticides tested were mostly organophosphates and carbamates, which poisoned the larval nervous system through both feeding and contact. However, chemical insecticides such as chlordimeform, which was synergistic with *B. t.* against cabbage pests (Jaques and Morris, 1981), had to be discontinued in several countries because they represented a serious health hazard. This chemical, as well as organophosphates and carbamates, has been replaced by pyrethroids and insect growth regulators (IGR). Early in the 1980s it was shown that the joint action of *B. t.* with sublethal levels of pyrethroids can be synergistic (Table 6.7). Diflubenzuron belongs to the IGR compounds which interfere with chitin synthesis; therefore larval mortality is not rapid but is postponed until the moult stage. The cessation of feeding caused by *B. t.* probably reduced the amount of IGR in the larval body which may explain the antagonistic effect (Mohamed *et al.*, 1983). Nevertheless, diflubenzuron persisted well on the crop and could control larvae recovering from *B. t.* intoxication. The combination of *B. t.* and the IGR has been used extensively for the control of the gypsy moth in California (Dreistadt and Dahlsten, 1989) and replaced the less selective chemical insecticide carbaryl. In

Table 6.7. Interactions of *Bacillus thuringiensis* with chemical insecticides

Target insect	Assay	Chemical insecticide	Interaction/ joint action	Ref.[a]
Operophtera brumata	Spray	Pyrethroid	Synergistic	(1)
Tortrix viridana	Spray	Pyrethroid	Synergistic	
Heliothis virescens	Cotton spray	Chlordimeform	Synergistic	(2)
Heliothis virescens	Cotton spray	Diflubenzuron	Antagonistic	(2)
Spodoptera littoralis	Diet/larvae	Pyrethroids	Synergistic	(3)
Spodoptera exigua	Diet	Neem	Antagonistic	(4)

[a] (1) Svestka and Vankova, 1980; (2) Mohamed *et al.*, 1983; (3) Salama *et al.*, 1984; (4) Moar and Trumble, 1987.

organic agriculture *B. t.* is often combined with the natural pyrethroid pyrethrum.

An extract of the Neem tree seed, *Azadirachta indica*, is a phagodeterrent as well as growth regulator, which may be the cause of its antagonistic interaction with *B. t.* (Table 6.7). More work is needed to evaluate the joint action of *B. t.* with other insect pathogens and with those new insecticides that adversely affect insect development.

Effect of *B. thuringiensis* on insect parasitoids and predators

The combination of microbial pathogens with parasite/predator control is one of the most effective strategies in integrated pest management (IPM) programmes used widely in organic agriculture. Much work in this area has been carried out on interactions between *B. t.* and the natural enemies of lepidopteran forest pests (Table 6.8). Enhancement of parasitism by natural populations of *Cotesia (Apanteles) melanoscelus* (Hymenoptera: Braconidae) in the gypsy moth was recorded by Ticehurst *et al.* (1982), Weseloh *et al.* (1983) and Webb *et al.* (1989). This was explained by the fact that the *B. t.* treated larvae were maintained at a life stage susceptible to parasitism for longer than the untreated insects, resulting in a higher rate of parasitism. The synergistic interaction of *B. t.* with the braconid parasitoids reflects a common aspect of *B. t.* control in which only a part of the insect population is controlled by *B. t.* and larvae surviving or recovering from treatment were further exposed to parasitism. Conversely, very few progeny of the larval parasitoid *Cotesia rubecula* emerged from *Anteogeia (Pieris) rapae*, imported cabbageworm, larvae treated with *B. t.*, because they could not develop in a host suffering acute intoxication (McDonald *et al.*, 1990). The lower the *B. t.* concentration fed to *P. rapae* larvae, the higher the parasitoid survival. Furthermore, parasitized larvae were less susceptible to *B. t.* because their ingestion of *B. t.* was reduced. These findings suggest that simultaneous application of *B. t.* with a release of larval parasitoids is undesirable in IPM programmes. Instead, the release of larval parasitoids at intervals between or after *B. t.* application can be a more effective biological control strategy.

Trichogramma spp. (Hymenoptera: Trichogrammatidae) egg parasitoids are one of the most effective entomophagous agents. They can be mass-produced and released in the IPM of several agricultural pests, such as *O. nubilalis*. It was reported that *B. t.* is compatible with *Trichogramma caoeciae* (Hassan *et al.*, 1983). Furthermore, feeding *Trichogramma platneri* with high *B. t.* concentrations did not affect their survival (Wysoki, 1989). The compatibility of *B. t.* with *Trichogramma* spp. thus seems to be high, because they are not affected by *B. t.* and the egg is not a target stage for *B. t.*

Carabid beetles (Cameron and Reeves, 1990) and shrews, which are

Table 6.8. Interaction of *Bacillus thuringiensis* with entomophagous parasitoids and predators

Target insect	B. t. strain	Host plant	Parasitoid/predator	Interaction	Ref.[a]
Boarmia selenaria	HD-1	Avocado	*Trichogramma platneri* (Trichogrammatidae)	No effect	(6)
Choristoneura occidentalis	HD-1 NRD12	Forest	*Apanteles fumiferanae* (Braconidae)	No effect	(3)
			Glypta fumiferanae (Ichneumonidae)	No effect	(3)
Choristoneura pinus	HD-1	Forest	*Sorex* spp. (shrews)	No effect on small mammals	(4)
Lymantria dispar	HD-1	Forest	*Cotesia melanoscela*[b] (Braconidae)	Synergistic	(5)
Lymantria dispar	HD-1 HD-263	Forest	*Apanteles melanoscelus*[b] (Braconidae)	Synergistic	(2)
Lymantria dispar	HD-1	Forest	*Apanteles melanoscelus*[b] (Braconidae)	Synergistic	(1)
Lymantria dispar	HD-1	Forest	*Callosoma frigidum* (Carabidae)	No effect	(7)
Pieris rapae	HD-1	Cabbage	*Cotesia rubecula* (Braconidae)	Antagonistic	(8)

[a] (1) Ticehurst *et al.*, 1982; (2) Weseloh *et al.*, 1983; (3) Niwa *et al.*, 1987; (4) Innes and Bendell, 1989; (5) Webb *et al.*, 1989; (6) Wysoki, 1989; (7) Cameron and Reeves, 1990; (8) McDonald *et al.*, 1990.
[b] *Cotesia melanoscela* = *Apanteles melanoscelus*.

probable predators of the jack-pine budworm, *Choristoneura pinus* (Innes and Bendell, 1988) were not affected by *B. t.* in treated forests. Several working groups of the International Organization for Biological Control (IOBC) recorded the compatibility of a commercial *B. t.* product (Dipel®) with important arthropod predators such as the lacewing, *Chrysopa carnea* (Neuroptera: Chrysopidae) as well as predacious mites (Hassan *et al.*, 1983)

Interactions among lepidopteran pests, *B. thuringiensis* and plant allelochemicals

The first investigations on the interaction between *B. t.* and plants reported *B. t.* spore inactivation by extracts from the leaves of trees (Jaques and Morris, 1981). The evaluation of interactions between specific allelochemicals, lepidopteran pests and *B. t.* has only been made in recent years (Table 6.9). An antagonistic interaction between *B. t.* and tobacco in *M. sexta* was caused by nicotine reducing larval feeding (Krischik *et al.*, 1988). Dietary cinnamic acid and *p*-coumarine enhanced the effect of *B. t.* in the sunflower moth, *Homoeosoma electellum* (Brewer and Anderson, 1990). Cotton condensed tannin antagonized the effect of *B. t.* strain HD-73 crystals on *H. virescens* neonates through a reduction in feeding and possibly through interaction with the toxin in the midgut (Navon, Hare and Federici, unpublished data). The antagonistic inter-action of tannin with the protoxin and the activated toxin *in vitro* (Lüthy *et al.*, 1985) suggests that postingestive interactions of tannin with the *B. t.* probably occur in the larval midgut. The enhancement of *B. t.* effects on *M. sexta* by dietary L-canavanine (Felton and Dahlman, 1984) could not be defined because the effect of this arginine analogue was exerted at several sites in the insect.

A synergistic effect was caused by the enhancement of the crystal pro-tein activity with polyphenol oxidase and chlorogenic acid (Ludlum *et al.*, 1991). Both allelochemicals serve as resistance factors against *H. zea* in tomato plants. *B. t.* toxicity was enhanced by the monophenolase activity on the tyrosil residues of the toxin. In addition, polyphenol oxidase produced *o*-quinones from chlorogenic acid and the quinones in turn alkylated the crystal protein. This alkylation was proposed as the mech-anism for increasing protein toxicity.

Whenever the efficacy of the *B. t.* in pest management is low the possi-bility of antagonistic interactions among the insect, the *B. t.* product and plant allelochemicals must not be excluded. An antagonistic effect on *B. t.* activity characterized by feeding reduction caused by deterrent allelochemicals will necessitate re-evaluation of the microbial control strategy. On the other hand, allelochemicals potentiating *B. t.* activity by poisoning the larva (without reducing feeding) could be used in

Table 6.9. Interaction among lepidopteran pests, *Bacillus thuringiensis* and plant allelochemicals

Target insects	B. t. preparation	Allelochemicals	Interaction	Ref.[a]
Manduca sexta	Spores + crystals	Nicotine	Antagonistic >0.4% reduced feeding	(3)
Manduca sexta	Spores + crystals	L-canavanine	Enhancement No specifications	(1)
Homoeosoma ellectellum	Spores + crystals	Cinnamic acid	Enhancement	(4)
Helicoverpa zea	Spores + crystals	p-Coumaric acid	Enhancement	
	Crystals	Polyphenol oxidase	Synergistic,	(5)
	Crystals	Polyphenol oxidase + chlorogenic acid	Enhancement of crystal toxicity	
Pieris brassicae	Crystals	Tannin	< Antagonistic	
	Protoxin	Tannin	Antagonistic	(2)
	Toxin	Tannin	Antagonistic	
Heliothis virescens	Crystals, HD-73	Cotton tannin	Antagonistic Reduced feeding	(6)

[a] (1) Felton and Dahlman, 1984; (2) Lüthy *et al.*, 1985; (3) Krischik *et al.*, 1988; (4) Brewer and Anderson, 1990; (5) Ludlum *et al.*, 1991; (6) Navon, Hare and Federici, (unpublished).

developing insect host plant resistance. In addition, the allelochemicals that potentiate the protein crystal might be useful in improving *B. t.* formulations.

CONCLUSIONS AND FUTURE PROSPECTS

An evaluation of the state-of-the-art in lepidopteran pest control with *B. t.* leads to the following conclusions and future prospects.

1. The choice of natural and genetically modified *B. t.* strains has increased significantly and the use of new strains and delivery systems instead of, or in addition to, the present approach will increase the potency and insecticidal spectrum of the microbial product.
2. Decision-making regarding the application of *B. t.* with other microbial pest control agents or chemical insecticides should be based on the type of interaction between *B. t.* (at acute intoxication) and other control means (baculovirus, entomophilic nematodes, larval parasitoids, IGRs, deterrent chemicals). Sequential rather than simultaneous applications of *B. t.* with other control agents should be considered. Compatibility between *Trichogramma* spp., other parasitoids and various important predators indicates use of *B. t.* with entomophagous arthropods is an effective substitute for chemical pesticides. The use of *B. t.* together with synergistic control agents, both at sublethal concentrations, could be another useful plant protection strategy.
3. Interactions among lepidopteran pests, *B. t.* and host plant allelochemicals cannot be ignored in developing microbial pest management methods. Research should be linked to programmes of crop breeding for insect host plant resistance and to the development of *B. t.* formulations.
4. It is expected that public awareness of environmental safety and health will increase. Nevertheless, the need for public education to accept *B. t.* as a useful substitute for harmful chemical pesticides should not be overlooked; for example, to convince the consumer that a *B. t.* treated agricultural product is preferable to a totally insect-free product which may be hazardous due to the presence of persistent chemical residues.
5. Although the area under organic agriculture is estimated in many countries at less than 1% of the total land available for agriculture, it is expected that consumer demands for organic vegetables, fruits and postharvest commodities will increase. This means that the flow of effective microbial pest management programmes from organic to conventional agriculture will be intensified. However, the main effort in *B. t.* application will in future be focused on the replacement of

chemical pesticides with new microbial products and delivery systems. This effort will be best achieved by adoption of more holistic approaches to *B. t.* interactions with biological and environmental systems.

REFERENCES

Andermatt, M., Mani, E., Wildbolz, Th. and Lüthy, P. (1988). 'Susceptibility of *Cydia pomonella* to *Bacillus thuringiensis* under laboratory and field conditions'. *Entomol. Exp. Appl.*, **49**, 291–295.

Beegle, C. C., Dulmage, H. T. and Wolfenberger, D. A. (1982). 'Relationships between laboratory bioassay-derived potencies and field efficacies of *Bacillus thuringiensis* isolated with different spectral activity'. *J. Invert. Pathol.*, **39**, 138–148.

Bell, M. R. and Romine, C. L. (1980). 'Tobacco budworm field evaluation of microbial control in cotton using *Bacillus thuringiensis* and nuclear polyhedrosis virus with a feeding adjuvant'. *J. Econ. Entomol.*, **73**, 427–430.

Brewer, G. J. and Anderson, M. D. (1990). 'Modification of the effect of *Bacillus thuringiensis* on sunflower moth (Lepidoptera: Pyralidae) by dietary phenols'. *J. Econ. Entomol.*, **83**, 2219–2224.

Broza, M., Sneh, B., Yawetz, A., Oron, U. and Honigman, A. (1984). 'Commercial application of *Bacillus thuringiensis* var. *entomocidus* to cotton fields for the control of *Spodoptera littoralis* Boisd. (Lepidoptera: Noctuidae)'. *J. Econ. Entomol.*, **77**, 1530–1533.

Burges, H. D. (1981). 'Strategy for the microbial control of pests in 1980 and beyond'. In *Microbial Control of Pests and Plant Diseases 1970–1980* (Ed. H. D. Burges), pp. 797–836, Academic Press, New York.

Cameron, E. A. and Reeves, R. M. (1990) 'Carabidae (Coleoptera) associated with gypsy moth, *Lymantria dispar* (L.) (Lepidoptera: Lymantriidae), populations subjected to *Bacillus thuringiensis* Berliner treatments in Pennsylvania'. *Can. Entomol.*, **122**, 123–129.

de Barjac, H. and Frachon, E. (1990). 'Classification of *Bacillus thuringiensis* strains'. *Entomophaga*, **35**, 233–240.

Delannay, X., LaVallee, B. J., Proksch, R. K., Fuchs, R. L., Sims, S. R., Greenplate, J. T., Marrone, P. G., Dodson, R. B., Augustine, J. J., Layton, J. G. and Frischoff, D. A. (1989). 'Field performance of transgenic tomato plants expressing the *Bacillus thuringiensis* var. *kurstaki* insect control protein'. *Bio/Technology*, **7**, 1265–1270.

Dreistadt, S. H. and Dahlsten, D. L. (1989). 'Gypsy moth eradication in Pacific coast states: History and evaluation'. *Bull. Entomol. Soc. Am.*, **35**, 13–19.

Dulmage, H. T., Boening, O. P., Rehnborg, C. S. and Hansen, G. D. (1971). 'A proposed standardized bioassay for formulations of *Bacillus thuringiensis* based on the international unit'. *J. Invert. Pathol.*, **18**, 240–245.

Dulmage, H. T., Graham, H. M. and Martinez, E. (1978). 'Interactions between the tobacco budworm, *Heliothis virescens*, and the δ-endotoxin produced by the HD-1 isolate of *Bacillus thuringiensis* var. *kurstaki*: relationship between length of exposure to the toxin and survival'. *J. Invert. Pathol.*, **32**, 40–50.

Dunkle, R. L. and Shasha, B.S. (1989). 'Response of starch-encapsulated *Bacillus thuringiensis* containing ultraviolet screens to sunlight'. *Env. Entomol.*, **18**, 1035–1041.

Entwistle, P. F. and Evans, H. F. (1985). 'Viral control'. In *Comprehensive Insect Physiology. Biochemistry and Pharmacology* (Eds. G. A. Kerkut and L. I. Gilbert) Vol. 12, pp. 347–412, Pergamon Press, Oxford.

Felton, G. W. and Dahlman, D. L. (1984). 'Allelochemical induced stress: effects of L-canavanine on the pathogenicity of *Bacillus thuringiensis* in *Manduca sexta*'. *J. Invert. Pathol.*, **44**, 187–191.

Fuxa, J. (1989). 'Fate of released entomopathogens with reference to risk assessment of genetically engineered microorganisms'. *Bull. Entomol. Soc. Am.*, **35**, 12–24.

Fuxa, J. R. and Richter, A. R. (1990). 'Response of nuclear polyhedrosis virus-resistant *Spodoptera frugiperda* larvae to other pathogens and to chemical insecticides'. *J. Invert. Pathol.*, **55**, 272–277.

Gardner, W. A. (1988). 'Enhanced activity of selected combinations of *Bacillus thuringiensis* and Beta-exotoxin against fall armyworm (Lepidoptera: Noctuidae) larvae'. *J. Econ. Entomol.*, **81**, 463–469.

Gelernter, W. D. (l990). 'Targeting insecticide-resistant markets: new developments in microbial based-products'. In *Managing Resistance to Agrochemicals: From Fundamental Research to Practical Strategies* (Eḍs. M. B. Green, W. K. Moberg and H. LeBaron), ACS Series 421, pp. 105–117.

Hassan, S. A., Bigler, F., Bogenschutz, H., Brown, J. U., Firth, P., Ledieu, M. S., Naton, E., Oomen, P. A., Overmeer, W. P. J., Rieckermann, W., Samsøe-Petersen, L., Viggiani, G. and van Zon, A. Q. (1983). 'Results of the second joint pesticide testing programs by IOBC/WPRS-working Group "Pesticides and Beneficial Arthropods"'. *Z. Ang. Entomol.*, **95**, 151–158.

Höfte, H. and Whiteley, H. R. (1989). 'Insecticidal crystal proteins of *Bacillus thuringiensis*'. *Microbiol. Rev.*, **53**, 242–255.

Innes, D. G. and Bendell, J. F. (1987). 'The effects on small-mammal populations of aerial application of *Bacillus thuringiensis*, fenitrothion, and Matacil® used against jack pine budworm in Ontario'. *Can. J. Zool.*, **67**, 1318–1323.

Jaques, R. P. and Morris, O. N. (1981). 'Compatibility of pathogens with other methods of pest control and with different crops'. In *Microbial Control of Pests and Plant Diseases 1970–1980* (Ed. H. D. Burges), pp. 695–715, Academic Press, New York.

Jaquet, F., Hütter, R. and Lüthy, P. (1987). 'Specificity of *Bacillus thuringiensis* delta-endotoxin'. *Appl. Environ. Microbiol.*, **53**, 500–504.

Jarrett, P. and Burges, H. D. (1986). 'Isolates of *Bacillus thuringiensis* active against *Mamestra brassica* and some other species: alternatives to the present commercial isolates HD-1'. *Biol. Agric. Hortic.*, **4**, 39–45.

Kaya, H. K. and Burlando, T. M. (1989). 'Development of *Steinernema feltiae* (Rhabditida: Steinernematidae) in diseased insect hosts'. *J. Invert. Pathol.*, **53**, 164–168.

Krieg, A. and Langenbruch, G. A. (1981). 'Susceptibility of arthropod species to *Bacillus thuringiensis*'. In *Microbial Control of Pests and Plant Diseases 1970–1980* (Ed. H. D. Burges), pp. 837–896, Academic Press, New York.

Krischik, V. A., Barbosa, P. and Reichelderfer, C. (1988). 'Three trophic level interactions: allelochemicals, *Manduca sexta* (L.), and *Bacillus thuringiensis* var. *kurstaki* Berliner'. *Env. Entomol.*, **17**, 476–482.

Lecadet, M.-M. and Martouret, D. (1987). 'Host specificity of the *Bacillus thuringiensis* δ-endotoxin toward lepidopteran species: *Spodoptera littoralis* Bdv. and *Pieris brassicae* L.'. *J. Invert. Pathol.*, **49**, 37–48.

Lethbridge, G. (1989). 'An industrial view of microbial inoculants for crop plants'. In *Microbial Inoculation of Crop Plants* (Eds. E. Campbell and R. M. MacDonald) *Special Publ. Soc. Gen. Microbiol.*, **25**, 11–28.

Li, R. S., Jarrett, R. and Burges, H. D. (1987). 'Importance of spores, crystals and δ endotoxins in the pathogenicity of different varieties of *Bacillus thuringiensis* in *Galleria mellonella* and *Pieris brassicae*'. *J. Invert. Pathol.*, **50**, 277–284.

Lipa, J. J. (1991). 'Microbial pesticides and their use in the EPRS-IOBC (Eastern Europe)'. *Proc. IInd European Meeting Microbial Control of Pests* (Ed. H. F. Evans), IOBC, WRPS Bulletin, Rome, 6–8 March, 1989, pp. 23–32.

Ludlum, C. T., Felton, G. W. and Duffy, S. S. (1991). 'Plant defences: chlorogenic acid and polyphenol oxidase enhance toxicity of *Bacillus thuringiensis* subsp. *kurstaki* to *Heliothis zea*'. *J. Chem. Ecol.*, **17**, 217–237.

Lüthy, P., Hofmann, C. and Jaquet, F. (1985). 'Inactivation of delta-endotoxin of *Bacillus thuringiensis* by tannin'. *FEMS Microbiol. Lett.*, **28**, 31–33.

Luttrell, R. G., Young, S. Y., Yearian, W. C. and Horton, D. L. (1982). 'Evaluation of *Bacillus thuringiensis*-spray, adjuvant-viral insecticide combinations against *Heliothis spp.* (Lepidoptera: Noctuidae)'. *Env. Entomol.*, **11**, 783–787.

MacIntosh, S. C., Stone, T. B., Sims, S. R., Hunt, P. L., Greenplate, J. T., Marrone, P. G., Perlak, F. J., Fischhoff, D. A. and Fuchs, R. L. (1990). 'Specificity and efficacy of purified *Bacillus thuringiensis* proteins against agronomically important insects'. *J. Invert. Pathol.*, **56**, 258–266.

McDonald, R. C., Kok, L. T. and Yousten, A. A. (1990). 'Response of fourth instar *Pieris rapae* parasitized by the braconid *Cotesia rubecula* to *Bacillus thuringiensis* subsp. *kurstaki* δ-endotoxin'. *J. Invert. Pathol.*, **56**, 422–423.

McGaughey, W. H. (1985). '*Bacillus thuringiensis*: A critical review'. In *Proc. 4th Int. Work. Conf. Stored-Products Protection* (Eds. E. Donahaye and S. Navarro), pp. 14–23, Tel Aviv, Israel.

McGaughey, W. M. H. and Johnson, D. E. (1987). 'Toxicity of different serotypes and toxins of *Bacillus thuringiensis* to resistant and susceptible Indianmeal Moths (Lepidoptera: Pyralidae)'. *J. Econ. Entomol.*, **80**, 1122–1126.

Moar, W. J. and Trumble, J. T. (1987). 'Toxicity, joint action, and mean time of mortality of Dipel 2X, avermectin B1, neem and thuringiensin against beet armyworms (Lepidoptera: Noctuidae)'. *J. Econ. Entomol.*, **80**, 588–592.

Moar, W. J., Osbrink, W. L. A. and Trumble, J. T. (1986). 'Potentiation of *Bacillus thuringiensis* var. *kurstaki* with thuringiensin on beet armyworm (Lepidoptera: Noctuidae)'. *J. Econ. Entomol.*, **79**, 1443–1446.

Mohamed, A. I., Young, S. Y. and Yearian, W. C. (1983). 'Susceptibility of *Heliothis virescens* (F.) (Lepidoptera: Noctuidae) larvae to microbial agent-chemical pesticide mixtures on cotton foliage'. *Env. Entomol.*, **12**, 1403–1405.

Morris, O. N. (1982). 'Bacteria as pesticides: forest application'. In *Microbial and Viral Pesticides* (Ed. E. Kurstak), pp. 239–287, Marcel Dekker, New York.

Morris, O. N. (1983) 'Protection of *Bacillus thuringiensis* from inactivation by sunlight'. *Can. Entomol.*, **115**, 1215–1227.

Nakamura, L. K. and Dulmage, H. T. (1988). *Bacillus thuringiensis* cultures available from the U.S. Department of Agriculture, *U. S. Department of Agriculture, Technical Bulletin No. 1738*, 38 pp.

Navon, A. (1989). 'Development of potency bioassays for selecting *Bacillus thuringiensis* preparations against agricultural pests in Israel'. *Israel J. Entomol.*, **23**, 115–118.

Navon, A., Wysoki, M. and Keren, S. (1983). 'Potency and effect of *Bacillus thuringiensis* preparations against larvae of *Spodoptera littoralis* and *Boarmia* (*Ascotis*) *selenaria*'. *Phytoparasitica*, **11**, 3–11.

Navon, A., Klein, M. and Braun, S. (1990). *Bacillus thuringiensis* potency bioassays against *Heliothis armigera*, *Earias insulana*, and *Spodoptera littoralis* larvae based on standardized diets'. *J. Invertebr. Pathol.*, **55**, 387–393.

Niwa, C. G., Stelzer, M. J. and Beckwith, R. C. (1987). 'Effects of *Bacillus thuringiensis* on parasites of western spruce budworm (Lepidoptera: Tortricidae)'. *J. Econ. Entomol.*, **80**, 750–753.

Perlak, F. J., Deaton, R. W., Armstrong, T. A., Fuchs, R. L., Sims, S. R., Greenplate, J. T. and Frischoff, D. A. (1990). 'Insect resistant cotton plants'. *Bio/Technology*, **8**, 939–943.

Salama, H. S., Foda, M. S., El-Sharaby, A., Matter, M. and Khalafallah (1981). 'Development of some lepidopterous cotton pests as affected by exposure to sublethal levels of endotoxins of *Bacillus thuringiensis* for different periods'. *J. Invert. Pathol.*, **38**, 220–229.

Salama, H. S., Foda, M. S., Zaki, F. N. and Moawad, S. (1984) 'Potency of combinations of *Bacillus thuringiensis* and chemical insecticides on *Spodoptera littoralis* (Lepidoptera: Noctuidae)'. *J. Econ. Entomol.*, **77**, 885–890.

Spies, A. G. and Spence. K. D. (1985) 'Effect of sublethal *Bacillus thuringiensis* crystal endotoxin treatment on the larval midgut of a moth, *Manduca*: SEM study'. *Tissue Cell*, **17**, 379–394.

Svestka, M. and Vankova, J. (1980). 'Über die wirkung von *Bacillus thuringiensis* in Kombination mit dem synthetischen Pyrethroid Ambusch auf *Operophtera brumata*, *Tortrix viridana* und die Insectenfauna eines Eichenbestandes'. *Anz. Schadlinsk. Pflanz. Unweltschutz*, **53**, 6–10.

Takaki, S. (1985). '*B. t.* preparations in Japan'. *Japan Pest. Inform.*, **25**, 23–26.

Thompson, W. T. (1989). *Agricultural Chemicals*, Book I, pp. 64–70, Thompson Publications, Fresno, CA.

Ticehurst, M., Fusco, R. A. and Blumenthal, E. M. (1982) 'Effects of reduced rates of Dipel 4L, Dylox 1.5 oil and Dimilin W-25 on *Lymantria dispar* (L.) (Lepidoptera: Lymantriidae), parasitism and defoliation'. *Env. Entomol.*, **11**, 1058–1062.

Tompkin, G. J., Linduska, J. J., Young, J. M. and Dougherty, E. M. (1986). 'Effectiveness of microbial and chemical insecticides for controlling cabbage looper (Lepidoptera: Noctuidae) and imported cabbageworm (Lepidoptera: Pieridae) on collard in Maryland'. *J. Econ. Entomol.*, **79**, 497–501.

Trottier, M. R., Morris, O. N. and Dulmage, H. T. (1988). 'Susceptibility of the Bertha armyworm, *Mamestra configurata* (Lepidoptera, Noctuidae), to sixty-one strains from ten varieties of *Bacillus thuringiensis*'. *J. Invert. Pathol.*, **51**, 242–249.

van Frankenhuyzen, K. (1990). 'Effects of temperature and exposure time on toxicity of *Bacillus thuringiensis* spray deposits to spruce budworm, *Choristoneura fumiferana* Clemens (Lepidoptera: Tortricidae)'. *Can. Entomol.*, **122**, 69–75.

van Frankenhuyzen, K. and Nystrom, C. (1989). 'Residual toxicity of a high-potency formulation of *Bacillus thuringiensis* to Spruce budworm (Lepidoptera: Tortricidae)'. *J. Econ. Entomol.*, **82**, 868–872.

Webb, R. E., Shapiro, M., Podgwaite, J. D., Reardon, R. C., Tatman, K. M., Venables, L. and Kolodny-Hirsch, D. M. (1989). 'Effect of aerial spraying with Dimilin, Dipel, or Gypchek on two natural enemies of the gypsy moth (Lepidoptera: Lymantriidae)'. *J. Econ. Entomol.*, **82**, 1695–1701.

Weseloh, R. M., Andreadis, T. G., Moore, R. E. B. (1983). 'Field confirmation of a mechanism causing synergism between *Bacillus thuringiensis* and the gypsy moth parasitoid, *Apanteles melanoscelus*'. *J. Invert. Pathol.*, **41**, 99–103.

Wysoki, M. (1989). '*Bacillus thuringiensis* preparations as a mean for control of lepidopteran avocado pests in Israel'. *Israel J. Entomol.*, **23**, 119–130.

Wysoki, M. and Scheepens, M. H. M. (1990). 'The pathogenicity of *Bacillus thuringiensis* strains HD-263 and HD-251 to the larvae of *Boarmia selenaria* (Lepidoptera: Geometridae)'. *Israel J. Entomol.*, **24**, 125–131.

7 Use of *Bacillus thuringiensis israelensis* against Mosquitoes and Blackflies

NORBERT BECKER[1] AND JOEL MARGALIT[2]

[1]*German Mosquito Control Association (KABS), Europaplatz 5, 6700 Ludwigshafen, Germany, and* [2]*Ben-Gurion University of the Negev, Laboratory for Biological Control, Department of Life Sciences, Beer-Sheva, Israel*

INTRODUCTION

Mosquitoes and blackflies are involved in human health problems in many parts of the world. As vectors of such debilitating conditions such as malaria, filariasis, encephalitis, dengue and yellow fever, they threaten more than three billion people in tropical and subtropical regions. Malaria, transmitted by anopheline mosquitoes, continues to be the most important vector-borne disease. It affects at least 102 tropical countries, placing more than half the world's population at risk, resulting in 100 million malaria infections each year and an estimated one million deaths (WHO, 1991). Lymphatic filarial diseases caused by the nematodes *Wuchereria bancrofti*, *Brugia malayi* and *Brugia timori* affect about 90 million people in Asia, Africa and South America. An estimated 905 million people are directly exposed to the infection by transmission through various genera of mosquitoes, the most important being *Culex quinquefasciatus* and *Mansonia* spp. Onchocerciasis, or 'river blindness', caused by the nematode *Onchocerca volvulus*, which is transmitted by blackflies of the genus *Simulium*, affects 18 million people in 28 countries. Eighty million people are exposed to the risk of infection.

Arbovirus diseases, such as dengue, primarily vectored by the mosquito *Aedes aegypti*, constitute an important burden to mankind in terms of morbidity and mortality. About 1.5 billion people in the tropics, mainly in Asia, the Western Pacific region, the Caribbean, as well as Central and South America, live under the risk of dengue infection (Halstead, 1980, 1982; Becker *et al.*, 1991). Yellow fever is enzootic in Africa and South

Bacillus thuringiensis, An Environmental Biopesticide: Theory and Practice. Edited by P. F. Entwistle, J. S. Cory, M. J. Bailey and S. Higgs
© 1993 John Wiley & Sons Ltd

America. Although no urban outbreaks have occurred in the Americas for many decades, the constant threat of jungle yellow fever remains.

Mosquitoes and blackflies can also cause problems in temperate zones. Due to the massive numbers which can occur along the larger river systems, such as the Upper Rhine Valley, mosquitoes, especially various *Aedes* species, can be a major nuisance and reduce the quality of life dramatically. This also applies to blackflies, which torment people as well as grazing livestock near certain rivers in Central Europe and North America. Occasionally, cattle are killed by the bites of these insects.

Despite considerable national and international efforts aimed at suppressing these vector-borne diseases, they still obstruct improvement in health and socioeconomic development in many tropical countries. Prophylactic solutions based on inexpensive vaccines remain unavailable for protection against many of these diseases. In addition to chemotherapy, vector control is an important and effective protective measure which attacks the problem at its root. Vector control programmes, however, still rely to a great extent on chemicals. The annual demand amounts to more than 50,000 tons, with DDT being by far the most commonly used insecticide. Besides their life-saving benefits, insecticides very often have undesirable effects; for example, vectors become resistant to them and their non-selectivity frequently causes ecological damage.

The discovery of *Bacillus thuringiensis* subspecies *israelensis* (*B. t. israelensis*) in 1976 inaugurated a new chapter in vector control (Goldberg and Margalit, 1977; Margalit and Dean, 1985). Not only did it mark a breakthrough in the biological control of mosquito and blackfly larvae but it was also the first time that a new pathotype (B)—specific only to the Nematocera—had been found in the 65 years that had elapsed since the discovery of the first *B. t.* pathotype (A), specific to Lepidoptera, by Berliner (1911). The newly isolated pathotype was characterized as serotype H-14. In the years since its discovery searches have been carried out worldwide for further dipteran-active isolates of *B. t.*, with the support of the United Nations Development Programme/World Bank/World Health Organization Special Programme for Research and Training in Tropical Diseases (WHO/TDR). Over a hundred local strains of *B. t.* H-14 have been isolated in many countries. However, no *B. t.* isolate has yet been found that has a substantially better mosquitocidal effect than the original *B. t. israelensis* isolate. For example, *B. t.* subspecies *medellin* (serotype H-30) isolated in Colombia (Orduz *et al.*, 1992) and *B. t.* 187 (serotype H-14) isolated in China in 1982 have properties which are comparable with those of the original isolate; however, mention should be made of *B. t.* subspecies *morrisoni* (serotype H8a:8b) which was isolated in the Philippines by Padua *et al.* (1984). This isolate, abbreviated as PG-14, possesses an extra protein with a molecular weight of 144 kDa, in addition to the mosquitocidal parasporal protein bodies that are characteristic of *B. t.*

israelensis. Although PG-14 belongs to a different subspecies and to a different serotype, subsequent experiments have shown that it is as effective as *B. t. israelensis* against mosquito larvae.

A second spore-forming bacterium, *Bacillus sphaericus*, has become of increasing importance in recent years. A mosquitocidal strain of *B. sphaericus* first isolated by Kellen and Meyers (1964) showed little toxicity towards mosquito larvae; however, more effective strains, such as strain 2362 isolated from *Simulium* adults in Africa (Weiser, 1984) have since been isolated under the auspices of WHO/TDR. As more effective strains were isolated, so the interest of industry in *B. sphaericus* was aroused and formulations similar to those developed for *B. t. israelensis* are now available for large-scale trials. The high potential of *B. sphaericus* as a microbial control agent lies in its spectrum of efficacy and its ability, under certain conditions, to recycle or to persist in nature, which means that long-term control can be achieved. *B. sphaericus* kills only mosquito larvae, particularly those of the genus *Culex*. The ability to recycle in nature opens up the possibility of successful and cost-effective control of *Culex* species, particularly of *C. quinquefasciatus*, the most important vector of lymphatic filariasis, which breeds primarily in urban areas in highly polluted water-bodies.

PROPERTIES OF *B. THURINGIENSIS ISRAELENSIS*

The insecticidal effect of *B. t. israelensis* emanates from the parasporal body (PSB), which contains four major proteins with molecular weights of 27, 65, 128 and 135 kDa. Neither the spores nor the living bacilli appear to be involved in the insecticidal process. The more or less spherical PSB is formed at the end of sporulation and consists of three types of protein inclusion separated by thin layers. The largest inclusion is round, of slight electron density, occupies approximately 50% of the total volume of the PSB (Federici *et al.*, 1990) and probably contains the 27 kDa protein. The second type of inclusion (about 20% of the volume of the PSB) is moderately electron dense, rod-shaped, and consists almost entirely of a 65 kDa protein (Ibarra and Federici, 1986a). The third type of inclusion (about 25% of the PSB) is spherical, highly electron dense and is thought to contain the 128–135 kDa protein(s).

Tests on the single, solubilized and purified proteins have shown that each type is mosquitocidal, but alone are not as toxic as the intact PSB. This high toxicity is due to a synergistic interaction between the 25 kDa protein (proteolytic product of the 27 kDa protein) and one or more of the higher molecular weight proteins (Ibarra and Federici, 1986b; Chilcott and Ellar, 1988). The PSB is toxic to many mosquito and blackfly species. Other Nematocera are only affected at significantly higher doses. Because of this high toxicity to many of the most important vectors of human disease, *B. t. israelensis* was developed rapidly with support of the WHO/TDR and, following extensive safety-tests and environmental impact studies, was quickly put into practice. This rapid exploitation of *B. t. israelensis* was aided by the relative ease with which it can be mass produced, its high efficacy, environmental safety, ease of handling, storage stability, cost-effectiveness, lack of resistance and suitability for integrated control programmes based on community participation. Furthermore, the costs for development and registration of *B. t. israelensis* (about US$ 500 000) are many times lower than those for a conventional chemical insecticide (about US$20 million).

Formulations

The basic requirement for the successful use of *B. t. israelensis* is the development of effective formulations suited to the biology and habitats of the target organisms. For use against larval blackflies in turbulent and flowing waters, formulations had to be developed that floated for as long as possible (particularly highly concentrated liquid formulations) to minimize the number of treatments a given stretch of river required. Although preparations with short-term effectiveness were sufficient for interventions against larvae of floodwater mosquitoes (either liquid concentrates or wettable powders), preparations effective over a longer period of time (such as briquets) were developed for mosquitoes with a continual succession of generations (such as *Culex* species). In this way, the length of time between treatments could be extended and personnel costs reduced.

The development of effective formulations requires that the feeding behaviour of the different target species be taken into account. Preparations which persist at the bottom of a water container are suited for the control of bottom-feeding larvae. The larvae of anophelines, on the other hand, are controlled most effectively by granules which float on the water surface and release the toxicant slowly. *B. t. israelensis* preparations can now be obtained as wettable powders, fluid concentrates, granules, pellets or briquets (Table 7.1).

A few hundred grams of powder, or 0.5–2.0 litres of liquid concentrate per hectare, are enough to kill all mosquito larvae. In field tests, wettable powders such as Teknar TC® proved to be particularly effective (Table 7.2). In addition to commercially available granules, based on ground corn-cobs, sand granules can also serve as a carrier for wettable powder formulations: a mixture of 50 kg fire-dried quartz sand with 0.8–1.4 litres of vegetable oil (as a binding agent) and 0.9–1.8 kg of *B. t. israelensis* powders should be sufficient for treating 2–3 hectares.

One limitation to the use of *B. t. israelensis* lies in its relatively brief

Table 7.1. Different *Bacillus thuringiensis israelensis* products used in mosquito and blackfly control programmes

Product	Formulation
Teknar TC	Powder
Teknar HP-D	Fluid
Teknar G	Granules
Vectobac TP	Powder
Vectobac 12 AS	Fluid
Vectobac G	Granules
Bactimos WP	Powder
Bactimos G	Granules
Bactimos	Briquets
Skeetal FC	Fluid

activity after it has been applied to water. Expensive retreatments are frequently necessary so that better formulations are needed to provide persistent control. In recent years, progress has been made in this direction with the production of briquets, pellets and tablet formulations. For example, a single briquet per rain barrel (surface area 0.16 m^2 and water depth 0.5 m) produces satisfactory control of larval *Culex pipiens* for more than 1 month as a result of the slow release effects. Sustained-release, floating granules are also being developed. New tablet formulations, based on an asporogenous *B. t. israelensis* strain or *B. t. israelensis* material sterilized by radiation to prevent contamination of drinking water with spores, can successfully be used for the control of container breeding mosquitoes such as *A. aegypti* (Becker *et al.*, 1991) or *C. pipiens* in breeding sites close to human settlements.

When appropriately stored, most *B. t. israelensis* preparations can be kept for long periods of time without losing activity. But preparations should be retested in bioassays according to WHO guidelines (de Barjac, 1983) after they have been stored for more than a year in temperate climates and after 6 months in tropical regions. The activity of particular *B. t. israelensis* products, particularly fluid concentrates, may be more labile than others. Experience has shown that powder formulations lose little of their activity even after many years in storage.

Standardized methods have been developed to determine the LC$_{50}$ values using standard formulations for comparative purposes (e.g. IPS-82 produced by Institut Pasteur) (de Barjac, 1983; Dulmage *et al.*, 1990). The potency is expressed in ITU (International Toxic Units) and has to be calculated according to the following formula:

$$\text{potency sample (ITU/mg)} = \frac{\text{LC}_{50}\ \text{standard}}{\text{LC}_{50}\ \text{sample}} \times \text{potency standard (ITU/mg)}$$

Table 7.2. Field evaluation of various wettable powder formulations of *Bacillus thuringiensis israelensis* against larvae of *Aedes vexans* and *A. cantans*

Product	Treatment rate	Habitat depth	Species	No. larvae/10 dips pre- and post-treatment				
				Pre-treatment	1 Day	2 Days	3 Days	4 Days
Teknar TC (HP-D)	0.1 kg/ha	Swampy woodlands	A. cantans L2/L3	866	28 (96.8)	14 (98.4)	6 (99.3)	0 (100)
Bactimos (1986)	0.2 kg/ha	Swampy woodlands	A. cantans L3/L4	62	3 (95.2)	0 (100)	0 (100)	0 (100)
Bactimos (1989)	0.25 kg/ha	Swampy woodlands	A. cantans L3/L4	168	29 (82.7)	33 (80.4)	46 (72.6)	21 (87.5)
Bactimos (1986)	0.2 kg/ha	Floodlands	A. vexans L1–L3	759	14 (98.2)	2 (99.7)	0 (100)	0 (100)
Vectobac TP (1989)	0.2 kg/ha	Swampy woodlands	A. cantans L4	723	–	45 (93.8)	–	37 (94.9)

Figures in parentheses are % reduction; L1–L4 = larval instars; volume of one dip = 1 litre.

Environmental safety

The environmental safety of *B. t. israelensis* has been confirmed in numerous tests. The US Environmental Protection Agency approved the use of *B. t. israelensis* preparations as early as 1981. Before large-scale use against mosquitoes, such as in the Upper Rhine Valley in the early 1980s, safety tests on representative aquatic organisms were carried out. None of the tested taxa appeared to be affected when exposed in water containing large amounts of *B. t. israelensis* (Schnetter *et al.*, 1981, Morawcsik, 1983; Margalit *et al.*, 1985; Jäger, 1990) (Table 7.3). Within the Diptera, the toxicity *B. t. israelensis* is restricted to a few nematocerous families. Apart from larval mosquitoes and blackflies, only the closely related dixids are similarly sensitive to *B. t. israelensis*. Larval psychodids, chironomids,

Table 7.3. Taxa which are not affected by *Bacillus thuringiensis israelensis*

Taxa	Dosage (ppm)	Species
Cnidaria	100	*Hydra* sp.
Turbellaria	180	*Dugesia tigrina, Bothromesostoma personatum*
Rotatoria	100	*Brachionus calyciflorus*
Mollusca	180	*Physa acuta, Aplexa hypnorum, Galba palustris, Anisus leucostomus, Bathyomphalus contortus, Hippeutis complanatus, Pisidium* sp.
Annelida	180	*Tubifex* sp.*, Helobdella stagnalis*
Acari	180	*Hydrachnella* sp.
Crustacea	180	*Chirocephalus grubei, Daphnia pulex, Daphnia magna, Ostracoda, Cyclops strenuus, Gammarus pulex, Asellus aquaticus, Orconectes limosus*
Ephemeroptera	180	*Cloeon dipterum*
Odonata	180	*Ischnura elegans, Sympetrum striolatum, Orthetrum brunneum*
Heteroptera	180	*Micronecta meridionalis, Sigara striata, Sigara lateralis, Plea leachi, Notonecta glauca, Ilyocoris cimicoides, Anisops varia*
Coleoptera	180	*Hyphydrus ovatus, Guignotus pusillus, Coelambus impressopunctatus, Hygrotus inaequalis, Hydroporus palustris, Ilybius fuliginosus, Rhantus pulverosus, Rhantus consputus, Hydrobius fuscipes, Anacaena globulus, Hydrophilus caraboides, Berosus signaticollis*
Trichoptera	180	*Limnophilus* sp.*, Phryganea* sp.
Pisces	180	*Esox lucius, Cyprinus carpio, Perca fluviatilis*
Amphibia (larvae)	180	*Triturus alpestris, Triturus vulgaris, Triturus cristatus, Bombina variegata, Bufo bufo, Bufo viridis, Bufo calamita, Rana esculenta, Rana temporaria*

sciarids and tipulids generally are far less sensitive (Table 7.4, p. 165). Other flies, such as the housefly, *Musca domestica* (Muscidae), and syrphids, such as *Helophilus pendulus*, are insensitive to *B. t. israelensis* (Ali, 1981; Colbo and Undeen, 1980; Garcia *et al.*, 1981; Margalit *et al.*, 1985; Miura *et al.*, 1980; Molloy and Jamnback, 1981; Mulla *et al.*, 1982).

In toxicological tests on various mammals, *B. t. israelensis* given orally, subcutaneously and percutaneously, intraperitoneally, ocularly, through inhalation and scarification appeared to be innocuous even at high dosages (10^8 bacteria/animal) (WHO, 1982).

Ease of handling

No special equipment is required for applications of *B. t. israelensis* and generally, simple knapsack sprayers are adequate. Preparations can be applied without special safety precautions. Because of the rapid knockdown effect and the high level of efficiency of *B. t. israelensis* preparations, the success of the treatment can generally be monitored within a few hours of application.

Cost-effectiveness

Compared to conventional chemical insecticides, the application of *B. t. israelensis* can be cost-effective. For instance, the German Mosquito Control Association (KABS) abatement project effectively suppressed mosquitoes deriving from a catchment area of more than 500 km², involving 100 km² of existing breeding grounds. The total yearly budget of US$1 million protects more than 2 million residents in the area.

B. t. israelensis can be produced inexpensively in tropical and subtropical countries if local raw materials are used (see Salama and Morris, Chapter 11). A comparison of the relative costs of larvicides in the Onchocerciasis Control Programme demonstrates that the costs of *B. t. israelensis* treatment are more or less comparable to the costs of treatments using chemical insecticides such as temephos or chlorphoxim when the discharge of the streams is low. The costs for *B. t. israelensis* treatments are slightly higher when the discharge of the streams is several hundred m³/sec.

Lack of potential for development of resistance

The development of target pest resistance to chemical insecticides represents a serious problem. Microbial control agents, however, appear less likely to provoke resistance because their mode of action is more complex (Davidson, 1990). Resistance of a lepidopteran pest of stored grain, *Plodia interpunctella*, to *B. t.*, however, has been demonstrated in the laboratory (McGaughey, 1985). Recent studies show that the commercial use of *B. t.* preparations in agriculture can lead to resistance within a few years. For example, diamondback moth, *Plutella xylostella* populations, which were repeatedly treated with *B. t.* in farms in Hawaii, were found to be 41 times more resistant than populations that were only minimally exposed to *B. t.* (Tabashnik *et al.*, 1990). Such resistance phenomena have not yet been observed with *B. t. israelensis* (see Marrone and MacIntosh, Chapter 10).

Resistance studies have been carried out by the KABS with populations of *Aedes vexans*, which were constantly exposed to *B. t. israelensis* for a period of 10 years. These mosquitoes were compared to similar *Aedes vexans* taken from a remote location and thus never under selection pressure. No reduction in sensitivity to *B. t. israelensis* (Bactimos®) could be detected; in 1991, the LC_{50} values were 0.105 ± 0.011 ppm within the application area and 0.119 ± 0.017 ppm elsewhere (Ludwig, 1991; Becker and Ludwig, in press). Similar results were obtained by Kurtak *et al.* (1989) who found that, after 7 years of intensive *B. t. israelensis* application in West Africa, the susceptibility of the blackfly, *Simulium damnosum*, had not changed.

The complex mode of action of *B. t. israelensis* may partly explain the relative absence of resistance. The lethal changes in the midgut cells are induced only by the synergistic effects of the different toxin proteins present in the parasporal body of *B. t. israelensis*. This combination reduces the likelihood of resistance. However, when the gene encoding the single 27 kDa toxin protein was cloned into a microorganism and then fed to larval mosquitoes, resistance was induced within a few generations (Gill, personal communication). The sheer size and genetic variability of many mosquito and blackfly populations protects against the development of resistance because only a portion of the population is exposed to the toxicant. Both floodwater mosquitoes and most blackflies migrate considerably. This behaviour produces a constant gene flow within their populations, which should at least slow the onset of resistance.

Suitability for integrated control programmes with community participation

As *B. t. israelensis* does not affect predatory organisms such as fish, predators can be included as additional elements in an integrated control strategy. The effect of predators can continue after the *B. t. israelensis* has been applied producing a sustained, suppressive effect (Mulla, 1990). *B. t. israelensis* can be mixed with other larvicides, such as preparations that are applied to alter the surface film. This enhances diffusion over the water surface and destroys pupal mosquitoes, which are not affected by *B. t. israelensis* (Roberts, 1989). *B. t. israelensis* has a considerable safety advantage over synthetic insecticides because neither the operator nor the occupants of treated sites become exposed to potentially dangerous chemicals. For this reason, such preparations are particularly well suited for use by volunteers.

FACTORS INFLUENCING THE EFFICACY OF *B. THURINGIENSIS ISRAELENSIS* TREATMENTS

B. t. israelensis efficacy depends upon the developmental stage of the target organisms in the site, their feeding behaviour, the organic content of the breeding site, the filtration capacity of target larvae as well as that of other non-target organisms, photosensitivity and other abiotic factors such as water temperature, depth and the sedimentation rate, as well as the shelf-life of the *B. t. israelensis* formulation itself (Mulla *et al.*, 1990a; Becker *et al.*, 1992). It is important to understand the impact of these factors on routine treatment, particularly in the calculation of the optimal dosage, the selection of the right formulation under a given environmental situation and the optimal timing for application against various target species.

Mosquitoes

Instar sensitivity

Sensitivity to *B. t. israelensis* decreases as the larvae develop. For instance, second instar *A. vexans* are approximately 11 times more sensitive than fourth instar larvae at a water temperature of 25 °C (second instars: $LC_{90} = 0.014 \pm 0.007$ mg/l; fourth instars $LC_{90} = 0.149 \pm 0.004$ mg/l). It is therefore recommended that control measures commence while the larvae are in early developmental stages. Large differences in sensitivity are also found in different species of mosquitoes due to their feeding habits and ability to activate the protoxin and toxin binding to midgut cell receptors.

Temperature

The feeding rates of mosquito larvae are influenced by water temperature. For example, the feeding rate of *A. vexans* decreases as temperature decreases, which results in a reduction in the consumption of *B. t. israelensis* crystals. In bioassays, second instar larval *A. vexans* are more than 10 times less sensitive at 5 °C than at 25 °C (5 °C: $LC_{90} = 0.145 \pm 0.065$ mg/l; 25 °C: $LC_{90} = 0.0142 \pm 0.007$ mg/l). The same effect can be found in fourth instar larvae (5 °C: $LC_{90} = 1.338 \pm 1.1$ mg/l; 25 °C: $LC_{90} = 0.149 \pm 0.004$ mg/l).

Water body size

Because *B. t. israelensis* crystals diffuse throughout the entire body of water, deep water requires more *B. t. israelensis* than shallow water of the same surface area. Since larvae of many mosquito species feed near the surface, effectiveness depends on the concentration and durability of *B. t. israelensis* in the upper 10 cm of the water body.

Larval density

Bioassays with *A. vexans* have shown that with increasing numbers of larvae, the amounts of *B. t. israelensis* applied must be increased markedly (Zgomba *et al.*, 1991; Becker *et al.*, 1992). At a density of 10 fourth-instar larvae per 150 ml water, the LC_{50} value is 0.0162 ± 0.004 mg/l; with 75 larvae, the LC_{50} is about seven times higher ($LC_{50} = 0.1107 \pm 0.02$ mg/l) (Figure 7.1). The presence of other filter-feeding organisms causes similar effects.

State of nutrition

The state of nutrition and the amount of available food influence the sensitivity of mosquitoes to *B. t. israelensis*. In laboratory studies two or three times more *B. t. israelensis* was required for an equal level of kill in the presence of added food (or polluted water) as compared with pure water (Mulla *et al.*, 1990a).

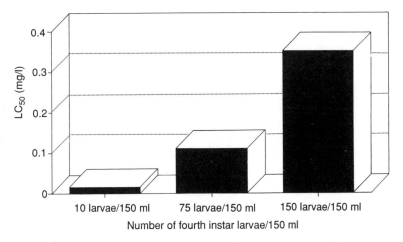

Figure 7.1. The effect of larval density on the efficacy of *B. t. israelensis* on *Aedes vexans*

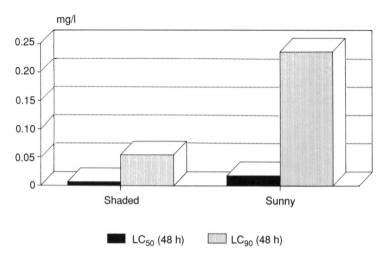

Figure 7.2 The effect of sunlight on the efficacy of *B. t. israelensis* on *Culex pipiens*

Sunlight

Strong sunshine appears to reduce *B. t. israelensis* activity. For example, the LC_{90} values obtained with Bactimos WP® in bioassays with third instar *C. pipiens* in sunny sites (sunlight: 6000 to 12 000 lux for 7 hours) and in shaded sites (< 150 lux) at the same time and under identical conditions (e.g. at same temperatures of $25 \pm 1\,°C$) were found to differ greatly (Ludwig, 1991; Becker *et al.*, 1992). Whereas the LC_{90} value is 0.054 ± 0.008 ppm under shaded conditions, it was approximately four times higher in sunny sites ($LC_{90} = 0.235 \pm 0.036$ ppm) (Figure 7.2). *B. sphaericus* preparations were active more than three times longer in shaded water than in water exposed to the sun (Sinegre, 1990).

Blackflies

All blackflies require fast-flowing water for their development. Larval Simuliidae pass through seven instars, and have a specialized filter apparatus with which to entrap food such as algae, bacteria and suspended matter that passes by in the flowing water. An insecticide that is to be ingested with food must therefore be used in an appropriate formulation and at the correct concentration. A number of abiotic and biotic influences affect the efficacy of *B. t. israelensis* formulations (Molloy, 1990).

Environmental factors

The speed of the current determines the time taken for the toxin to be transported. In slow-flowing waters, its efficacy is restricted to a relatively short stretch of the river, whereas in rapid-flowing waters the toxin can be transported for several kilometres. Additional physical factors or elements of the water flow morphology, such as water temperature, turbulence, or river profile, may also influence the efficacy of *B. t. israelensis*. For example, at lower temperatures the efficacy of *B. t. israelensis* is reduced. High levels of turbidity and polluted waters also have a negative effect.

Biotic factors

As with mosquitoes, the early larval stages of blackflies are substantially more sensitive to *B. t. israelensis* than the later ones. Additionally, changes in the filtration rate as, for example, during larval ecdysis, can influence the efficacy of *B. t. israelensis*. Overall, it has been shown that in programmes of routine application, the liquid formulations are more suitable for blackfly control because their sedimentation rates are lower than powder formulations with a relatively large particle size.

ORGANIZATION OF MOSQUITO CONTROL PROGRAMMES

In the preparative phase of the control programme a baseline collection of entomological and ecological data must be carried out, recording information such as the species present, density of larvae, density of biting adults, location of the breeding sites and abiotic and biotic conditions of the breeding sites. In the implementation phase it is necessary to evaluate the following: LC_{50}/LC_{90} values using different *B. t. israelensis* formulations against the main mosquito species; optimum effective dosage in small-scale field tests; testing of the most suitable application technique(s) and formulation(s). Depending on the outcome of this preparation phase, the cost-effectiveness of the use of *B. t. israelensis* should be subject to evaluation.

Various methods for collecting baseline data for a *B. t. israelensis* intervention have been standardized.

Abundance of mosquitoes

The density of larvae should be monitored biweekly in about 10–15 different breeding sites. At least 10 dips (each containing about 500 ml of water) from selected sites and from the centre of each plot should be taken. The first/second and third/fourth instar larvae and pupae should be separately recorded and a sample of the fourth instar larvae and emerged adults should be identified to species (WHO, 1990). Adult landing rates should also be determined biweekly. Biting rates must be assessed for different locations, for example, in swamps, at the outskirts of settlements, inside towns as well as inside houses, and at different times of the day and considering different abiotic conditions, especially humidity, temperature, wind and sunshine. Abiotic and biotic data from selected breeding sites should be collected giving special reference to the occurrence of predators and vegetation.

Assessment of the efficacy of different *B. t. israelensis* formulations

The susceptibility of the local field-collected larvae should be determined according to the WHO guidelines for bioassays (de Barjac and Larget-Thiery, 1984; Dulmage *et al.*, 1990). In preliminary small-scale field tests at selected trial sites, the optimum dosage of the most effective formulation must be determined. The evaluation should be carried out in well-defined and relatively small breeding sites. At least three replicates at each dosage (e.g. $2 \times LC_{90}$, $4 \times LC_{90}$ and $8 \times LC_{90}$) should be conducted and two similar plots must remain untreated as controls. Larval density (based on a sample of 10 dips per plot) must be monitored before treatment and 1, 2 and 4 days after treatment. Thereafter, plots should be monitored once each week. Monitoring should be continued until the larvae have been reduced by at least 70%. Treatments should be repeated when feasible.

Integration of the intervention

Additional control procedures for the development of an integrated control programme should be tested according to the specific circumstances in the region (e.g. environmental management, use of fish).

Organization

Cooperation between political decision-makers, public authorities, scientists and the general public is a necessary foundation for sustained success in dipteran control. The control programme must be well organized and be flexible enough to respond to individual situations with the most appropriate techniques and at the correct time.

SUCCESSFUL CONTROL PROGRAMMES USING *B. THURINGIENSIS ISRAELENSIS*

Mosquitoes

Mosquito control in Germany

In the Upper Rhine Valley mosquitoes can be a major public nuisance. The dominant species include the floodwater mosquitoes *A. vexans*,

A. *sticticus* and A. *rossicus* (Becker and Ludwig, 1981) which breed in large numbers in temporary water bodies resulting from elevated water levels on the river. C. *pipiens* is the main mosquito pest in houses in the Rhine Valley. Their breeding sites usually consist of rainwater containers near houses and other temporary water bodies occurring in summer. A control campaign, initiated in 1976, now covers 100 towns and villages along 300 km of river and an inundation area of 500 km^2 (Becker and Ludwig, 1983). Its main aim is to reduce mosquito abundance without disturbing ecologically sensitive river bank areas. On the basis of precise mapping of all mosquito breeding sites, a mosquito control strategy for each area is worked out together with the environmental protection authorities. By allowing the ecological conditions to direct the choice of application technique used, any disturbance of wild-life and the natural flora can be minimized.

There are normally about 300 trained people active in the KABS, supported by an annual budget of US$ 1 million. The scheme is organized along integrated control lines applying different but compatible methods to reduce the likelihood of the development of resistance. Methods which reduce breeding sites, promote fish and other natural predators and the use of B. t. *israelensis* are encouraged. Today the mosquitoes in more than 90% of the area are controlled exclusively with B. t. *israelensis*. B. *sphaericus* preparations are also increasingly being used against C. *pipiens molestus*. For young larvae and shallow water, 250 g of B. t. *israelensis* powder with an activity of 6000 ITU/mg are mixed with 10 litres of filtered pond water for each hectare treated and applied by knapsack sprayer. For deeper sites or when later instars are present, the dose is doubled. Equivalent dosages are applied for the liquid concentrates. Sand or corn-cob granules are used at high water levels or when vegetation is dense. Between 1981 and 1991 approximately 50 000 hectares were treated with 23 tons of B. t. *israelensis* wettable powder and 19 000 litres liquid concentrates. As a result, mosquito incidence was reduced by 90% each year.

Similar results against nuisance mosquitoes have been achieved in other parts of the world, for example in many mosquito control programmes in the USA and Europe. e.g., in Czechoslovakia, France, Hungary, Italy, Russia, Slovenia, Spain, Switzerland and Yugoslavia (Rettich, Sinegre, Mas, Gruffaz, Schaffner, Pfirsch, Bellini, Aranda, Eritja, Lüthy and Zgomba—personal communications).

Microbial control of mosquitoes in the People's Republic of China

The lakes along the Yangtze River, together with the high groundwater level and the widely distributed drainage system, provide not only optimal conditions for growing rice but also ample breeding conditions for mosquitoes. The 69 mosquito species recorded to date include major malaria vector species, such as *Anopheles sinensis*. More than 20 million people living along the Yangtze River are threatened by the malaria agent *Plasmodium vivax*. During 1985 about 82 000 cases of malaria were reported, which corresponds to an annual incidence of about 168 cases/100 000 people. The control of *Anopheles sinensis* with insecticides has recently become increasingly difficult due to developing resistance and ecological and toxicological risks. Other diseases transmitted by mosquitoes in Hubei include Japanese B encephalitis (transmitted by *Culex tritaeniorhynchus*), as well as bancroftian filariasis (transmitted by *C. pipiens quinquefasciatus*). Dengue is transmitted by *Aedes albopictus* but few cases have been reported.

For more than 30 years, the use of chemical insecticides against adults and larvae has allowed relatively simple control of almost all vector species in Hubei. Forty tons of DDT are now used annually for residual applications and about 3 tons of deltamethrin for bednet impregnation. The development of resistance, the high costs and the environmental risks have caused chemicals to become unacceptable.

Early on, the 'Institute of Parasitic Diseases', with WHO support, began investigations into microbial mosquito control working principally with *B. t. israelensis* and *B. sphaericus* (strain 2362), but also isolating agents locally, such as *B. t.* 187 isolated from a soil sample in 1982 and *B. sphaericus* C_{3-41} in 1986. Laboratory tests showed that while *B. t. israelensis* preparations were very effective against the larvae of *A. albopictus* ($LC_{50} = 0.025–0.058$ ppm) they were less active against larvae of *C. pipiens quinquefasciatus* ($LC_{50} = 0.037–0.28$ ppm). *B. sphaericus*, however, killed *Culex* larvae at relatively low dosages ($LC_{50} = 0.001–0.009$ ppm) whereas *Aedes* larvae were less sensitive ($LC_{50} = 0.54–0.72$ ppm). High concentrations (LC_{50} 1.9–10 ppm) of both preparations were necessary to control *Anopheles sinensis*. Large-scale field tests of powdered *B. t. israelensis* and *B. sphaericus* destroyed larval *C. pipiens quinquefasciatus* at concentrations of 0.5–1.0 ppm for about 1 week. Tests showed that more effective preparations needed to be developed to control *Anopheles* larvae. Therefore slow release floating granules were produced using elm bark as a carrier, wheat bran (inducer) and *B. t.* 187 or *B. sphaericus* C_{3-41} cultures (Xu *et al.*, 1991). These granules showed a sustained effect for more than 15 days in the field.

In routine treatment fluid formulations are applied using a high-pressure sprayer attached to a 600 litre tank pulled by a mini-tractor. In the summer treatment is required every 7 days and every 10 days in the autumn, depending on temperature, with a total of approximately 24 treatments per season. In the last few years about 10 tons of *B. t.* 187 (at US$1 per litre at 400 ITU/mg) and 14 tons of *B. sphaericus*, C_{3-41} (US$ 1.20 per litre at 270 ITU/mg) have been produced in Hubei province using local materials. This has been enough to treat 12 000 hectares.

Before application in 1986 there was a density of 8–9 mosquitoes/person/hour, whilst in 1987, 1988 and 1989 this had fallen to 3.7, 2.0 and 1.3 respectively, post treatment. Similarly, the incidence of malaria per 10 000 people fell from 5.6 (1986) to 2.14, 1.6 and 0.8 cases (1987, 1988 and 1989 respectively) in the operational area. The reduction in the number of malaria cases was confirmed by blood smear tests: before application in 1986 1.6% were positive; after large-scale routine treatments it was reduced to 0.24% (Xu et al., 1992).

Blackflies

The Onchocerciasis Control Programme

In the Onchocerciasis Control Programme (OCP) of West Africa, the organophosphate temephos has been used exclusively since 1975 against the larval *Simulium damnosum* (*sensu lato*), the vector of *Onchocerca volvulus*, the agent of onchocerciasis (WHO, 1985; Leveque et al., 1988). The programme now operates in 11 West African countries, protecting 2–3 million people living along 50 000 km of the Upper Volta River system. About 18 000 km of this system are treated regularly, requiring about 10 000 hours of flight time and 800 000 litres of insecticides per year, at a cost of some $US29 million (Kurtak et al., 1989). The programme provides a model for successful international cooperation under the guidance of the United Nations (UN). Based on concerted efforts by four sponsoring international agencies (UNDP, FAO, World Bank and WHO), an important step towards improved socioeconomic development in the countries involved has been realized (WHO, 1985).

By 1979 the first signs of resistance of *S. damnosum* against temephos began to appear (Guillet et al., 1980; Kurtak, 1986). Therefore, in 1981, experiments with *B. t. israelensis* (Teknar®) were conducted under the guidance of the WHO in the areas most affected by this developing resistance. After early difficulties and consequent improvement in the potency of the *B. t. israelensis* product used, satisfactory results were obtained. Large-scale application of *B. t. israelensis* proved effective at application rates of 0.4–1.2 mg/l per 10 minutes.

In order to avoid additional increases in resistance, an 'integrated' OCP effort was developed in 1985. In those regions where no resistance had developed, temephos continued to be used. Where resistance had occurred, *B. t. israelensis* was applied if the river discharge rate was less than 75 m³/sec. Above this level, the strategy was to alternate the use of insecticides such as chlorphoxim, permethrin and carbosulfan (Leveque et al., 1988; Kurtak et al., 1989). As a result of this strategy, *B. t. israelensis* (Teknar HP-D® and Vectobac 12 AS®) is used in increasing amounts in weekly intervals during the dry season. Whereas only 8000 litres of *B. t. israelensis* were applied, out of a total of 222 000 litres of insecticides (about 3.6%), in 1981, about 750 000 litres from a total of 923 000 litres of insecticides were applied in 1988, resulting in 81% of the region being protected by B. t. israelensis (Kurtak et al., 1989; Davidson, 1989).

The substitution of temephos by B. t. israelensis presents certain technical difficulties (Guillet et al., 1990). B. t. israelensis sprays must be applied more carefully, covering the entire width of the river, rather than at a single point, as in the case of synthetic chemicals. Non-corrosive tanks and nozzles must be used. In addition, the presence of algae at a density of 1500–3000 cells per ml may require a doubling of the operational dose of B. t. israelensis.

Since the inception of the OCP, few new cases of infection with O. volvulus have been recorded. Prevalence of this infection among the people protected by the OCP is rapidly approaching zero. The three million or more children born within the first 10 years of the programme are free from onchocercal infection and 70% of the patients originally suffering from onchocercal eye infections have recovered. The very fertile low-lying valleys can now once again be used by the human population, permitting enhanced socioeconomic development of the region.

Use of B. thuringiensis israelensis against blackflies in temperate climates

In temperate parts of the world, where blackflies do not transmit pathogens that affect human health, these insects can still have considerable nuisance value (Molloy, 1990). Cattle are also severely affected. Temperate regions subject to these problems generally are ecologically sensitive, for example, mountain resorts. Particular problems affect the use of B. t. israelensis in mountain streams. The quantity of B. t. israelensis to be applied varies with: discharge rate of the water, profile of the stream, turbidity, presence of pollutants, water temperature, pH, degree of vertical mixing, settling due to presence of pools and characteristics of the substrate. Although the ideal particle size seems to be about 35 μm, the rate of application varies several-fold, depending on these characteristics (Molloy et al., 1984). Larger particles settle faster than smaller particles. This provides liquid formulations with a distinct advantage over powders; the small particles in a liquid formulation carry better, a crucial factor in the treatment of fast-moving streams. Some 5–30 ppm of B. t. israelensis per minute generally provide satisfactory control of larval blackflies over a span of 50–250 m in moderate sized streams (Knutti and Beck, 1987).

The non-target effects of B. t. israelensis in these sites are minimal. Of the wide variety of organisms present, only filter-feeding chironomids were found to be sensitive to B. t. israelensis, but at very high rates of application. B. t. israelensis is the sole larvicide that can be used in such sensitive ecosystems.

Control of other diptera

Due to the increasing eutrophication of many rivers and lakes, aquatic midges such as *Chironomus* species often find optimal developmental conditions in aquatic sediments. Although the large-scale occurrence of flying insects can be very unpleasant (Ali, 1980; Mulla *et al.*, 1990b) chironomids occupy an important position in the food chain because of their large biomass. Furthermore, they play a critical role in the self-cleaning properties of waterways, leading to a reduction in the degree of eutrophication. Chironomids are neither vectors of diseases nor biting pests. Their ecological value far outweighs the disadvantages of their occurrence and, therefore, one should generally refrain from their control.

In Germany studies have shown that larval *Chironomus plumosus* and *C. annularius*, which occur in aquatic sediments, are only affected at doses of Bactimos® above 4 ppm, approximately 10 times the dosage used in mosquito control. Mulla *et al.* (1990b) achieved similar results in California where they found that *C. decorus* could be controlled successfully with an application dosage of 5 kg Vectobac TP® or 10 litres Vectobac 12 AS® per hectare.

Because of their ecological value chironomids and similar insects are protected in Germany. Due to their low sensitivity towards *B. t. israelensis* and to their occurrence mostly in permanent waters with silty sediments, in which no *Aedes* mosquitoes occur, such protection is easy. Additionally, other species of midge, such as *Smittia* or *Procladius* and *Tanypus*, are unaffected by *B. t. israelensis* (Table 7.4). As a result of the selective

Table 7.4. Susceptibility of larvae of other Nematocera and Brachycera species to *Bacillus thuringiensis israelensis*

Insects	Dosage (ppm)	Results
Psychoda alternata	1	100% mortality
Chaoborus crystallinus	180	No ill effects
Mochlonyx culiciformis	1.8	No ill effects
Dixa spp.	2	100% mortality
Procladius sp.	1.8	No ill effects
Tanypus spp.	0.5	No ill effects
Psectrotanypus varius	0.5	No ill effects
Orthocladius sp.	1.8	No ill effects
Smittia sp.	18	No ill effects
Chironomus sp.	1.8	90% mortality
Ceratopogonidae	180	No ill effects
Tipula sp.	30	50% mortality
Helophilus pendulus	180	No ill effects
Sciaridae	10^4	79% mortality

action of *B. t. israelensis*, enough insects remain as prey for birds and other predators after application.

Psychodidae, particularly *Psychoda alternata*, can occasionally become troublesome in sewage treatment plants. In laboratory experiments, a dose of 1 ppm of Bactimos® was sufficient to kill larval *P. alternata*. In the field, however, a dose of 100 ppm was required to achieve nearly 100% mortality. Economic control of *Psychoda* therefore seems not feasible. Even if *B. t. israelensis* were active against the related phlebotomine vectors of leishmaniasis, similar problems in the control of these vectors could be anticipated (de Barjac and Larget, 1981).

Tipulid larvae (e.g. *Tipula paludosa* and *T. oleracea*) are the main insect pests in grasslands in Europe. Experiments in the laboratory, greenhouses and in the field showed that first instar tipulid larvae could be successfully controlled by spraying *B. t. israelensis* (Smits and Vlug, 1990). To avoid the high dosages of *B. t. israelensis* which are usually necessary for tipulid control, food baits containing high concentrations of *B. t. israelensis* can be used as an alternative method (Langenbruch, personal communication).

FUTURE PROSPECTS FOR THE USE OF *B. THURINGIENSIS ISRAELENSIS*

With the development of appropriate formulations effective and economic control of mosquitoes and blackflies is now generally possible. Interest in *B. t. israelensis* is increasing worldwide year by year; about 1000 tons of *B. t. israelensis* preparations are now being used annually. So far there have been no negative effects. *B. t. israelensis* and *B. sphaericus* are promising agents in the battle against dangerous diseases such as malaria, filariasis and arbovirus infections. In suitable formulations, these microbial agents are useful supplements to, or replacements for, broad-spectrum chemicals. Further improvements, particularly to extend their long-term effect and to enhance *Anopheles* control, will accelerate this process still further. In temperate regions *B. t. israelensis* offers an ecologically defensible compromise between the desire of man to protect himself from troublesome mosquitoes or blackflies and the requirements of current environmental policies to protect sensitive ecosystems by the use of non-selective methods.

ACKNOWLEDGEMENTS

We especially acknowledge the assistance of Dr Andrew Spielman, Harvard School of Public Health, in producing this paper. Furthermore the cooperation of the following persons and/or institutions is gratefully recognized: Dr Frantizek Rettich and Dr Georg Ryba, Institute of Hygiene and Epidemiology, Prague; Dr Marija Zgomba and Dr Dusan Petric, University of Novi Sad, Yugoslavia; Professor Dr Peter Lüthy, Institute of Microbiology, Zürich, Switzerland; Professor Dr Herbert W. Ludwig, Dr Wolfgang Schnetter, Dr Mario Ludwig, Achim Kaiser, Ulrich Jäger and Michael Riffel, University of Heidelberg; Professor Dr Xu Bozhao and Dr Xiao Xianq, Institute for Parasitic Diseases, Wuhan, PR-China; Dr Robert Rose; Dr Muhamad Masyhur, Regional Health Office in Jakarta and Dr S Djakaria, Department of Parasitology; also David Russell, University of Heidelberg, and Adrian C. Pont for their help in preparing the manuscript. We are especially indebted to the German Mosquito Control Association, for providing financial support, and the UNDP/World Bank/WHO Special Programme for Research and Training in Tropical Diseases (TDR).

REFERENCES

Ali, A. (1980). 'Nuisance chironomids and their control: a review'. *Bull. Entomol. Soc. Am.*, **26**, 3–16.

Ali, A. (1981). '*Bacillus thuringiensis* serovar. *israelensis* (ABG-6108) against chironomids and some nontarget aquatic invertebrates'. *J. Invert. Pathol.*, **38**, 264–272.

Becker, N. and Ludwig, H. W. (1981). 'Untersuchungen zur Faunistik und Ökologie der Culicinae und ihrer Pathogene im Oberrheingebiet'. *Mitt. Dtsch. Ges. Allg. Ang. Entomol.*, **2**, 186–194.

Becker, N. and Ludwig, H. W. (1983). 'Mosquito control in West Germany'. *Bull. Soc. Vector Ecol.*, **8**, 85–93.

Becker, N. and Ludwig, M. 'Investigations on resistance of *Aedes vexans* populations after 10 years of applications of *Bacillus thuringiensis* var. *israelensis*'. *Bull. Soc. Vector Ecol.* (in press).

Becker, N., Djakaria, S., Kaiser, A. and Zuhasril (1991). 'Efficacy of a new tablet formulation of an asporogenous strain of *Bacillus thuringiensis israelensis* against larvae of *Aedes aegypti*'. *Bull. Soc. Vector Ecol.*, **16**(1): 176–182.

Becker, N., Zgomba, M., Ludwig, M., Petric, D. and Rettich, F. (1992). 'Factors influencing the efficacy of the microbial control agent *Bacillus thuringiensis israelensis*'. *J. Am. Mosq. Control Ass.* **8**, 285–289.

Berliner, E. (1911). 'Uber die Schlaffsucht der Mehlmottenraupe'. *Ztschr. Ges. Getreidewesen* (Berlin), **3**: 63–70.

Chilcott, C. N. and Ellar, D. J. (1988). 'Comparative toxicity of *Bacillus thuringiensis* var. *israelensis* crystal proteins in vivo and in vitro'. *J. Gen. Microbiol.*, **134**, 2551–2558.

Colbo, M. H. and Undeen, A. H. (1980). 'Effect of *Bacillus thuringiensis* var. *israelensis* on non-target insects in stream trials for control of Simuliidae'. *Mosq. News*, **40**, 368–371.

Davidson, E. W. (1989). *Microbial Control of Vector Insects*. UCLA Symposia on Molecular and Cellular Biology. Vol. 112, p. 14.

Davidson, E. W. (1990). Development of insect resistance to biopesticides. Proc. Second Sympos. on Biocontrol, Brasilia. Oct. 1990, p. 19.

de Barjac, H. (1983). 'Bioassay procedure for samples of Bacillus thuringiensis israelensis using IPS-82 standard'. WHO Report TDR/VED/SWG (5)(81.3).

de Barjac, H. and Larget, J. (1981). 'Toxicité de Bacillus thuringiensis var. israelensis sérotype H-14 pour les larves de Phlebotomes, vecteurs de Leishmanioses'. Bull. de la Société de Pathol Exotic, 74, 485–489.

de Barjac, H. and Larget-Thiery, I. (1984). 'Characteristics of IPS 82 as standard for biological assays of Bacillus thuringiensis H-14 preparations'. WHO/VBC/84.892.

Dulmage, H. T., Correa, J. A. and Gallegos-Morales, G. (1990). 'Potential for improved formulations of Bacillus thuringiensis israelensis through standardization and fermentation development'. In Bacterial Control of Mosquitoes and Blackflies, pp. 110–133, Rutgers University Press.

Federici, B. A., Lüthy, P. and Ibarra, J. E. (1990). 'Parasporal body of Bacillus thuringiensis israelensis'. In Bacterial Control of Mosquitoes and Blackflies, pp. 16–44, Rutgers University Press.

Garcia, R., DesRochers, B. and Tozer, W. (1981). 'Studies on Bacillus thuringiensis var. israelensis against mosquito larvae and other organisms'. Proc. Calif. Mosq. Vector Control Ass., 49, 25–29.

Goldberg, L. H. and Margalit, J. (1977). 'A bacterial spore demonstrating rapid larvicidal activity against Anopheles sergentii, Uranotaenia unguiculata, Culex univattatus, Aedes aegypti and Culex pipiens'. Mosq. News, 37, 355–358.

Guillet, P., Escaffre, H., Quedrago, M. and Quillevere, D. (1980). 'Mise en evidence d'une resistance au temephos dans le complexe Simulium damnosum (S. sanctipauli et. S. soubrense) en Côte d'Ivoire'. Cah. ORSTOM. ser. Ent. Med. Parasit, 18, 291–298.

Guillet, P., Kurtak, D. C., Philippon, B. and Meyer, R. (1990). 'Use of Bacillus thuringiensis israelensis for Onchocerciasis Control in West Africa'. In Bacterial Control of Mosquitoes and Blackflies, pp. 187–201, Rutgers University Press.

Halstead, S. B. (1980). 'Dengue haemorrhagic fever—a public health problem and a field for research'. Bull. WHO 58, 1, 1–21.

Halstead, S. B. (1982). 'WHO fights dengue haemorrhagic fever'. WHO 38, 2, 65–67.

Ibarra, J. E. and Federici, B. A. (1986a). 'Isolation of a relatively nontoxic 65-kilodalton protein inclusion from the parasporal body of Bacillus thuringiensis subsp. israelensis'. J. Bacteriol., 165(2), 527–533.

Ibarra, J. E. and Federici, B. A. (1986b). 'Parasporal bodies of Bacillus thuringiensis subsp. morrisoni (PG-14) and Bacillus thuringiensis, subsp. israelensis, are similar in protein composition and toxicity'. FEMS Microbiol. Lett., 34(1), 79–84.

Jäger, U. (1990). Ökotoxikologische Untersuchungen zur Anwendung von Bacillus thuringiensis var. israelensis-Präparaten. Diplomarbeit. University of Heidelberg, 71 pp.

Kellen, W. R. and Meyers, C. M. (1964). 'Bacillus sphaericus Neide as a pathogen of mosquitoes'. J. Invert. Pathol., 7, 442–448.

Knutti, H. J. and Beck, W. R. (1987). 'The control of blackfly larvae with Teknar'. In Blackfly Ecology. Population Management, An Annotated World List (Eds. K. C. Kim and R. W. Merritt), pp. 409–418, University Park, Penn; Pennsylvania State University.

Kurtak, D. (1986). 'Insecticide resistance in the Onchocerciasis Control Programme'. Parasitol. Today, 2, 20–21.

Kurtak, D., Back, C., Chalifour, A., Doannio, J., Dossou-Yovo, J., Duval, J., Guillet, P., Meyer, R., Ocran, M. and Wahle, B. (1989). 'Impact of B. t. israelensis on blackfly control in the Onchocerciasis Control Programme in West Africa'. Israel J. Entomol., 23, 21–38.

Leveque, C., Fairhurst, C. P., Abban, K., Paugy, D., Curtis, M. S. and Traore, K.

(1988). 'Onchocerciasis control programme in West Africa: Ten years monitoring of fish populations'. *Chemosphere*, **17**, 421–440.

Ludwig, M. (1991). Untersuchungen zum natürlichen Auftreten von Resistenzerscheinungen bei *Aedes vexans* (Culicinae) nach 10 Jahren *B. t. israelensis* Applikationen. KABS-Report. 37 pp.

McGaughey (1985). 'Insect resistance to the biological insecticide *Bacillus thuringiensis*'. *Science*, **229**, 193–195.

Margalit, J. and Dean, D. (1985). 'The story of *Bacillus thuringiensis israelensis* (B.ti.)', *J. Am. Mosq. Control Ass.*, **1**, 1–7.

Margalit, J., Lahkim-Tsror, L., Bobroglo, H., Paskar, C. and Barak, Z. (1985). 'Biological control of mosquitoes in Israel'. In *Integrated Control of Vectors* (Eds. M. Laird and J. W. Miles), pp. 361–374, Academic Press.

Miura, T., Takahashi, R. M. and Mulligan, F. S. (1980). 'Effects of the bacterial mosquito larvicide, *Bacillus thuringiensis* serotype H-14 on selected aquatic organisms'. *Mosq. News*, **40**, 619–622.

Molloy, D. (1990). 'Progress in the biological control of blackflies with *Bacillus thuringiensis*, with emphasis on temperate climates'. In *Bacterial Control of Mosquitoes and Blackflies*, pp. 161–186, Rutgers University Press.

Molloy, D. and Jamnback, H. (1981). 'Field evaluation of *Bacillus thuringiensis* var. *israelensis* as a blackfly biocontrol agent and its effect on nontarget stream insects'. *J. Econ. Entomol.*, **74**, 314–318.

Molloy, D., Wraight, S. P., Kaplan, B., Gerardi, J. and Peterson, P. (1984). 'Laboratory evaluation of commercial formulations of *Bacillus thuringiensis* var. *israelensis* against mosquito and blackfly larvae'. *J. Agric. Entomol.*, **1**, 161–168.

Morawcsik, J. (1983). Untersuchungen zur Wirkung von *Bacillus thuringiensis israelensis* auf aquatische Nontarget-Organismen. Doktorarbeit, University of Heidelberg, 114 pp.

Mulla, M. S. (1990). 'Activity, field efficacy, and use of *Bacillus thuringiensis israelensis* against mosquitoes'. In *Bacterial Control of Mosquitoes and Blackflies*, pp. 134–160, Rutgers University Press.

Mulla, M. S., Federici, B. A. and Darwazeh, H. A. (1982). 'Larvicidal efficacy of *Bacillus thuringiensis* serotype H-14 against stagnant water mosquitoes and its effects on nontarget-organisms'. *Env. Entomol.*, **11**, 788–795.

Mulla, M. S., Darwazeh, H. A. and Zgomba, M. (1990a). 'Effect of some environmental factors on the efficacy of *Bacillus sphaericus* 2362 and *Bacillus thuringiensis* (H-14) against mosquitoes'. *Bull. Soc. Vector Ecol.*, **15**, 166–175.

Mulla, M. S., Chaney, J. D. and Rodcharoen, J. (1990b). 'Control of nuisance aquatic midges (Diptera: Chironomidae) with the microbial larvicide *Bacillus thuringiensis* var. *israelensis* in a man-made lake in Southern California'. *Bull. Soc. Vector Ecol.*, **15**, 176–184.

Orduz, S., Rojas, W., Correa, M. M., Montoya, A. E. and de Barjac, H. (1992). 'A new serotype of *Bacillus thuringiensis* from Colombia toxic to mosquito larvae'. *J. Invert. Pathol.*, **59**, 99–103.

Padua, L. E., Ohba, M. and Aizawa, K. (1984). 'Isolation of *Bacillus thuringiensis* strain (serotype 8a:8b) highly and selectively toxic against mosquito larvae'. *J. Invert. Pathol.*, **44** (1), 12–17.

Roberts, G. M. (1989). 'The combination of *Bacillus thuringiensis* var. *israelensis* with a monomolecular film'. *Israel J. Entomol.*, **23**, 95–97.

Schnetter, W., Engler, S., Morawcsik, J. and Becker, N. (1981). 'Wirksamkeit von *Bacillus thuringiensis* var. *israelensis* gegen Stechmucken und Nontarget-Organismen'. *Mittlg. Dtsch. Ges. Allg. Ang. Entomol.*, **2**, 195–202.

Sinegre, G. (1990). 'Utilisation de *Bacillus sphaericus* dans la lutte contre les moustiques-présentation, performances et limites d'emploi. Conseil Scientifique et Technique de l'E.I.D. Rhone-Alpes, February 1990, 1–9.

Smits, P. H. and Vlug, H. J. (1990). Control of Tipulid larvae with *Bacillus thuringiensis* var. *israelensis*. *Proceedings of the Vth International Colloquium on Invertebrate Pathology and Microbial Control*, Adelaide, p. 343.

Tabashnik, B. E., Cushing, N. L., Finson, N. and Jonson, M. W. (1990). 'Development of resistance to *Bacillus thuringiensis* in field populations of *Plutella xylostella* in Hawaii'. *J. Econ. Entomol.* (in press).

Weiser, J. (1984). 'A mosquito-virulent *Bacillus sphaericus* in adult *Simulium damnosum* from Northern Nigeria'. *Zbl. Mikrobiol.*, **139**, 57–60.

WHO (1982) 'Data sheet on the biological control agent *Bacillus thuringiensis* serotype H-14 (de Barjac 1978)'. Document WHO/VBC/79.750 Rev.1.

WHO (1985). 'Ten years of onchocerciasis control in West Africa'. Document OCP/GVA/85.1B.

WHO (1990). 'Report on the TDR/CTD Information Consultation on Large-scale Field Trials with *Bacillus sphaericus* against Filariasis Vectors'. Geneva, p. 5.

WHO (1991). *Tropical Diseases. Progress in Research, 1989–1990*, Tenth Programme Report. UNDP/WORLD BANK/WHO, p. 135.

Xu, B. Z., Fangyu, X., Xianqi, X. and Becker, N. (1991). 'Evaluation of *Bacillus thuringiensis* 187 and *Bacillus sphaericus* preparations against *Anopheles sinensis* and *Culex quinquefasciatus* in laboratory and field'. KABS-Report 1991.

Xu, B. Z., Becker, N., Xiao, X. Q. and Ludwig, H. W. (1992). Microbial Control of Mosquitoes in Hubei Province, People's Republic of China. *Bull. Soc. Vector Ecol.*, **17**, 1–10.

Zgomba, M., Petric, D. and Ludwig, M. (1991). 'Effect of temperature and larval density on the efficacy of *B. t. israelensis*'. KABS-Report, 10 pp.

8 Control of Coleopteran Pests by *Bacillus thuringiensis*

BRIGITTE KELLER AND GUSTAV-ADOLF LANGENBRUCH
Federal Biological Research Centre for Agriculture and Forestry, Institute for Biological Control, Heinrichstrasse 243, D-6100 Darmstadt, Germany

INTRODUCTION

In 1982, Huger isolated and Krieg identified a novel *Bacillus thuringiensis* (*B. t.*) strain (BI 256-82) from a dead pupa of the yellow mealworm *Tenebrio molitor* (Coleoptera, Tenebrionidae) which was shown by Schnetter to be active against Coleoptera. It was described as *Bacillus thuringiensis* subspecies *tenebrionis* and represented a new *B. t.* pathotype (C) (Krieg *et al.*, 1983). Its δ-endotoxin is especially active against leaf beetles (Chrysomelidae) but is ineffective against Lepidoptera, Diptera or other insect orders. Up until this time only *B. t.* strains producing δ-endotoxins active against Lepidoptera and Diptera had been described, although pests belonging to the Coleoptera and other Orders could be controlled by the β-exotoxin of some *B. t.* subspecies, e.g. *thuringiensis*, *morrisoni*, *tolworthi*, *darmstadiensis* and *kumamotoensis* (Burgerjon and Biache, 1966; Bond *et al.*, 1971; Krieg, 1986). As an exception to this rule, Sharpe (1976) reported the toxicity of the crystal of a strain of *B. t.* subspecies *galleriae* to larvae of Japanese beetle, *Popillia japonica*. Additionally, Ignoffo *et al.* (1982) reported that the HD-1 strain of *B. t.* subspecies *kurstaki* was effective against the larvae of the Colorado potato beetle (CPB), *Leptinotarsa decemlineata*. However, this could not be verified by Krieg *et al.* (1984). It is possible that a thermolabile α-exotoxin could have caused the effect.

In the following years, details of further strains with coleopteran activity appeared, e.g. *B. t.* subspecies *san diego* (Herrnstadt *et al.*, 1986), which was later found to be identical to *B. t. tenebrionis* (Krieg *et al.*, 1987a), as was strain EG 2158 (Ecogen company) (Donovan *et al.*, 1988). One other strain from the Ecogen company (EG 2838) also showed coleopteran activity though it possesses different δ-endotoxin genes to *B. t. tenebrionis* (Sick *et al.*, 1990). Payne (1988) (Mycogen Corporation)

Bacillus thuringiensis, An Environmental Biopesticide: Theory and Practice. Edited by P. F. Entwistle, J. S. Cory, M. J. Bailey and S. Higgs
© 1993 John Wiley & Sons Ltd

reported the isolation of approximately 50 coleopteran active strains without giving details. Recently in Italy a further *B. t.* strain (*B. t.* subspecies *donegani*) with coleopteran activity has been patented (Cidaria *et al.*, 1990).

This chapter will primarily concentrate on *B. t. tenebrionis* as this is currently the most important commercial subspecies used in the control of coleopteran pests.

B. THURINGIENSIS TENEBRIONIS PRODUCTS AND THEIR REGISTRATION

Four companies have placed *B. t. tenebrionis* products on the market: M-One® (Mycogen), Trident® (Sandoz), Ditera® (Abbott) and Novodor® (Novo Biokontrol). Ecogen also offers the product Foil®, which is a combined strain of pathotypes A and C for the control of European corn borer, *Ostrinia nubilalis*, a lepidopteran and *L. decemlineata*, which are pests of potatoes in parts of the USA. The preparations M-One®, Trident®, Ditera®, Foil® and Novodor® were registered in the USA in 1991. Novodor® is also registered in Bulgaria, Czechoslovakia, Poland, Switzerland, Yugoslavia and, conditionally, in Russia.

CHARACTERIZATION OF *B. THURINGIENSIS TENEBRIONIS* AND OTHER COLEOPTERAN-ACTIVE SUBSPECIES

Morphological and structural characterization

In common with other *B. t.* pathotypes, *B. t. tenebrionis* produces an active proteinaceous, parasporal crystal which is located outside the exosporium and is formed during the sporulation process. In contrast to the majority of lepidopteran-active crystals which have typical bipyramidal shapes and dipteran-active crystals which are spherical, the parasporal crystals of *B. t. tenebrionis* appear as flat plates with square, rectangular or rhomboidal outlines. In general, their axes measure $0.7–2.4 \times 0.7$ μm and the thickness of the plates ranges from 0.15 to 0.25 μm with pointed or roof-shaped edges. Sporangia frequently harbour two identical crystals of similar size in either parallel or serial locations (Huger *et al.*, 1986).

On nutrient rich media especially, *B. t. tenebrionis* is able to produce an additional parasporal crystal, the toxicity of which has yet to be determined, that is characteristically spindle-, spheroidal- or plate-shaped with a size range of 0.5–1.4 by 0.35–0.77 μm (Huger and Krieg, 1989). Both types of crystal are frequently closely associated (Figure 8.1A). The second crystal type typically displays a distinct hexagonal lattice with a periodicity of about 7.8 nm in negatively stained preparations, as compared to 5.8 nm for the lattice of the toxic crystal plates. Under the same culture conditions the second type of crystal is also synthesized by *B. t. san diego* (Figure 8.1B) and *B. t.* strain EG 2158 (Donovan *et al.*, 1988), further confirming their identity with *B. t. tenebrionis.*

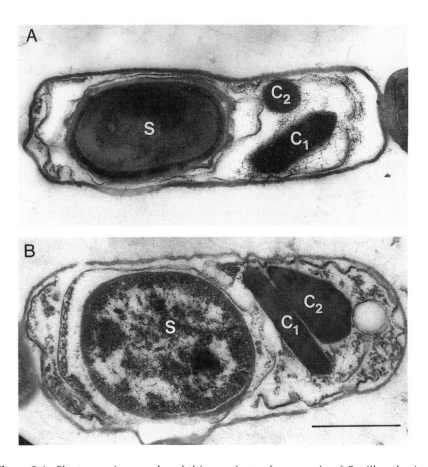

Figure 8.1. Electron micrographs of thin sections of sporangia of *Bacillus thuringiensis* subspecies *tenebrionis* (A) and *Bacillus thuringiensis* subspecies *san diego* (B); each sporangium additionally harbours, besides the spore (S) and the plate-shaped parasporal crystal (C_1), the second type of crystal (C_2). Scaling bar for A and B = 500 nm

Serological and biochemical characterization

Serological classification by flagellar antigens (H-serotype) (de Barjac, 1981) shows that *B. t. tenebrionis* belongs to the H-serotype 8a8b, as does the identical *B. t. san diego* (de Barjac and Frachon, 1990). *B. t. tenebrionis* shares the same H-serotype as *B. t. morrisoni* (pathotype A) and the *B. t.* subspecies *morrisoni* PG 14 (pathotype B). There is no immunological relationship between the crystal toxin of *B. t. tenebrionis* and the δ-endotoxins of the other strains in the H-serotype 8a8b (Huger *et al.*, 1986). Furthermore, *B. t. tenebrionis* belongs to the O-serotype IX (in contrast to PG 14) (Krieg *et al.*, 1987b). In view of the practical use of *B. t. tenebrionis* in biological control it is important to note that it does not produce a β-exotoxin (Krieg *et al.*, 1983, 1987b).

In contrast to the δ-endotoxins of *B. t.* subspecies with activity against Lepidoptera and Diptera, the crystals of *B. t. tenebrionis* are soluble in concentrated solutions of NaBr, KBr or NaI at pH 10–12.5 (Bernhard, 1986). The solubilized δ-endotoxin readily reforms active crystals after removal of the salt or neutralization.

Existing methods for the purification of the *B. t. tenebrionis* crystals from the spores include separation using discontinuous density gradient centrifugation in CsCl solutions (Huger and Krieg, 1989) and Ludox gradient centrifugation (Zhu *et al.*, 1989). From SDS–PAGE analysis it was found that the proteinaceous parasporal crystals can be separated into subunits of 68 and 55 kDa (Bernhard, 1986) or 74 and 68 kDa (Krieg *et al.*, 1987b). Other investigations showed that the insecticidal δ-endotoxin crystals of *B. t. tenebrionis* contain, in addition to the major polypeptide of 67 kDa, minor polypeptides of 73, 72, 55 and 46 kDa, depending on the growth stage monitored (Carroll *et al.*, 1989). Only the 73 kDa polypeptide was formed during sporulation at stage I, whilst the 67 kDa polypeptide was first detected at stage II, its concentration increasing throughout the later stages of sporulation and after crystal release (a discussion of the *B. t.* sporulation stages can be found in Fast, 1981). There was a concomitant decrease in the 73 kDa polypeptide. This transition from the 73 kDa to the 67 kDa polypeptide can be prevented by the addition of proteinase inhibitors.

B. t. tenebrionis δ-endotoxin crystals can be solubilized by treatment with trypsin or insect gut extract, resulting in a 55 kDa cleavage product. Asparagine was detected at position 159 as the N-terminus of the amino acid sequence of the toxin (Höfte *et al.*, 1987; Sekar *et al.*, 1987; McPherson *et al.*, 1988). This polypeptide was found to be as toxic *in vivo* as the native δ-endotoxin.

Genetic characterization

The *cryIIIA* gene encoding the coleopteran-active crystal protein has a regional homology with the toxin-encoding domain of the lepidopteran (*cryI*) and dipteran (*cryIV*) genes but lacks an area corresponding to the 3′ half of these molecules (Herrnstadt *et al.*, 1987). When expressed in *Escherichia coli* the *cryIIIA* gene directs the synthesis of a 72 kDa protein toxic for CPB (Jahn *et al.*, 1987; Rhim *et al.*, 1990; McIntosh *et al.*, 1990). The conversion of this protein into a 66 kDa toxin results from spore-associated proteases which remove 57 N-terminal amino acids (McPherson *et al.*, 1988).

In addition to the *B. t. tenebrionis cryIIIA* gene, another coleopteran active gene has been described, i.e. *cryIIIB* (Sick *et al.*, 1990). The *cryIIIB* gene was identified in *B. t.* strain EG 2838 and subsequently shown to encode a 74.2 kDa protein with insecticidal activity against CPB. This strain was classified by flagellar serotype as *B. t.* subspecies *tolworthi* (H9). The *cryIIIB* gene shares only 69% homology with the *cryIIIA* gene.

HOST RANGE OF *B. THURINGIENSIS TENEBRIONIS* AND SUSCEPTIBILITY LEVELS OF SOME PEST SPECIES

Since the first isolation of *B. t. tenebrionis* and the discovery of its effect on Coleoptera many species have been tested to assess susceptibility (Table 8.1). Most of the known susceptible species belong to the chrysomelid family. Only six species from other families have been reported as being susceptible, often at low levels (Table 8.1). So far no species have been identified in which the adults are susceptible to *B. t. tenebrionis* but the larvae are not and there is only one case (*Galerucella viburni*) where the adults seem to be more susceptible than the larvae.

Leptinotarsa decemlineata, Colorado potato beetle

One of the most important pests in the Coleoptera, which causes extensive damage on potatoes, tomatoes and egg plants, is the Colorado potato beetle (CPB), *Leptinotarsa decemlineata* (Chrysomelidae). This beetle was first described in 1824 and named after Colorado, USA, where it was found. The native home is central Mexico, where the beetle feeds on *Solanum rostratum* and related species (Casagrande, 1985). CPB was found in Nebraska, USA, for the first time in 1859, after successful adaptation to potato plants. From there it spread East and West in the USA and was accidentally imported to Europe in 1922. Today potato plants are infested throughout most of Europe with the exception of the UK. Historically, chemicals have been the major control measure for CPB (Gauthier *et al.*,

Table 8.1. Susceptibility of larvae (L) and adults (A) of some coleopteran species to *Bacillus thuringiensis* subspecies *tenebrionis*

Species	Susceptibility Laboratory		Field	Literature
	L	A		
Chrysomelidae				
[a] *Agelastica alni*	+ + +			Krieg *et al.* 1983
		+		Meyer (1989)
			+ + +	G. A. Langenbruch, unpublished
Calligrapha scalaris	+ + +	+ +		W. Gelernter, personal communication
[a] *Chrysomela scripta*	+ + +	+ +	+ + +	Bauer (1990)
Chrysophtharta bimaculatus	+ +			O. Skovmand, personal communication
[a] *Galerucella (Pyrrhalta) viburni*	+ +	+ + +		Meyer (1989)
			+ +	Riethmüller (1990)
[a] *Leptinotarsa decemlineata*	+ + +	(+)	+ + +	Krieg *et al.* (1984)
Melasoma vigintipunctata	+ + +			Riethmüller (1990)
[a] *Paropsis charybdis*	+ + +	−		Jackson and Poinar (1989)
Plagiodera versicolora	+ + +	+ +		Meyer (1989)
[a] *Xanthogaleruca (Pyrrhalta) luteola*	+ + +	+ + +		Herrnstadt *et al.* (1986)
exclamalionis			+ + +	Cranshaw *et al.* (1989)
zygogramma	+ + +	+ +		W. Gelernter, personal communication
Chrysomela herbacea	+ +			Riethmüller (1990)
Gastrophys (Gastroidea) viridula	+ +			C. Balser, personal communication
		+		Meyer (1989)
Haltica tombacina	+ +	+ +		Herrnstadt and Soares (1986)
Phaedon cochleariae	+ +	(+)		Meyer (1989)
Phyllodecta vulgatissima		+ +		Riethmüller (1990)
[a] *Crioceris asparagi*	+			Riethmüller (1990)
[a] *Crioceris duodecimpunctata*	+			Riethmüller (1990)
[a] *Diabrotica undecimpunctata*	+	−		Herrnstadt *et al.* (1986)
Chrysolima (Chrysomela) fastuosa		(+)		Meyer (1989)
Galeruca tanaceti	+			Riethmüller (1990)
[a] *Lilioceris lilii*	+			Riethmüller (1990)
Diabrotica longicornis	−			Gelernter (1990)
Haltica oleracea	−			W. Schnetter, personal communication
Oulema melanopa	−	−		Meyer (1989)
Oulema spp.	+			G. A. Langenbruch, unpublished

Table 8.1. (*Continued*)

Species	Susceptibility Laboratory Field		Literature
	L	A	
Phyllotreta atra		−	Krieg *et al.* (1983)
Phyllotreta undulata		−	Krieg *et al.* 1983
Podagrica fuscicornis		−	Riethmüller (1990)
Other Families			
Carabidae			
Zabrus tenebrioides		−	Meyer (1989)
Cerambycidae			
Strangalia maculata		−	Meyer (1989)
Curculionidae			
[a]*Anthonomus grandis*	+	+	Herrnstadt *et al.* (1986)
Cylas puncticollis		−	Riethmüller (1990)
[a]*Hypera brunneipennis*	+ + +		Herrnstadt and Soares (1986)
[a]*Otiorrhynchus sulcatus*	+ +	(+)	Herrnstadt *et al.* (1986) Riethmüller (1990)
Zacladus affinis		−	Meyer (1989)
Coccinellidae			
[a]*Epilachna varivestis*	+		C. Balser, personal communication
Dermestidae			
Attagenus (*megatoma, piceus*)	−		Herrnstadt and Soares (1986)
Attagenus unicolor	−		Herrnstadt *et al.* (1986)
Trogoderma granarium	−		O. Skovmand, personal communication
Elateridae			
Agriotes lineatus		−	Riethmüller (1990)
Nitidulidae			
Meligethes aeneus		−	Meyer (1989)
Scarabaeidae			
Melolontha hippocastani	−		Meyer (1989)
Melolontha melolontha			Riethmüller (1990)
Tenebrionidae			
[a]*Tenebrio molitor*	+		Krieg *et al.* (1983)
Tribolium castaneum	−	−	Herrnstadt *et al.* (1986)

+ + +, High susceptibility, probably sufficient for control; + +, moderate susceptibility; +, low susceptibility; (+), no mortality, but feeding inhibition; −, no significant reaction.
[a] Susceptible species of importance in plant protection.

1981) which has responded to this pressure by evolving resistance to most of the presently available insecticides; this is especially so in the north-eastern USA (Forgash, 1985; Boiteau *et al.*, 1987). The lack of successful control of CPB with registered chemical insecticides has stimulated interest in the testing of alternative control measures.

Previous attempts to control CPB with natural enemies, notably the predator *Perillus bioculatus* (Hemiptera, Pentatomidae) and the parasite *Myiopharus doryphorae* (Diptera, Tachinidae), had only limited success (Casagrande, 1987). The use of the egg parasite, *Edovum puttleri* (Hymenoptera, Eulophidae) (Schroder and Athanas, 1989), Neem seed extract (Zehnder and Warthen, 1988), a juvenile hormone analogue (Koopmanshap *et al.*, 1989) and the fungal pathogen, *Beauveria bassiana* (Anderson *et al.*, 1988) also gave poor control of CPB. In comparison, *B. t. tenebrionis* has produced the best results.

Bioassay methods

Several bioassay methods have been developed for CPB larvae. In order to compare different preparations using an LC_{50}, value potato leaflets are dipped in *B. t. tenebrionis* suspensions and compared to a standard preparation. Riethmüller and Langenbruch (1989) used this technique for first instar larvae and, using an unformulated *B. t. tenebrionis* standard (a freeze-dried powder BI XXI/3), obtained an LC_{50} of 0.0038–0.0076%, indicating $1–2 \times 10^6$ spore equivalents/ml (assuming that the relationship between spores and crystals is approximately 1:1). Ferro and Gelernter (1989) used a similar technique for second instar larvae where the dipped leaflets were placed in small containers supplied with nutrient solution. The LC_{50} was estimated in units of a standard preparation (a spray-dried technical powder assigned a potency of 50 000 Colorado potato beetle International Units (CPB IU/mg)) and not in spore equivalents. The LC_{50} was 10 200 CPB IU/ml of suspension. Because no information regarding the number of spores or crystals/ml was supplied it is not possible to compare these data without conducting comparative bioassays. Riethmüller and Langenbruch (1989) also developed a second method to ascertain the LD_{50} by applying defined *B. t. tenebrionis* doses to small leaf discs which are then completely consumed by the larvae. Using the above-mentioned standard an LD_{50} of 2×10^3 spore equivalents/larva was calculated for first instar larvae.

Sensitivity and efficacy

With increasing age CPB larvae become less sensitive to *B. t. tenebrionis*. The LC_{50} of third instar larvae from Darmstadt, Germany, is about 10 times higher than that of the first instar. This rises to between 100 and

1000 times for fourth instar larvae (Langenbruch and Hommel, 1990). First instar larvae from Heidelberg (Germany) had an LC_{50} of 39 ng of a *B. t. tenebrionis* standard/cm^2 leaf area ($= 2.3 \times 10^4$ spore equivalents/cm^2), 44 ng/cm^2 for the second instar and 680 ng/cm^2 for the third instar. The fourth instar was so insensitive that the LC_{50} could not be measured: 15 minutes after ingestion of a high *B. t. tenebrionis* dose the larvae stopped feeding and vomited. A week later all the larvae were still alive (Meyer, 1989). It is thus considered uneconomic to control fourth instar CPB larvae in the field, though there may be local strains of CPB with different levels of susceptibility.

When adults are exposed to contaminated leaves they stop feeding and egg-laying; however, this behaviour is reversed as soon as they find untreated leaves (Krieg *et al.*, 1984). Zehnder and Gelernter (1989) had similar results with adults and also second and third instar CPB larvae. Meyer (1989) found no adult mortality even after an application of 0.3 mg *B. t. tenebrionis* standard/cm^2 leaf area.

Several bioassays with first instar CPB larvae showed that *B. t. tenebrionis* spores were not involved in the efficacy of spore-crystal preparations (Riethmüller and Langenbruch, 1989). This was confirmed by testing high doses of *B. t. tenebrionis* mutants which produced either no spores or no crystals, or neither spores nor crystals, in CPB larvae. All strains without crystals were ineffective whereas strains with crystals and without spores caused high mortality (Meyer, 1989). If other important pests could also be controlled by *B. t. tenebrionis* without living spores, it would be of practical advantage to inactivate the spores by γ-irradiation. In this way replication of sprayed *B. t. tenebrionis* could be prevented and the restriction on its use in water-catchment areas in countries such as Germany might be avoided.

Little work has been carried out on adjuvants, although Riethmüller (1990) tested 17 additives (spreader-sticker, feeding stimulants, oils) for enhancement of the efficacy of *B. t. tenebrionis* for CPB larvae but found that only the feeding stimulant Coax® (Trader Oil Mill Company, USA) increased the effectiveness.

Agelastica alni, alder leaf beetle

Agelastica alni damages alders in Europe and is therefore of importance in forest management, parks and urban areas. With CPB, *A. alni* was the first chrysomelid species to be tested in *B. t. tenebrionis* bioassays (Krieg *et al.*, 1983). After eating *B. t. tenebrionis*-contaminated leaflets the larvae ceased feeding. The LD_{50} of third instar larvae was estimated as 0.5×10^7 spore equivalents/ml. Ultraviolet (UV) irradiation of the spores reduced the efficacy of the preparation. According to Meyer (1989), the LC_{50} of the second instars was about 10 times higher than the LC_{50} of CPB second instars. No more than 50% of the adults died after feeding on 0.3 mg *B. t. tenebrionis*/cm^2 leaf area.

Chrysomela scripta, cottonwood leaf beetle

Bauer (1990) described a standardized bioassay method for *Chrysomela scripta* in which immature leaves of hybrid poplars were dipped in suspensions of a commercial *B. t. tenebrionis* product (M-One®) and air dried. Second and third (final) instar larvae and adults were allowed to feed for 96 h on the leaves. The LC$_{50}$ was estimated at 10 000 CPB IU/ml for second instars, approximately 20 times higher for third instars and 30–40 times higher for adults. Hence the susceptibility of second instar *C. scripta* is similar to *L. decemlineata*. Adults and larvae placed on treated foliage stopped feeding and began to wander. This behaviour may increase the opportunity for predators and parasitoids to exert control in addition to *B. t. tenebrionis*.

Galerucella (Pyrrhalta) viburni, cranberry tree leaf beetle

Meyer (1989) found that second and third instar larvae and adults were significantly less sensitive than CPB larvae and required more than 1 μg *B. t. tenebrionis* standard/cm^2 to achieve a LC$_{50}$. The adults appeared more susceptible than the larvae as when fed with 1 μg *B. t. tenebrionis*/cm^2 they died after 6 days. However, interpretation of the assays depends on a knowledge of the relative feeding rates and hence the quantity of *B. t. tenebrionis* ingested by the larvae and the adults.

Paropsis charybdis, eucalyptus tortoise beetle

Jackson and Poinar (1989) tested the eucalyptus tortoise beetle, a major chrysomelid pest of eucalyptus in New Zealand, for susceptibility to *B. t. tenebrionis*. They found that larvae were susceptible with a first instar LC$_{50}$ of approximately 2 CPB IU/mg, a sensitivity similar to that of CPB. Adults were found to be insensitive, although their food intake may have decreased.

Xanthogaleruca (Pyrrhalta) luteola, elm leaf beetle

Because CPB is a quarantine species in California, Mycogen Corporation initially used *Xanthogaleruca luteola* for bioassays with *B. t. tenebrionis*. Both adults and larvae showed more than 90% mortality when exposed to leaves treated with a spore-crystal preparation at 10^4 spore equivalents/cm^2 (Herrnstadt *et al.*, 1986).

Susceptibility of other coleopteran species

Anthonomus grandis, the cotton boll weevil, is a pest of great economic importance in the USA and Mexico. The susceptibility of *A. grandis* to *B. t. tenebrionis* was disclosed in a patent (Herrnstadt and Soares, 1986) and a publication in the same year (Herrnstadt *et al.*, 1986). There are, however, no further publications either on *B. t. tenebrionis* bioassays or field trials to control cotton boll weevil.

The chrysomelids *Crioceris asparagi*, *C. duodecimpunctata* (both severe pests on asparagus in Europe) and *Lilioceris lilii*, which damages lilies in Europe, are so insensitive to *B. t. tenebrionis* that it seems unlikely that they can be controlled by this pathogen. However, Skovmand (personal communication) of Novo Nordisk (Denmark) has reported better results with *L. lilii*.

Some other important species, *Diabrotica undecimpunctata howardi*, the Western spotted cucumber beetle, *Otiorhynchus sulcatus*, the black vine weevil, and *Epilachna varivestis*, the Mexican bean beetle, show only slight susceptibility. Their economic control with *B. t. tenebrionis* therefore seems unlikely.

Surprisingly, larvae of the yellow mealworm, *Tenebrio molitor*, the original 'host' of *B. t. tenebrionis*, showed a remarkable resistance to *B. t. tenebrionis* spores and crystals (Huger *et al.*, 1986; Herrnstadt and Soares, 1986; Herrnstadt *et al.*, 1986). The effect of *B. t. tenebrionis* on adult *T. molitor* has not been tested.

THE IMPACT OF *B. THURINGIENSIS TENEBRIONIS* ON NON-TARGET ANIMALS

Before using a pathogen in pest control its host range and possible side-effects must be investigated. Table 8.2 shows the animals, other than Coleoptera, tested with *B. t. tenebrionis* and found to be insensitive. The impact of *B. t. tenebrionis* on beneficial arthropods is important but has only been studied in a small number of species. A lacewing, *Chrysoperla carnea*, was insensitive (Table 8.2) but in a preliminary bioassay of *B. t. tenebrionis* on the seven spot ladybird, *Coccinella septempunctata*, 63% mortality was observed when first instar larvae were placed for 6 days on potato shoots with aphids, dipped in a solution of 1% Novodor® (G. A. Langenbruch, unpublished data). This result needs to be verified in further trials. It should also be noted that one eighty-third of the concentration used in the bioassays causes more than 80% mortality in first instar CPB in comparable bioassays. It is therefore considered unlikely

Table 8.2. Species insensitive to *Bacillus thuringiensis* subspecies *tenebrionis*

CLASS/**Order**/Family	Species	Tested Stages L	A	Reference
CRUSTACEA				
Isopoda:				
Porcellionidae:	*Porcellio scaber*		x	Meyer (1989)
INSECTA				
Ensifera:				
Gryllidae	*Acheta domestica*	x		Meyer (1989)
Phaneropteridae	*Leptophyes punctatiss*	x		Meyer (1989)
Caelifera:				
Acrididae	*Locusta migratoria*	x		Meyer (1989)
Dermaptera:				
Forficulidae	*Forficula auricularia*	x		Meyer (1989)
Blattodea:				
Blattidae	*Periplaneta americana*	x		Krieg *et al.* (1983)
Planipennia:				
Chrysopidae	*Chrysoperla carnea*	x		G. A. Langenbruch, unpublished
Hymenoptera:				
Tenthredinidae	*Athalia rosae*	x		Krieg *et al.* (1983)
Lepidoptera:				
Hyponomeutidae	*Hyponomeuta malinellus*	x		Meyer (1989)
Lymantriidae	*Euproctis chrysorrhoea*	x		Meyer (1989)
Noctuidae	*Spodoptera exigua*	x		Herrnstadt *et al.* (1986)
	Trichoplusia ni	x		Herrnstadt *et al.* (1986)
Nymphalidae	*Inachis io*	x		Meyer (1989)
Phycitidae	*Ephestia kuehniella*	x		Krieg *et al.* (1983)
Diptera:				
Culicidae	*Aedes aegypti*	x		Krieg *et al.* (1983)
	Culex pipiens	x		Meyer (1989)
Tipulidae	*Tipula* sp.	x		Meyer (1989)
ARACHNIDA				
Acari:				
Tetranychidae	*Tetranychus urticae*		x	Chapman *et al.* (1991)
PISCES				
Atheriniformes:				
Poeciliidae	*Lebistes reticulata*		x	Meyer (1989)
AMPHIBIA				
Anura:				
Pipidae	*Xenopus laevis*	x		Meyer (1989)
MAMMALIA				
Rodentia:				
Muridae	*Mus musculus* (*per os* and intraperitoneal)		x	Meyer (1989)

L, larvae; A, adults.

that *B. t. tenebrionis* will be harmful to coccinellids at doses recommended for field application.

Chapman and Hoy (1991) compared different *B. t. tenebrionis* preparations and Dipel 2X® (*B. t. kurstaki*) for their effect on the spider mite pest *Tetranychus urticae* and its mite predator *Metaseiulus occidentalis*. In laboratory bioassays they applied rates of 0.1, 0.5 and 1.0 times the field rate of Abbott preparations and comparable doses of M-One® and Trident® to leaf discs of the French bean, *Phaseolus vulgaris*. Abbott formulations reduced the survival of *M. occidentalis* adult females, but not adult females of the phytophagous *T. urticae*. The egg hatch of both species was not significantly affected but the survival of *M. occidentalis* immatures was reduced. Trident®, and especially M-One® were less noxious but still produced significant mortality. It was concluded that the impact of *B. t. tenebrionis* in the field would be difficult to predict from laboratory studies and that field trials are necessary. Both these sets of data indicate that more detailed investigations are needed of the effects of *B. t. tenebrionis* on beneficial arthropods in the field.

CONTROL OF COLEOPTERA BY β-EXOTOXIN

Fifty-five species, belonging to ten Orders of insects, have been found to be susceptible to the β-exotoxin, including three species of beetle, CPB, *Phaedon cochleariae* (Krieg and Langenbruch, 1981) and the Mexican bean beetle, *Epilachna varivestis* (Cantwell *et al.*, 1985). The β-exotoxin is noxious for many pest species such as *Pieris brassicae* (Lepidoptera), largewhite cabbage butterfly; *Locusta migratoria* (Orthoptera) migratory locust; *Musca domestica* (Diptera); common house fly; *Tetranychus urticae* (Acarina); *Meloidogyne incognita* (Nematoda). However, it is also harmful to beneficial insects like the honey bee, *Apis mellifera* (Hymenoptera) (Krieg and Langenbruch, 1981). Vertebrates are also sensitive. By injection the intraperitoneal LD_{50} for mice amounts to 10–20 $\mu g/g$ body-weight and the *per os* toxicity of some salts of the β-exotoxin in hens is considerable, even with low doses (3 $\mu g/g$ body-weight) (Sebesta *et al.*, 1981).

From these results it is obvious that the β-exotoxin is less selective than the δ-endotoxins and it is therefore not used for plant protection in most countries. However, in the former Soviet Union and some countries of Eastern Europe, combined *B. t.* preparations with both the δ-endotoxin and the β-exotoxin, especially of the *B. t.* subspecies *thuringiensis*, are commercialized. These products are used to control early instars of CPB (Sebesta *et al.*, 1981). In the USA β-exotoxin preparations have also been tested as an alternative control measure against CPB. In field trials on tomatoes a 98–100% reduction of older larvae and adults was observed after two experimental preparations were applied at weekly intervals (Cantwell and Cantelo, 1984).

There is no doubt that it is possible to control many coleopteran species with *B. t.* β-exotoxin. However, due to its unselective mode of action we have to consider if it is more hazardous than alternative measures such as chemical insecticides.

FIELD TRIALS WITH *B. THURINGIENSIS TENEBRIONIS*

Leptinotarsa decemlineata, Colorado potato beetle

To be effective *B. t. tenebrionis* crystals must be ingested by the pest and thus all parts of the plants which are eaten by the larvae should be treated with the toxin preparation. A special boom and nozzle arrangement can be employed to contaminate the underside of the leaves (Langenbruch *et al.*, 1985) or this can be carried out using an additional air stream in the sprayer.

In agreement with the bioassay results, field trials demonstrated that only first and second instar larvae could be controlled by an economically acceptable dose of *B. t. tenebrionis* (Langenbruch and Riethmüller, 1990). Mortality of third and fourth instar larvae was below 70%, which is unacceptable in terms of field control. Exact determination of the time of egg-laying and hatching must be used to time spray applications accurately. It is not possible to use the same concept of an economic threshold pest population density, as employed for chemical insecticides, as *B. t. tenebrionis* is less effective against older larvae and adults.

To evaluate the persistence of *B. t. tenebrionis*, Meyer (1989) treated leaves of potted potato plants in the laboratory with 1 μg of a *B. t. tenebrionis* standard per cm^2 leaf area. A leaf was cut off daily and fed to first instar larvae. The leaves cut on the first 2 days caused 100% mortality within 2 days and leaves cut 7 days after contamination resulted in 70% mortality within 2 days and 100% mortality within 4 days. Plants were then dosed with 0.1 μg or 1 μg *B. t. tenebrionis* standard/cm^2 leaf area and left outdoors in August. When leaves were sampled on the day of application, 0.1 μg/cm^2 gave 80% mortality after 4 days and 1 μg/cm^2 100% mortality after 4 days. Three days later there was no mortality in either treatment.

In another trial, plants were sprayed with three commercial *B. t. tenebrionis* preparations and leaves were cut off at 0, 1, 2, 3, 5, 7 and 9 days after the application and bioassayed with first instar larvae (Langenbruch and Riethmüller, 1990) (Figure 8.2). After 3 days, two emulsion preparations caused less than 50% mortality. After 5 days without rain these products caused 10% and 40% mortality respectively. A third product, a wettable powder, caused more than 50% mortality 9 days after application, in spite of 4 mm precipitation on the 6th and 8th days. Laboratory

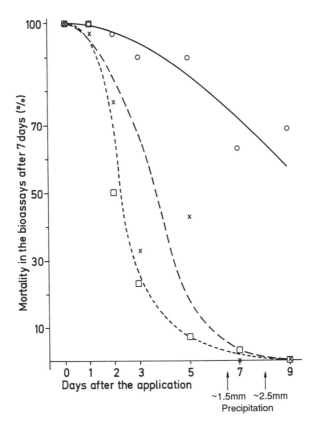

Figure 8.2. Foliar persistence of three *B. t. tenebrionis* preparations (wettable powder S, liquid formulations A and D). Tests were carried out using first instar Colorado potato beetle, *Leptinotarsa decemlineata*, larvae. Key, preparations S (———○), D (— — — — —×) and A (- - - - -□)

and field trials indicate that apart from rainfall, solar irradiation reduces the foliar persistence of *B. t. tenebrionis* crystals (Riethmüller and Langenbruch, unpublished data). *B. t. tenebrionis* preparations thus appear to have brief foliar persistence and applications need to be repeated at 7–10 day intervals if hatching of larvae continues. In field trials, one to three applications were needed to control the first generation of CPB (Langenbruch and Hommel, 1991).

In an attempt to overcome the short field persistence of *B. t. tenebrionis* δ-endotoxin Mycogen Corporation has produced the crystals in *Pseudomonas fluorescens* cells. The bacterium is then killed, but the crystal is left protected by the bacterial cell wall. The activity of this product, MYX 1806®, is reported to be significantly longer than M-One®.

Following these strictures on spray methodology, timing and frequency of application, good control of CPB (mortality between 73% and 91%) was achieved on potatoes in Germany with 3×10^{10} spore equivalents/m^2 of an unformulated, freeze-dried technical *B. t. tenebrionis* powder (spreader-sticker added), 2 kg/ha of a wettable powder and 3–5 litres/ha of a commercial liquid formulation (Langenbruch et al., 1985; Langenbruch and Riethmüller, 1990; Langenbruch and Hommel, 1991). Jaques and Laing (1989) obtained good results when they field tested a commercial formulation of *B. t. tenebrionis* applied with a spreader-sticker and a sunlight protectant at 220×10^9 CPB IU/ha sprayed on four dates at intervals of 6–8 days in potatoes and on five dates, with different intervals on tomatoes. There are no data on the instars present at the time of applications.

Ferro and Gelernter (1989) sprayed the manufacturer's recommended rate of 157×10^9 CPB IU/ha (= 7 litres M-One®) in a volume of 525 litres/ha five times at intervals of 7–10 days, as long as CPB larvae were present. The applications began when economic threshold levels were reached. Excellent control was obtained (a rate of 53×10^9 CPB IU was slightly less effective). Zehnder and Gelernter (1989) emphasize that the *B. t. tenebrionis* applications must be timed to coincide with CPB egg hatch or when early instars are the predominant life stage.

As these results show *B. t. tenebrionis* may be an important step towards integrated pest management in potatoes, particularly when the price of *B. t. tenebrionis* products becomes competitive with non-selective chemical alternatives.

Agelastica alni, alder leaf beetle

When alders were sprayed with Novodor® (1%) at egg hatch, *A. alni* mortality of more than 70% was noted after 7 days. During succeeding days 84% of larvae died on treated and 4% on untreated twigs held in the laboratory. Adults did not feed or lay eggs as long as the leaves were contaminated. However, if egg-laying continues, two or more applications may be necessary (G. A. Langenbruch, unpublished data).

Chrysomela scripta, cottonwood leaf beetle

Bauer (1990) cites a field trial conducted by Bystrak in 1988 where *Populus* sp. were sprayed with M-One® 9.3 l/ha (2.1×10^{11} CPB IU/ha) by air at York, Pennsylvania (USA), with successful suppression.

Galerucella viburni, cranberry tree leaf beetle

Viburnum opulus, guelder rose, is sporadically damaged by *G. viburni* in Germany; it often attacks the same bushes repeatedly, resulting eventually in their death. Riethmüller (1990) sprayed *V. opulus* bushes with unformulated *B. t. tenebrionis* (1%) to control the young larvae and observed a high mortality. The treated bushes remained green while the branches of the untreated bushes dried up.

Xanthogaleruca luteola, Elm leaf beetle

Cranshaw *et al.* (1989) sprayed 2.9–7.6 litres M-One® per 379 litres of water (spreader-sticker added) on Siberian elm, *Ulmus pumila*, when Elm leaf beetle larvae were predominantly in the first and second instars. While 2.9 litres gave inadequate mortality, 3.8 and 7.6 litres M-One® resulted in mortality of 88–96%. A second *B. t. tenebrionis* preparation (ABG 6263) showed even higher efficacy in glasshouse tests. High mortality rates were also observed for adults on potted elm trees treated with *B. t. tenebrionis* in the glasshouse. The persistence of both preparations was short and 1 or 2 days after the application mortality levels dropped from 79% to 1.5%. Applications therefore should be accurately timed to coincide with the end of peak egg hatch. Because the elms are often located in urban areas, the authors see great advantages in using an environmentally acceptable biological agent.

FUTURE PROSPECTS

Payne (1988) reported about 50 coleopteran-active isolates belonging to five serotypes with different crystal morphologies from 33 locations throughout the USA. Such progress suggests that we may now be standing at the beginning of a new chapter in the biological control of Coleoptera by *B. t.*, if new isolates are to be used commercially.

REFERENCES

Anderson, T. E., Roberts, D. W. and Soper, R. S. (1988). 'Use of *Beauveria bassiana* for suppression of Colorado potato beetle populations in New York State (Coleoptera: Chrysomelidae)'. *Env. Entomol.*, **17**, 140–145.

Bauer, L. S. (1990). 'Response of the Cottonwood leaf beetle (Col.: Chrysomelidae) to *Bacillus thuringiensis* var. *san diego*'. *Env. Entomol.*, **19**, 428–431.

Bernhard, K. (1986). 'Studies on the delta-endotoxin of *Bacillus thuringiensis* var. *tenebrionis*'. *FEMS Microbiol. Lett.*, **33**, 261–265.

Boiteau, G., Parry, R. H. and Harris, C. R. (1987). 'Insecticide resistance in New Brunswick populations of the Colorado potato beetle (Coleoptera: Chrysomelidae)'. *Can. Entomol.*, **119**, 459–463.

Bond, R. P. M., Boyce, C. B. C., Rogoff, M. H. and Shieh, T. R. (1971). 'The thermostable exotoxin of *Bacillus thuringiensis*'. In *Microbial Control of Insects and Mites* (Eds H. D. Burges and N. W. Hussey), pp. 275–302, Academic Press, London, New York.

Burgerjon, A. and Biache, G. (1966). 'Alimentation au laboratoire de *Perillus bioculatus* Fabr. avec des Larves de *Leptinotarsa decemlineata* Say Intoxiquées par la Toxine Thermostable de *Bacillus thuringiensis* Berliner', *Entomophaga*, **11**, 279–284.

Cantwell, G. E. and Cantelo, W. W. (1984). 'Control of the Colorado potato beetle (Coleoptera: Chrysomelidae) on tomatoes with *Bacillus thuringiensis* subsp. *thuringiensis*'. *The Great Lakes Entomologist*, **17**, 145.

Cantwell, G. E., Cantelo, W. W. and Schroder, R. F. W. (1985). 'The integration of a bacterium and parasites to control the Colorado potato beetle (*Leptinotarsa decemlineata*) and the Mexican bean beetle (*Epilachna varivestis*)'. *J. Entomol. Sci.*, **20**, 96–103.

Carroll, J., Li, J. and Ellar, D. J. (1989). 'Proteolytic processing of a coleopteran-specific δ-endotoxin produced by *Bacillus thuringiensis* var. *tenebrionis*'. *Biochem. J.*, **261**, 99–105.

Casagrande, R. A. (1985). 'The "Iowa" Potato Beetle, its discovey and spread to potatoes'. *Bull. Entomol. Soc. Am.*, **31**, 27–29.

Casagrande, R. A. (1987). 'The Colorado potato beetle: 125 years of mismanagement'. *Bull. Entomol. Soc. Am.*, **33**, 142–150.

Chapman, M. H. and Hoy, M. A. (1991). 'Relative toxicity of *Bacillus thuringiensis* subsp. *tenebrionis* to the Two-spotted spider mite (*Tetranychus urticae* Koch) and its predator *Metaseiulus occidentalis* (Nesbitt) (Acari, Tetranychidae and Phytoseiidae)'. *J. Appl. Entomol.*, **111**, 147–154.

Cidaria, D., Cappai, A., Vallesi, A., Caprioli, V. and Pirali, G. (1990). '*Bacillus thuringiensis* var. *donegani* and preparation or toxin obtained therefrom, endowed with insecticidal activity against Coleoptera'. *European Patent Application No. 90114708.2*.

Cranshaw, W. S., Day S. J., Gritzmacher, T. J. and Zimmermann, R. J. (1989). 'Field and laboratory evaluations of *Bacillus thuringiensis* strains for control of the Elm leaf beetle'. *J. Arboric.*, **15**, 31–34.

de Barjac, H. (1981). 'Identification of H-Serotypes of *Bacillus thuringiensis*'. In *Microbial Control of Pests and Plant Diseases 1970–1980* (Ed. H. D. Burges), pp. 35–43, Academic Press, London, New York.

de Barjac, H. and Frachon, E. (1990). 'Classification of *Bacillus thuringiensis* strains'. *Entomophaga*, **35**, 233–240.

Donovan, W. P., Gonzalez, J. M. Jr, Gilbert, M. P. and Dankocsik, C. (1988). 'Isolation and characterization of EG 2158, a new strain of *Bacillus thuringiensis* toxic to coleopteran larvae and nucleotide sequence of the toxin gene'. *Mol. Gen. Genet.*, **214**, 365–372.

Fast, P. G. (1981). 'The crystal toxin of *Bacillus thuringiensis*'. In *Microbial Control of Pests and Plant Diseases 1970–1980* (Ed. H. D. Burges), pp. 223–248, Academic Press, London, New York.

Ferro, D. N. and Gelernter, W. D. (1989). 'Toxicity of a new strain of *Bacillus thuringiensis* to Colorado potato beetle'. *J. Econ. Entomol.*, **82**, 750–755.

Forgash, A. J. (1985). 'Insecticide resistance in the Colorado potato beetle'. In *Proc. of the Symp. on the Colorado Potato Beetle XVII Int. Congress of Entomology, Massachusetts*. (Eds. D. N. Ferro and R. H. Voss), pp. 33–52.

Gauthier, N. L., Hofmaster, R. N. and Semel, M. (1981). 'History of Colorado potato beetle control'. In *Advances in Potato Pest Management* (Eds. J. H. Lashomb and R. Casagrande), pp. 13–33.

Gelernter, W. D. (1990). '*Bacillus thuringiensis*, bioengineering and the future of bioinsecticides.' Brighton Crop Protection Conference — Pests and Diseases. Vol. 2, 617–624.

Herrnstadt, C., Gilroy, T. E., Sobieski, D. A., Benett, B. D. and Gaertner, F. A. (1987).
'Nucleotide sequence and deduced amino acid sequence of a coleopteran-
active delta endotoxin gene from *Bacillus thuringiensis* subsp. *san diego*.' *Gene*,
57, 37–46.

Herrnstadt, C. and Soares, G. G. (1986). 'Cotton boll weevil, Alfalfa weevil, and
Corn rootworm via contact with a strain of *Bacillus thuringiensis*'. *US Patent No.
4,797,276* Jan 10, 1989 (filed: Mar. 21, 1986).

Herrnstadt, C., Soares, G. G., Wilcox, E. R. and Edwards, D. L. (1986). 'A new strain
of *Bacillus thuringiensis* with activity against coleopteran insects'. *Bio/Tech-
nology*, **4**, 305–308.

Höfte, H., Seurinck, J., Van Houtven, A. and Vaeck, M. (1987). 'Nucleotide
sequence of a gene encoding an insecticidal protein of *Bacillus thuringiensis*
var. *tenebrionis* toxic against Coleoptera'. *Nucl. Acids Res.*, **15**, 7183.

Huger, A. M. and Krieg, A. (1989). 'Über zwei Typen parasporaler Kristalle beim
käferwirksamen Stamm BI 256-82 von *Bacillus thuringiensis* subsp. *tenebrionis*'.
J. Appl. Entomol., **108**, 490–497.

Huger, A. M., Krieg, A., Langenbruch, G. A. and Schnetter, W. (1986). 'Discovery
of a new strain of *Bacillus thuringiensis* effective against Coleoptera'. In
*Symposium in Memoriam Dr. Ernst Berliner anlaßlich des 75. Jahrestages der
Erstbeschreibung von Bacillus thuringiensis* (Eds. A. Krieg and A. M. Huger) Mitt.
Biol. Budesanstalt Land-u. Forstwirtschaft, Berlin-Dahlem, 333, 83–96.

Ignoffo, C. M., Garcia, C. and Kroha, M. (1982). 'Susceptibility of the Colorado
potato beetle *Leptinotarsa decemlineata* to *Bacillus thuringiensis*'. *J. Econ.
Entomol.*, **39**, 244–246.

Jackson, T. A. and Poinar, G. O. Jr (1989). 'Susceptibility of Eucalyptus tortoise
beetle (*Paropsis charybdis*) to *Bacillus thuringiensis* var. *san diego*'. In *Proc.
42nd N.Z. Weed and Pest Control Conf.* pp. 140–142.

Jahn, N., Schnetter, W. and Geider, K. (1987). 'Cloning of an insecticidal toxin gene
of *Bacillus thuringiensis* subsp. *tenebrionis* and its expression in *Escherichia coli*
cells'. *FEMS Microbiol. Lett.*, **48**, 311–315.

Jaques, R. P. and Laing, D. R. (1989). 'Effectiveness of microbial and chemical
insecticides in control of the Colorado potato beetle on potatoes and tomatoes'.
Can. Entomol., **121**, 1123–1131.

Koopmanshap, A. B., Couchi, H. and de Kort, C. A. D. (1989). 'Effects of a juvenile
hormone analogue on the eggs, post-embryonic development, metamorphosis
and diapause induction of the Colorado potato beetle, *Leptinotarsa decemlineata*'. *Entomol. Exp. App.*, **50**, 255–263.

Krieg, A. (1986). *Bacillus thuringiensis—ein mikrobiologisches Insektizid.* (Acta
Phytomedica 10), Parey, Berlin and Hamburg.

Krieg, A., Huger, A. M., Langenbruch, G. A. and Schnetter, W. (1983). '*Bacillus
thuringiensis* var. *tenebrionis*: ein neuer, gegenüber Larven von Coleopteren
wirksamer Pathotyp'. *Z. Ang. Entomol.*, **96**, 500–508.

Krieg, A., Huger, A. M., Langenbruch, G. A. and Schnetter, W. (1984). 'Neue Ergeb-
nisse über *Bacillus thuringiensis* var. *tenebrionis* unter besonderer Berücksichtigung selner Wirkung auf den Kartoffelkäfer (*Leptinotarsa decemlineata*)'.
Anz. Schädlingskde., Pflanzenschutz, Umweltschutz, **57**, 145–150.

Krieg, A., Huger, A. M. and Schnetter, W. (1987a). '*Bacillus thuringiensis* var. *san
diego* M-7 ist identisch mit dem zuvor in Deutschland isolierten käferwirksamen
B. thuringiensis subsp. *tenebrionis* Stamm BI 256-82'. *J. Appl. Entomol.*, **104**, 417–424.

Krieg, A. and Langenbruch, G. A. (1981). 'Susceptibility of arthropod species to
Bacillus thuringiensis'. In *Microbial Control of Pests and Plant Diseases
1970–1980* (Ed. H. D. Burges), pp. 837–899, Academic Press, London, New York.

Krieg, A., Schnetter, W., Huger, A. M. and Langenbruch, G. A. (1987b). '*Bacillus thuringiensis* subsp. *tenebrionis*, strain Bl 256-82: a third pathotype within the H-serotype 8a8b'. *System. Appl. Microbiol.*, **9**, 138–141.

Langenbruch, G. A. and Hommel, B. (1990). 'Zur Bekämpfung des Kartoffelkäfers mit *Bacillus thuringiensis* subsp. *tenebrionis*'. *Mitt. Biol. Bundesanst. Land-Forstwirtsch. Berlin-Dahlem.*, **266**, 294.

Langenbruch, G. A. and Hommel, B. (1991). 'Zur Bekampfung des Kartoffelkafers (*Leptinotarsa decemlineata*) mit *Bacillus thuringiensis* ssp. *tenebrionis*'. *Gesunde Pflanzen*, **43**, 193–196.

Langenbruch, G. A., Krieg, A., Huger, A. M. and Schnetter, W. (1985). 'Erste Feldver-suche zur Bekämpfung der Larven des Kartoffelkäfers (*Leptinotarsa decem-lineata*) mit *Bacillus thuringiensis* var. *tenebrionis*'. *Med. Fac. Landbouww. Rijksuniv. Gent.*, **50/2a**, 441–449.

Langenbruch, G. A. and Riethmüller, U. (1990). 'Kartoffelkäferbekämpfung mit *Bacillus thuringiensis* subsp. *tenebrionis*'. *Nachrichtenbl. Deut. Pflanzen-schutzd.*, **42**, 65–69.

McIntosh, S. C., Pherson, S. L., Perlak, J. J., Marrone, P. G. and Fuchs, R. L. (1990). 'Purification and characterization of *Bacillus thuringiensis* var. *tenebrionis* insec-ticidal proteins produced in *E. coli*'. *Biochem. Biophys. Res. Commun.*, **170**, 665–672.

McPherson, S. A., Perlak, F. J., Fuchs, R. L., Marrone, P. G., Lavrik, P. B. and Fisch-hoff, D. A. (1988). 'Characterization of the coleopteran-specific protein gene of *Bacillus thuringiensis* var. *tenebrionis*'. *Bio/Technology*, **6**, 61–66.

Meyer, D. (1989). 'Untersuchungen zum Wirtsspektrum und Wirkungsmechan-ismus von *Bacillus thuringiensis* subsp. *tenebrionis*'. Dissertation, University of Heidelberg (Germany), Faculty of Natural Sciences and Mathematics.

Payne, J. (1988). 'New isolates of *Bacillus thuringiensis* with activity against Coleop-tera'. In *Abstr. XXI Annu. Meet. Soc. for Invertebrate Pathology*, San Diego, p. 180.

Rhim, S. L., Jahn, N., Schnetter, W. and Geider, K. (1990). 'Heterologous expression of a mutated toxin gene from *Bacillus thuringiensis* subsp. *tenebrionis*'. *FEMS Microbiol. Lett.*, **66**, 95–100.

Riethmüller, U. (1990). 'Labor-und Freilanduntersuchungen über den Einsatz von *Bacillus thuringiensis* subspecies *tenebrionis* gegen Coleopteren unter besonderer Berücksichtigung der Larven des Kartoffelkäfers (*Leptinotarsa decemlineata* Say)'. Dissertation, College of Darmstadt (Germany), Biological Division.

Riethmüller, U. and Langenbruch, G. A. (1989). 'Zwei Biotestmethoden zur Prüfung von *Bacillus thuringiensis* subsp. *tenebrionis* gegen Larven des Kartoffelkäfers (*Leptinotarsa decemlineata*)'. *Entomophaga*, **34**, 237–245.

Schroder, R. F. W. and Athanas, M. M. (1989). 'Potential for the biological control of *Leptinotarsa decemlineata* (Col.: Chrysomelidae) by the egg parasite, *Edovum puttleri* (Hym.; Eulophidae) in Maryland, 1981–84'. *Entomophaga*, **34**, 135–141.

Šebesta, K., Farkaš, J., Horska, K. and Vánková, J. (1981). 'Thuringiensin, the beta-exotoxin of *Bacillus thuringiensis*'. In *Microbial Control of Pests and Plant Diseases 1970–1980* (Ed. H. D. Burges), pp. 249–281, Academic Press, London, New York.

Sekar, V., Thompson, D. V., Maroney, M. J., Bookland, R. G., Adang, M. J. (1987). 'Molecular cloning and characterization of the insecticidal crystal protein gene of *Bacillus thuringiensis* var. *tenebrionis*'. *Proc. Natl Acad., Sci.*, **84**, 7036–7040.

Sharpe, E. S. (1976). 'Toxicity of the parasporal crystal of *Bacillus thuringiensis* to Japanese beetle larvae'. *J. Invert. Pathol.*, **27**, 421–422.

Sick, A., Gaertner, F. and Wong, A. (1990). 'Nucleotide sequence of a coleopteran-active toxin gene from a new isolate of *Bacillus thuringiensis* subsp. *tolworthi'*. *Nucl. Acids Res.*, **18**, 1305.
Zehnder, G. W. and Gelernter, W. D. (1989). 'Activity of the M-One formulation of a new strain of *Bacillus thuringiensis* against the Colorado potato beetle: relationship between susceptibility and insect life stage'. *J. Econ. Entomol.*, **82**, 756–761.
Zehnder, G. and Warthen, J. D. (1988). 'Feeding inhibition and mortality effects of Neem-seed extract on the Colorado potato beetle (Coleoptera: Chrysomelidae)'. *J. Econ. Entomol.*, **81**, 1040–1044.
Zhu, Y. S., Brookes, A., Carlson, K. and Filner, P. (1989). 'Separation of protein crystals from spores of *Bacillus thuringiensis* by Ludox gradient centrifugation'. *Appl. Env. Microbiol.*, **55**, 1279–1281.

9 *Bacillus thuringiensis* in the Environment: Ecology and Risk Assessment

MARTIN P. MEADOWS

Microbiology and Crop Protection Department, Horticulture Research International, Worthing Road, Littlehampton, Sussex BN17 6LP, and Energy Technology Support Unit, Harwell, Oxfordshire OX11 0RA, UK

INTRODUCTION

Use of microbial pest control agents, in particular *Bacillus thuringiensis* (*B. t.*), appears likely to increase dramatically in the future as farmers move towards more environmentally acceptable agricultural practices. Sales of *B. t.* products are estimated to rise at least 20% per annum in the next few years (Rigby, 1991). However, the application of *B. t.* could be even greater if some of its commercial limitations, such as rapid inactivation in the field and a highly specific host spectrum, could be overcome. To this end, there has been much effort to isolate new *B. t.* strains with increased potency against target pest insect species and/or a wider host range. In addition, application of recombinant DNA technology provides the potential to create new strains of *B. t.* and an opportunity to transfer *B. t.* toxins to alternative expression systems such as transgenic plants, other bacteria and baculoviruses (see Gelernter and Schwab, Chapter 4; Ely, Chapter 5 and Marrone and MacIntosh, Chapter 10). *B. t.* strains with new combinations of δ-endotoxins can be tailored to particular pest complexes by conjugation (Burges and Jarrett, 1990), transformation (Lereclus *et al.*, 1989) and recombinant DNA techniques (Baum *et al.*, 1990). Such approaches, allied with exploitation of the diversity of wild-type isolates, will increase the scope for the application of *B. t.* The potential benefits to agriculture are considerable, but the possibility of adverse environmental impact caused by large-scale application of novel wild-type *B. t.* strains, recombinant *B. t.* strains or *B. t.* δ-endotoxin genes expressed in other genetically modified organisms (GMOs) needs to be considered. There is a well-documented history of the safe application of *B. t.* in the environment: a small number of wild-type strains, formulated as commercial products,

Bacillus thuringiensis, An Environmental Biopesticide: Theory and Practice. Edited by P. F. Entwistle, J. S. Cory, M. J. Bailey and S. Higgs
© 1993 John Wiley & Sons Ltd

have been applied in increasing quantities as insecticides for over three decades to food crops, ornamentals, forest trees and stored grains without incident of harm (Meadows, 1992). In addition, the decades of pure and applied research that have accompanied the exploitation and study of *B. t.* have provided some limited knowledge of its behaviour in the environment.

ROLE IN THE ENVIRONMENT

To be able to estimate the risk of releasing any microorganism into the environment it is important to understand the way that it interacts with its surroundings and the other biota. In the case of *B. t.*, there has been extensive study of its toxicology and safety, but there has been only limited research on its role in the environment. Because of the predicted increase in use of *B. t.* and, in particular, the potential application of recombinant strains, further study of its ecology is desirable for risk assessment.

The life cycle of *B. t.* consists of two distinct stages: vegetative cell division and sporulation, with the production of δ-endotoxin taking place during sporulation. The spore exhibits little or no metabolic activity and is able to remain viable under conditions of environmental stress, such as extremely low nutrient availability or desiccation, to which the vegetative cells are more sensitive. Germination of spores, and subsequent cell division, occur in response to environmental changes usually associated with conditions favourable for multiplication, such as increased nutrient levels. Studies indicate that *B. t.* persists in soil predominately as spores, with only limited multiplication of vegetative cells (Akiba *et al.*, 1980; West *et al.*, 1984, 1985a,b; Petras and Casida, 1985; and Akiba, 1986), but the relative roles of spores and vegetative cells in persistence, and possible spread of *B. t.* in the environment are not known. *B. t.* can be readily isolated from insects, soil, stored product dusts, deciduous and conifer leaves and also aquatic environments. It is usual to find relatively small numbers of crystalliferous spore-forming bacteria in most samples of natural field material (Martin and Travers, 1989; Meadows *et al.*, 1990), and so *B. t.* seems to be an indigenous bacterium in many environments.

The reason(s) for the widespread distribution of *B. t.* are unknown and this section uses currently available knowledge to describe three hypotheses concerning the possible role of *B. t.* in the environment.

B. thuringiensis as an entomopathogen

Many strains of *B. t.* have been isolated from insects but epizootics (widespread, rapidly spreading infections) are extremely rare in the outdoor environment (Burges, 1973). Occasionally, *B. t.* has been found to be enzootic (i.e. ever-present but not at outbreak proportions) in insects in the field while epizootics usually occur only indoors where insects are kept in dry, confined environments in which spores survive almost indefinitely (Burges, 1973). Indeed, one of the present major commercial strains, HD-1, was isolated from within a mass reared colony of the pink bollworm, *Pectinophora gossypiella* (Dulmage, 1970). An example of an outdoor epizootic was the discovery of a crystalliferous spore-former from a diseased, very dense population of *Culex pipiens* in a small pond in north-central Negev Desert, Israel (Goldberg and Margalit, 1977; Margalit and Dean, 1985). The isolate, highly toxic to certain mosquito and blackfly species, was reported by de Barjac (1978) as a new *B. t.* serotype (H-14), and identified as subspecies *israelensis*.

It seems unlikely that *B. t.* would expend energy and nutrients during sporulation (a time of nutrient limitation or physiological stress) to produce a proteinaceous crystal occupying up to 35% of the cell dry weight without this conferring a selective advantage. Ellar (1990) has argued the following. The spore and crystal form at the same time and are released together when the cell lyses. The crystal is inactive at non-alkaline pH and, hence, of no apparent value to the spore under non-alkaline conditions. However, when the spore and crystal are ingested by an insect larva, the gut pH is too high to allow vegetative growth but the alkaline conditions and gut enzymes activate the crystal toxin. The toxin causes cytolysis of the insect gut epithelial cells which upsets osmotic balance and control resulting in a lowering of pH. The spore is maintained in the alimentary canal due to gut paralysis caused by the toxin. It is being held in a nutrient-rich environment where germination, and more importantly, subsequent growth can take place. Without a crystal, *B. t.* would not be able to colonize the gut because of high pH, rapid food transit time and possible lack of available germinants in a healthy larval gut. *B. t.* therefore gains an advantage over non-toxic spore-formers in the environment of an insect gut. Further evidence for *B. t.*'s evolutionary relationship with insects is found with the ability of certain strains to inhibit their host's immune system. For example, *B. t.* subspecies *alesti* produces two extracellular inhibitors of the immune system of saturniid pupae (Edlund *et al.*, 1976).

An example of an ecosystem where *B. t.* and insects appear in intimate association is the stored product warehouse (Burges and Hurst, 1977; DeLucca *et al.*, 1982; Meadows *et al.*, 1990, 1992). New larval infestations frequently encounter insect cadavers containing *B. t.* spores. Additionally, liberated spores and crystals may persist for a long time in the dry environment of a warehouse. Thus, there may be natural enrichment in this closed system leading to new infections. There is also the probable transfer of plasmids between different strains of *B. t.* during growth within the insect (Jarrett and Stephenson, 1990). Thus, the stored product ware-house provides an ideal environment for bacterial growth, the creation of strains with new toxin combinations and also the long-term survival of spores.

Although *B. t.* is well adapted to multiply in the insect gut and thereby survive in environments populated by a wide range of insects, it is perhaps difficult to reconcile the apparently close adaptation of *B. t.* to this niche with the widespread distribution of the bacterium in the environment.

B. *thuringiensis* and the phylloplane

An important piece of work has recently demonstrated that *B. t.* can be readily isolated from the leaves of deciduous and coniferous trees in North America (Smith and Couche, 1991). *B. t.* was found to comprise between 0 and 95% of total spore-forming colony-forming units (CFUs) recovered from broadleaved trees (depending on the species) and *B. t.* iso-lates were obtained from 11 of 14 common species of deciduous tree and shrub. The numbers of isolates and recovery frequency were considered to be higher than would be expected of casual isolates that may simply have been transferred to the phylloplane by wind dissemination or soil splash. Isolates were also recovered from the phylloplane of coniferous trees throughout winter. The isolates were naturally occurring rather than the residues from commercial applications of *B. t.* as there was no cross reactivity with antibodies specific for any crystal proteins formulated into commercial products. *B. t.* was recovered in high enough numbers to be considered an epiphyte by Smith and Couche (1991), who hypothesized that conifer needles and not the soil serve as a functional reservoir for *B. t.* spores during the winter in temperate climates. Wind dissemination to deciduous leaves takes place in spring where there may be some growth of *B. t.* on the leaf exudates leading to protection from insect attack by causing low level mortality or by acting as an antifeedant. (*B. t.* may be pro-tected from inactivation by ultraviolet (UV) radiation by being sheltered by leaf structures or by being on the underside of the leaf.) Soil simply func-tions as a sink for *B. t.* spores when the leaves fall in the autumn whereas the true ecological niche of *B. t.* may be the phylloplane. This hypothesis is an important contribution in the ecology of *B. t.* but is based on a single study (Smith and Couche, 1991) and further research is required before it can be confirmed.

B. *thuringiensis* as a soil microorganism

In early studies on soil, *B. t.* was found only at low concentrations in some soils. For example, in the USA, only 0.5% of the isolates from the *Bacillus cereus*/*B. t.* colony morphological group were classified as *B. t.* (Delucca *et al.*, 1981). A study of Japanese soils found 2.7% of the same grouping to be *B. t.* (Ohba and Aizawa, 1986b). In the Philippines, only 12 out of 54 soil samples harboured *B. t.* (Padua *et al.*, 1982). However, these studies may have underestimated the distribution of *B. t.* in soil due to the problem of distinguishing *B. t.* from soil samples containing a high background level of other microorganisms.

Development of an acetate selection technique (Travers *et al.*, 1987) allowed a more accurate determination of the distribution of *B. t.* This simple technique increases the efficiency of isolation of crystalliferous spore-forming bacteria in the presence of other spore-formers. In a study of soils mainly from the USA, and from 29 other countries representative of five continents, *B. t.* was isolated from 785 of 1115 soil samples, a rate demonstrating that it is a common soil microorganism (Martin and Travers, 1989). The percentage of samples that contained *B. t.* varied depending on the continent from which the sample came. *B. t.* was found in 94% of soil samples from Asia and Central and South Africa and in 84% of the samples from Europe. The lowest recoveries were from New Zealand (56%) and the USA (60%). *B. t.* was most frequently found in samples from savannah, desert and agricultural land, but was also isolated from urban areas, forest, Arctic tundra and steppe. It was less frequently isolated from beaches, caves, some desert locations and rainforest. Perhaps surprisingly, there appeared to be no correlation between the preponderance of insects and the occurrence of *B. t.*

There are four possible explanations for the presence of *B. t.* in soil. The first is that *B. t.* rarely grows in soil, but is deposited there from insect cadavers, leaves, or larvae descending to pupate in soil. Soil acts as a reservoir or sink for *B. t.* spores which may then spread long distances by wind dissemination.

Some root- or tuber-feeding insects are known to be susceptible to *B. t.* but, alone, these are too few to account for the frequency of *B. t.* in soil. A second explanation could be that *B. t.* may have as yet undiscovered pathogenicity to common soil-inhabiting insects and other animals. Interestingly, three separate studies have shown that 40–64% of *B. t.* isolates from soil were not highly pathogenic to a limited number of test insect species from the Lepidoptera, Coleoptera and Diptera (Ohba and Aizawa, 1986b; Martin and Travers, 1989; Meadows *et al.*, 1990), and such isolates may be pathogenic to other insect or animal species.

A third explanation is that *B. t.* may grow in soil when nutrients become available, for example, from decaying plant or animal matter (West and Burges, 1985) or after release of nutrients into the soil (West *et al.*, 1985a), such as after rainfall or frost.

A fourth possible explanation for the presence of *B. t.* in soil is based on the affinity between *B. t.* and the very similar, more common, soil inhabiting *B. cereus*. *B. t.* has been co-isolated with *B. cereus* from soil in a wide range of geographical areas (DeLucca *et al.*, 1981; Ohba and Aizawa, 1986b,c; Salama *et al.*, 1986; Meadows *et al.*, 1990). However, in most soils, *B cereus* greatly outnumbers *B. t.* (DeLucca *et al.*, 1981; Ohba and Aizawa, 1986; Meadows *et al.*, unpublished data) and it appears that it may be able to grow under conditions that do not permit growth of *B. t.* (West and Burges, 1985; West *et al.*, 1985a,b). The reasons for this difference are not yet known. To date, the only consistent characteristic in which *B. t.* and *B. cereus* have been found to differ is the production, by *B. t.*, of parasporal δ-endotoxin crystals (Baumann *et al.*, 1984; Priest *et al.*, 1988; Zahner *et al.*, 1989). It is not known if possession of these crystals has an effect on growth characteristics, although some evidence suggests that the ability to produce δ-endotoxin crystals may be linked to the conditions required for spore germination. Thus, loss of plasmids encoding for the crystal can simultaneously alter the requirements for spore germination (Jarrett, 1985). Acrystalliferous or small crystal mutants have been found to respond to lower concentrations of germination-stimulating compounds and to germinate much faster than the crystal-producing parent strains (Jarrett, 1985).

Therefore, the fourth explanation for the presence of *B. t.* in soil may be as follows. *B. t.* strains do not readily germinate in soil but persist until favourable conditions for growth occur. The control of germination may not be absolute throughout a population of dormant spores, and a proportion of the *B. t.* spores may germinate in the soil under unfavourable conditions for growth. If this were to happen, cells which spontaneously lose plasmids may gain a selective advantage in a relatively nutrient-poor environment and outgrow cells harbouring a greater number of plasmids. Studies with *Escherichia coli* growing in continuous culture have demonstrated that there can be selective disadvantages associated with the maintenance of plasmids in a population. For example, there is a maintenance cost to the cell as a result of the need to reproduce the plasmid, to synthesize RNA and protein from plasmid-borne genes and to carry out other functions determined by the plasmid (Godwin and Slater, 1979). When populations of *E. coli* containing a small, non-conjugative plasmid were grown under carbon-limited conditions, the presence of the plasmid lowered the growth rate of the host bacterium and the proportion of plasmid-containing organisms in the total population declined (Helling *et al.*, 1981). Thus, in the case of *B. t.*, dividing cells which lose genetic material, such as δ-endotoxin-coding plasmids, might have a selective advantage in the soil niche over *B. t.* cells carrying a larger plasmid burden. A *B. t.* sporangium without a parasporal crystal is indistinguishable microscopically from *B. cereus* and colonies produced from such spores would be classified as *B. cereus*. δ-Endotoxin-coding genes will not be completely lost because a proportion of the original population of spores that did not germinate may persist, explaining the regular, if sparse, incidence of *B. t.* recovered from soil. This argument, involving loss and/or gain of plasmids, is highly simplified. The population dynamics of plasmid maintenance and loss in microbial populations is complex, involving interactions between host strains, the plasmids themselves and environmental factors (e.g. Caulcott *et al.*, 1987), and further studies are required in order to confirm or reject such an explanation.

The explanation of a proportion of *B. cereus* soil strains possibly being derived from *B. t.* is supported by some experimental evidence, suggesting that, in soil, *B. cereus* greatly outnumbers *B. t.* and that *B. cereus* bearing *B. t.* antigens greatly outnumber *B. t.* crystal formers. Ohba and Aizawa (1986b) reported that, out of 6910 soil isolates from Japan, morphologically referrable to the *B. cereus/B. t.* group, most were acrystalliferous (6724), of which, 29.6% (1990) reacted with known *B. t.* H (flagellar) antisera (Ohba and Aizawa, 1986c). This number is much greater than the number of crystal formers (186), of which, 62 reacted with the same antisera. Also Krieg (1969) recorded a number of *B. cereus* strains possessing *B. t.* flagellar antigens. The explanation is also supported indirectly by records of *B. cereus* isolates, not specifically of soil origin, in which *B. t.* antigens were uncommon, possibly because the strains had evolved in a different way and had not been derived from *B. t.* as may be the case in soil. Thus, only a few *B. cereus* isolates among 150 of unspecific origin (de Barjac, 1981) and 33 of insect origin (Lemille *et al.*, 1969) possessed *B. t.* H-antigens. Also, 137 *B. cereus* isolates from food poisoning incidences and 66 isolates from uncooked rice were divided between 18 *B. cereus* H-serotypes, of which only serotype No. 18 reacted with *B. t.* antigens (IVA and IVB) (Taylor and Gilbert, 1975). Interestingly, insect-pathogenic strains of *B. cereus*—albeit of low potency—have also been described (Kumari and Neelgund, 1985; Rahmet-Alla and Rowley, 1989).

The firm delineation between *B. t.* and *B. cereus* in soil may be an artifact caused by using a 'snapshot' of a dynamic process in which proportions of *B. t.* or *B. cereus* within *B. t./B. cereus* populations alter as a result of loss and/or transfer of δ-endotoxin-coding plasmids (see also section below on transfer of genetic information). Very small differences in growth characteristics between members of populations may result in large differences in numbers of different strains found in soil.

The taxonomic relationship between *B. t.* and *B. cereus* has been an area for much discussion in the past (Lysenko, 1983). After review of the taxonomic characteristics of *B. t.* and *B. cereus*, Gordon (1977) recommended *B. t.* should be described as a variety of *B. cereus*. On the basis of classical microbial taxonomy, bacteriologists may regard *B. t.* as a variety of *B. cereus* due to the very close relatedness of these two bacteria, demonstrated using well-established bacteriological characters but, from the point of view of insect pathologists and applied microbiologists involved in biological control, the possession of δ-endotoxin crystals is a major functional characteristic that argues for maintenance of *B. t.* as a species. In addition, although certain strains of *B. cereus* are responsible for mild diarrhoeal and emetic types of food poisoning (Gilbert, 1979; Johnson, 1984) and others for secondary medical and veterinary conditions (Parry *et al.*, 1983), no strains of *B. t.* have been associated with food poisoning or conclusively implicated with any human or other mammalian health incidents. This perhaps further argues for the maintenance of *B. t.* as a separate species in order to avoid possible confusion in the eyes of the public and regulatory authorities. At present, *B. t.* and *B. cereus* are recognized as separate species in the Second Edition of *Bergey's Manual of Systematic Bacteriology* (Claus and Berkeley, 1986).

RISK ASSESSMENT

Risk is an estimate of the probability and severity of harm or a prediction of the likelihood of something going wrong in a certain set of circumstances (King, 1985). Risk assessment is the process of obtaining quantitative and qualitative measures of risk levels (Ficksell and Covello, 1986; Fuxa, 1989) There have been few published quantitative studies of the environmental impact of *B. t.* insecticides, whereas their toxicology and safety have been studied extensively (see Burges, 1981; Siegel *et al.*, 1987).

Recently, core thinking about the safety of released microorganisms has switched from the subject of wild-type strains to GMOs. In turn, in North America, the focus of study on possible environmental impact of GMOs has shifted. Now, attention is directed to the properties of the organism to be released and the ecology of its target environment, rather than the method by which it was created (National Academy of Sciences, 1987). Over 15 years' experience of genetic manipulation in the laboratory indicates that GMOs are not inherently unsafe as a result of the means of their manipulation. Alternatively, the European approach remains that GMOs must be assessed independently from related wild-type species because the modified organisms may present an additional risk to the ecosystem over and above that of introduced wild-type species. At present, there is a great reluctance on the part of European regulatory authorities to allow even small-scale, highly controlled releases of GMOs for experimental purposes.

Research and development has brought us to the point where the identification of specific *B. t.* genes that code for insecticidal toxins has allowed these genes to be cloned and expressed in other *B. t.* strains (Klier *et al.*, 1983; Garduno *et al.*, 1988; Bone and Ellar, 1989, Schurter *et al.*, 1989); other bacteria (Adang *et al.*, 1985; Obukowicz *et al.*, 1986; Shivakumar *et al.*, 1986); crop plants such as cotton, tomatoes and soya (Barton *et al.*, 1987; Fischoff *et al.*, 1987; Vaeck *et al.*, 1987, 1989, see Ely,

Chapter 5) and a baculovirus in insect cells (Martens et al., 1990, see Gelernter and Schwab, Chapter 4). Other genetic studies include the creation of a strain of B. t. with combined activity against certain Spodoptera and Heliothis species using the natural process of plasmid exchange by conjugation (Burges and Jarrett, 1990). Thus, a series of methods have been used to effect genetic change, from advanced genetic engineering grading progressively into natural processes.

A dividing line has been established to determine which techniques fall within the GMO regulations. Conjugation is regarded as a natural process allowed to function in the laboratory. Conjugated strains have been registered for release in both North America and Europe and are not considered as GMOs for regulatory purposes and risk assessment. Directed mutagenesis of selected regions of a B. t. δ-endotoxin protein to alter receptor binding (Ahmad and Ellar, 1990) and translational fusion of different δ-endotoxin genes to enlarge the insecticidal spectrum (Honée et al., 1990) may have to be considered as separate studies for risk assessment.

There are different properties of the gene expression systems in different GMOs that have a bearing on risk assessment. Releases of wild-type B. t. strains for commercial use have demonstrated that B. t. does not develop into a naturally spreading epizootic, neither does it persist or spread in the land environment, indicating that recombinant genes expressed in B. t. strains will remain within, or close to, the area of release. Such relative confinement also applies to B. t. added directly to water courses for mosquito and blackfly control. To date, limited carry of B. t. has been obtained only in high discharge streams (Lacey et al., 1982), but extremely poor carry was obtained in small mountain streams (Undeen et al., 1981) as the spores soon become trapped in various ways, e.g. on plant and humic matter, and cannot be detected more than a few miles downstream. However, the possibility of dissemination of GMOs may be greater if other expression systems are chosen. For example, a baculovirus may multiply and spread by epizootic, and pollen from transgenic plants may be widely disseminated by insects or wind.

To date, there have been no deliberate releases of GMOs involving B. t. toxins in the United Kingdom, whereas, in the USA, a number of strictly regulated field trials using B. t. toxins inserted into other host systems, such as crop plants (see Ely, Chapter 5) and the natural endophyte, Clavibacter xyli subsp. cynodontis (Kostka, 1991) have taken place. No adverse environmental impact has been reported from such trials.

Concerns relating to the introduction of genetically modified entomopathogens

Fuxa (1989) summarized three major environmental concerns relating to the introduction of GMOs in general and genetically modified entomopathogens in particular. However, certain of these concerns could be extended to any introductions of species, wild-type or recombinant, into environments where they are not indigenous or in numbers which may disturb the ecological balance.

The first concern is that a well characterized organism in the laboratory may have unexpected properties in the field. Such properties could be unpredictably harmful or beneficial. In the case of an entomopathogen such as *B. t.*, there could be unexpected effects on virulence, host range and survival. Genetic modification may alter the evolutionary potential of the organism, such as by the bacterium becoming pathogenic to non-target organisms.

The second area of concern is the possibility of disturbing the balance of an ecosystem by the intentional release of an organism, in particular the change of an organism intended to be beneficial to the status of a pest species.

The third area of concern is the unintended transfer of genetic information between organisms. This compounds the two earlier points by increasing the likelihood of unpredictable behaviour in the field. For example, pathogenicity could be transferred to previously non-pathogenic organisms, or the introduced organism could receive genetic traits altering its spectrum of pathogenicity.

There are additional points that need to be considered when assessing the possible uses of new and modified *B. t.* strains. For example, what are the risks to susceptible, non-target species? In general, the *B. t.* bacterium represents much less risk to non-target species than most synthetic, chemical contact insecticides because of its high specificity, short persistence and poor dispersal ability. *B. t.* strains used as insecticides and genes selected for expression on transgenic plants are commercially viable because of their highly potent action on target pest species. Because of this selection it is considered unlikely, even if non-target species were susceptible, that they would be as highly susceptible as the target species. Additionally, when *B. t.* is applied, the short half-life of the spore/crystal mixture requires that it is sprayed when the susceptible, early instars of the target species are present: again, it is unlikely that the most susceptible instars of non-target species will be present at exactly the same time. Even with sensitive insects, only larval stages that can ingest enough toxin are killed; eggs, pupae and usually adults are unharmed.

There may be slightly greater risk to non-target species caused by the continuous presence of *B. t.* toxins constitutively expressed by transgenic plants. However, the risk would apply only to those species that are both *susceptible* to the toxin and that actually feed upon the same plant as the pest species. Arthropod susceptibility to *B. t.* has been studied in some detail: for a comprehensive review see Krieg and Langenbruch (1981). In all cases, the very low risk to non-target species represented by the use of *B. t.* needs to be considered alongside the possible risks of using some of the less specific, more recalcitrant synthetic chemical insecticides.

Development of resistance by the target pest species to *B. t.* applied as a spore/crystal mixture or as cloned δ-endotoxin genes in bacteria, baculoviruses or plants is another factor for consideration in risk assessment (Meadows, 1992 and see Marrone and MacIntosh, Chapter 10). The very low incidence of resistance to wild-type *B. t.* applied as an insecticide in the field has been one of the proven advantages of its use, and, despite widespread application of *B. t.* insecticides for more than 20 years, there has been only one report of substantial resistance in open field populations (Tabashnik *et al.*, 1990). This good record may, in part, be due to wild-type strains of *B. t.* often containing more than one δ-endotoxin gene (Dulmage *et al.*, 1981; Jarrett 1985; Aronson *et al.*, 1986; Aronson and Beckman, 1987; Höfte and Whiteley, 1989; Yong-Man Yu *et al.*, 1991). There may be specific receptors in the insect gut for each protein expressed by different δ-endotoxin genes (Hofmann *et al.*, 1988). The resistance mechanism to *B. t.* appears to be a change in the insect's receptor and not a non-specific effect, such as proteolytic cleavage of the active toxin by gut enzymes (Johnson *et al.*, 1990). If each of the toxins in a wild strain acts on a different receptor, then the chances of resistance developing to all the toxins simultaneously should be slight. There should be a greater chance of selecting for resistance if recombinant *B. t.* strains were produced that expressed only a single toxin gene. Thus, if genetically modified strains of *B. t.* are used in pest control, it should be better to use strains with more than one cloned δ-endotoxin gene.

Stone *et al.* (1989), using a genetically engineered *Pseudomonas fluorescens* strain containing the 134 000 mol. wt. δ-endotoxin of *B. t.* HD-1, were able, under laboratory conditions, to select for resistance to this endotoxin in *Heliothis virescens*. Resistance was 13- to 20-fold over non-selected lines in seven generations. The selected insects were also resistant to a commercial product of HD-1 (Dipel®). After 23 generations, resistance was 75-fold to *P. fluorescens* and 50-fold to Dipel®. Attempts to select for resistance using wild-type *B. t.* strains in the laboratory, which incorporate three to five toxins in the crystals, have produced, at most, only low levels of resistance (see Meadows, 1992). Evidence to date suggests there is no cross resistance to different *B. t.* toxins (McGaughey and Johnson, 1987; Van Rie *et al.*, 1990).

Transgenic plants constitutively expressing *B. t.* toxin protein in all tissues might increase selective pressure on the insect. All the insect population feeding on the crop in question would be exposed for a long time at all stages of growth. Any resulting resistant insect populations might not be controlled by the transgenic plant, or by applications of *B. t.* as an insecticide containing the same toxins. Selective pressure on insect populations may be reduced by using transgenic plants producing δ-endotoxins only in specific tissues, or by the use of non-constitutive expression systems that respond to, for example, tissue damage. It will also be important for farmers to take advantage of resistance management practices. Resistance management, a very important aspect of the ecology and application of *B. t.*, is reviewed by Marrone and MacIntosh, Chapter 10.

Ecology and risk assessment

A limitation to the rate of progress in introduction of many potentially beneficial microbial agents in agriculture is the relative lack of knowledge about their ecology, and more generally about the fate and effects of applied microorganisms in the field. Mode of application, the persistence of introduced microorganisms, their reproductive rate (multiplication), their rate of gene transfer to indigenous organisms, their movement away from the site of application (dissemination) and the effects on the balance and functioning of the exposed ecosystem (safety/benefit/harm) are of prime importance and must be assessed before a release can be considered (Levin, 1982; Trevors *et al.*, 1986). In the case of *B. t.* its lack of adverse environmental impact is well documented. An understanding of its ecology will aid the prediction of environmental impact of the application of new *B. t.* strains, both recombinant and wild-type, which could provide criteria for risk assessment of other introduced microorganisms.

Application

B. t. has been applied to the environment since 1933, but it was not successfully commercialized until the 1950s, when the then new technology of deep tank aerobic liquid fermentation was used to produce spore and crystal preparations. Commercial development has involved the large-scale use of a few strains although others have been used in smaller quantities for experimental purposes and field trials.

Major applications of *B. t.* have taken place in North America for the control of over 40 pest species in field, forest, orchard, vineyard, parkland and gardens (Burges and Daoust, 1986). A search for safe, effective insecticides for forest regions surrounding residential areas provided an ideal opportunity for the application of a microbial insecticide. Lepidopteran-active varieties of *B. t.* have been used in increasingly large quantities on forests for the last 30 years. The first aerial spray applications were in 1961 when spruce budworm, *Choristoneura fumiferana*, was treated in New Brunswick (Mott *et al.*, 1961) and Western black-headed budworm, *Acleris gloverana*, in British Columbia (Kinghorn *et al.*, 1961). Use of *B. t.* in North America has increased greatly over recent years. For example, between

1979 and 1983, 1–4% of the sprayed forest was treated with *B. t.* in eastern Canada. By 1988, the figure had increased to 63%. Between 1985 and 1988, a total of 1 856 548 ha were treated to control spruce budworm. In Ontario, *B. t.* was used on a total of 836 171 ha to control jack pine budworm, *Choristoneura pinus*, between 1985 and 1987 and on a total of 35 756 ha to control hemlock looper, *Lambdina fiscellaria fiscellaria*, in Newfoundland between 1985 and 1988 (Cunningham, 1990). An approximation of the number of viable spores released into the environment during such applications can be made as follows. *B. t.* is applied at a rate of, for example, 30 billion international units (IU) per hectare to control spruce budworm (Morris, 1982). There may be approximately $3–6 \times 10^4$ viable spores per IU (taking a 1974 formulation of Dipel®, Abbott Laboratories, as an example). That represents a release of approximately 10^{15} viable spores per hectare.

   B. t. israelensis has also been applied in large quantities for disease vector control. Its first major release was in the river blindness eradication programme against blackflies (*Simulium* spp.) in West Africa (Guillet, 1984). In the first year, 40 tons of *B. t. israelensis* was released, resulting in successful control. In other application programmes, *B. t. israelensis* was released on a large scale in mosquito-infested areas in southern Switzerland (Lüthy, 1989) and along the Rhine in Germany (see Becker and Margalit, Chapter 7).

Survival

The impact of wild-type and recombinant *B. t.* strains on the environment will depend partly on their ability to survive and remain potent in the ecosystem. *B. t.* is sensitive to the UV component of sunlight, although the spores and crystals may be deactivated at different rates (Jarrett, 1980). The half-life of *B. t.* spores applied to foliage may be only a few days (Pinnock *et al.*, 1971) but below the soil surface, spores are protected from UV light and are able to survive for much longer.

   Although *B. t.* is not capable of germination and growth in most soils, a proportion of the spores can persist and remain viable for some time. Pruett *et al.* (1980) inoculated soil in the laboratory with spores and crystals of *B. t.* subspecies *galleriae*. The number of spores (estimated as colony-forming units or CFUs) fell slowly to 24% of the initial number over 135 days. Pathogenicity of the soil mixed into artificial insect food and fed to larvae of the wax moth, *Galleria mellonella*, fell rapidly to 12% of the original value in 15 days and to < 1% in 135 days, indicating that the crystals were degraded more rapidly than spores. The soil was not pasteurized during the isolation process so presumably any vegetative cells present as the result of growth would have been detected.

   West *et al.* (1985a) compared the effect of pH, nutrient availability and moisture content on the survival of streptomycin-resistant *B. t.* and *B. cereus* spore inocula in soil stored at constant temperature in the laboratory. The most important factor was nutrient availability. In unamended soil, *B. t.* did not grow under most conditions. Numbers per gram remained stable except in soil saturated with moisture, when a slow 10-fold increase was detected, and in dry soil where a slow decrease of 60% over 64 days occurred. In contrast, *B. cereus* increased approximately 10- to 100-fold in all but the driest soil, including unamended soils. When the nutrient availability of the soil was increased, either by autoclaving or by addition of nutrients, growth of both bacilli was stimulated.

In another study, viable log phase vegetative cells of *B. t.* added to untreated soil rapidly disappeared, with a 91% loss occurring in 24 h. No spores were detected following this loss. The initial rapid disappearance was followed by an exponential loss between 1 and 10 days, after which cells were no longer detectable by immunofluorescent techniques (West *et al.*, 1984). In contrast, when spores were added, their number remained unaltered for 91 days and, during this period, no germination took place. Scanning electron micrographs showed the possible degradation of *B. t.* vegetative cells by an actinomocyte-like microorganism. Casida (1988) has identified bacteria that degrade both *B. t.* spores and crystals so it is possible that the reduction in *B. t.* numbers in soil may, at least in part, be due to predation by other microorganisms.

Martin and Reichelderfer (1989) studied field persistence using antibiotic-resistant, marked strains. Spores and crystals were applied at a rate of 10^{12} spores per acre on potato and corn. Recovery from leaves declined to 10% in 1 week and 1% in 3 weeks on potato and to 10% in 3 weeks on corn, with 5% retained throughout winter on the remains of foliage. In soil, spores declined to 10% in 3 weeks—in good agreement with laboratory studies of West *et al.* (1985a,b) and Petras and Casida (1985)—and then remained on a plateau for 15 months. Some increase in spores was noted after frost, probably due to release of nutrients from the soil, which confirms the findings of West *et al.* (1985a, b) that *B. t.* can grow in soil under nutrient-enriched conditions. Petras and Casida (1985) found that when viable spores of a commercial preparation of *B. t.* were added to soil in the field they decreased by approximately 1 log in 2 weeks and then remained constant for at least 8 months. This indicated that the population of spores consisted of a proportion that lost viability rapidly in soil and a proportion that did not lose viability for a long period. Spores produced in soil through multiplication of added vegetative cells survived only for a short time. Soil pH had little effect on spore survival beyond 2 weeks for those that survived the first 2 weeks of incubation.

Multiplication

Some studies have been performed on possible growth of *B. t.* in the environment outside an insect (Akiba *et al.*, 1980; West *et al.*, 1984, 1985a, b; West and Burges, 1985; Petras and Casida, 1985; Martin and Reichelderfer, 1989). These indicate that little or no multiplication takes place in natural soils unless increased nutrients become available. In contrast, depending on the strain of *B. t.* and the host insect, *B. t.* is able to grow well in insect cadavers (Yamvrias, 1962; Aizawa and Fujiyoshi, 1964; Prasertphon *et al.*, 1973; Khawaled *et al.*, 1990). Burges and Hurst (1977) found $6.6–42.2 \times 10^8$ spores per gram dried insect in diseased larval cadavers of the stored product moths *Plodia interpunctella*, *Ephestia cautella*, *Anagasta kuehniella*, *Ephestia elutella* and *G. mellonella*, bred and infected in the laboratory. In another study larvae of *G. mellonella* (on artificial diet) and the Egyptian cotton worm, *Spodoptera littoralis* (on leaf material), were fed spores and crystals of *B. t.*, and microscopic examination of disease-killed larvae showed that *B. t.* was able to grow and sporulate in insect cadavers. The number of *B. t.* spores isolated from dead *S. littoralis* larvae ranged from 5.0×10^5 to 9.2×10^7 per larva (Jarrett and Stephenson, 1990). Many strains of *B. t.* produce factors (thuricins) antagonistic to some Gram-positive bacteria (Krieg, 1970; de Barjac and Lajudie, 1974) and some strains of *B. t.* are able to inhibit the growth of other bacteria by the production of antibiotics (Pendleton, 1969), which may give them an advantage when competing for nutrients with other bacteria in the insect body. Although a recent survey of the phylloplane (Smith and Couche, 1991) indicated that *B. t.* may be a natural epiphyte of some deciduous and coniferous trees (see above), to date, there is no evidence of growth of *B. t.* on leaf material.

Dissemination

An important factor that may influence possible environmental impact is the movement of an organism away from the site of application. Outdoors, *B. t.* rarely spreads by epizootics either naturally, or when applied as an insecticide (Burges, 1973). Laboratory studies have indicated that such infectious spread of *B. t.* occurs only when the susceptible insect population is at a high density and conditions favour long-term spore survival, such as a very dry indoor environment (Burges and Hurst, 1977). This has some disadvantages for pest control but the potential for environmental perturbation may be lower than that of rapidly spreading organisms. Although persistence of *B. t.* in the field following application has been investigated in detail, there has been limited study of its dispersal. Martin and Reichelderfer (1989) studied dispersal using antibiotic-resistant, marked strains. No vertical movement through soil deeper than

6 cm was observed and movement outside the plot was less than 10 m, even along drainage courses. No evidence of genetic exchange was noted, although limited potential for this to take place might exist in soil under circumstances where *B. t.* can multiply (see above) (Jarrett and Stephenson, 1990).

Non-susceptible, non-target organisms may play a role in the dissemination of *B. t.* For example, when fathead minnows, *Pimephales promelas*, were exposed in the laboratory to *B. t. israelensis* (2.2×10^5 CFU ml^{-1}), they rapidly accumulated spores within 1 h of exposure, producing 4.0×10^6 CFU per fish (Snarski, 1990). The major route of entry was by ingestion. There were no toxic effects on the fish. When the fish were transferred to clean water the bacterial count of the whole fish dropped rapidly, by 1000-fold in 1 day, but spores were detected in faeces for over 2 weeks. It is also likely that *B. t.* applied on land could be dispersed by non-target organisms such as scavenging or predatory birds and mammals ingesting infected insects: spores will survive passage through the vertebrate gut, e.g. Smirnoff and Macleod (1961). *B. t.* may also be spread by invertebrate predators and parasitoids.

Transfer of genetic information

This is one of the least understood areas of *B. t.* biology in the environment, but perhaps the one with most potential for environmental impact. *B. t.* makes a good model for studying transfer of DNA as there is already much information on its genetics and molecular biology, although this has not yet been extended to field studies.

Plasmid transfer between strains of *B. t.* growing in infected lepidopterous larvae in the laboratory has been reported by Jarrett and Stephenson (1990). Transfer rates of plasmids coding for crystal production and antibiotic resistance were high, reaching levels similar to those obtained in laboratory broth cultures, in which rates of plasmid transfer of up to 60% were detected. The ability of *B. t.* to multiply to populations of 10^8 cells per larva probably contributes to the high levels of plasmid transfer observed. In addition, in broth cultures, *B. t.* transferred plasmids into other species of spore-forming bacteria from soil samples (Jarrett and Stephenson, 1990). Thus, the potential for plasmid transfer between strains of *B. t.* and soil-inhabiting bacteria in nature has been demonstrated. However, it is unlikely to take place at high rates because of the relatively low levels of *B. t.* multiplication in soil. It has not yet been shown whether non-insect-associated plasmid transfer takes place between *B. t.* strains in the field, although there are increasing numbers of reports of plasmid transfer in soil microcosms by other Gram-positive and by Gram-negative bacteria, including *B. cereus* and *B. subtilis*, e.g., Van Elsas *et al.* (1987). These authors demonstrated the transfer of a 2.8 MDa plasmid, pFT30, at a frequency of 0.7×10^{-7} from *B. cereus* to *B. subtilis* incubated together in a sterile, nutrient-amended loamy sand. High minimum inocula of 5×10^7 CFU g^{-1} dry soil were required for the detection of transfer.

In addition to intra-specific plasmid transfer, inter-specific transfer of *B. t.* plasmids can also take place. Plasmids coding for δ-endotoxin genes transfer at a high frequency between strains of *B. t.* and *B. cereus* by a conjugation-like process when grown in mixed culture (Gonzalez *et al.*, 1982). Plasmids from *B. t.* have been transferred into cultured *B. subtilis* and the soil inhabiting, mammalian pathogen *Bacillus anthracis* (Klier *et al.*, 1983). Under laboratory conditions, plasmids can be shuttled between strains of *B. cereus*, *B. anthracis* and *B. t.* (Battisti *et al.*, 1985) and self-transmissible, cryptic plasmids in four *B. t.* subspecies have been identified (Reddy *et al.*, 1987). Conceivably, cross-species movement of *B. t.* plasmids could take place in the environment allowing for horizontal spread of a new genotype. However, in the laboratory, the rates of plasmid transfer from *B. t.* to *B. subtilis*, *B. cereus* and *B. anthracis* of between 10^{-3} and 10^{-7} recipients per donor cell (Klier *et al.*, 1983; Battisti *et al.*, 1985; Reddy *et al.*, 1987) are much lower than the rates of transfer between *B. t.* strains (up to 60%) (Jarrett and Stephenson, 1990). Additionally, there are environmental barriers that may restrict the rate of plasmid transfer, e.g. temperature and moisture levels (Miller, 1988). Thus, the possible transfer of plasmids between *B. t.* and other *Bacillus* species appears to represent an acceptable risk. Despite the widespread application of *B. t.*, there have been no reports of the isolation of strains of *B. t.*, *B. cereus* or *B. anthracis* with unexpected pathogenic properties.

Possible transfer of genetic information from and to *B. t.* strains may not be limited to the mechanism of conjugative plasmid transfer. Transduction and transformation might also play a role in the movement of DNA, but recorded rates of transfer have been consistently much lower than with conjugation. Transformation, whereby naked pieces of double-stranded DNA are taken up by competent cells and integrated into the genome, has been demonstrated in the laboratory in many common soil microorganisms (*Rhizobium*, *Bacillus*, *Pseudomonas*) (Reanney *et al.*, 1982). It is therefore possible that transformation occurs in soil if free DNA is present. If this is indeed the case, the death and lysis of a potential donor cell would not prevent gene transfer. Graham and Istock (1978, 1979) proposed transformation as the mechanism for the observed gene exchange by *B. subtilis* incubated in sterile soil since the strain did not contain plasmids or transducing phages. Transformation recorded when purified DNA was added to the sterile soil supports this suggestion.

Safety/benefit/harm

The safety record of *B. t.* in use is excellent (Meadows, 1992). By 1981, it had been registered in products by four different companies, giving four-fold replication of safety tests (Burges, 1981) and further sets of registration tests on new commercial strains have been performed since then. One of only two reservations concerning use of *B. t.* is the production, by some strains, of a soluble toxin, β-exotoxin, which has a less specific range of activity and is toxic to some mammals (Sebesta *et al.*, 1981). Thus, as a precaution, in many countries the β-exotoxin is prohibited by law in commercial products using spores and crystals as the active components. This has not limited the use of *B. t.* because it is relatively easy to select strains that do not produce the β-exotoxin. The second reservation is that solubilized δ-endotoxin of *B. t. israelensis* is cytolytic to a wide range of cells and is toxic to mammals when administered by injection. However, the unsolubilized toxin produces no cytolytic effects (Thomas and Ellar, 1983; Armstrong *et al.*, 1985) and the toxin is solubilized only under alkaline conditions such as those found in many insect midguts: non-target animals, including birds and mammals, challenged with high doses of *B. t. israelensis* spores and crystals by all routes have not been harmed.

To the author's knowledge there have been no reports of adverse environmental effects caused by the application of *B. t.* The apparent lack of environmental impact may, in part, be due to the widespread existing distribution of *B. t.* and consequent ecological balance with the environment, although the survival and growth characteristics of *B. t.* may also play a role. In its lack of effect on predators, parasites and scavengers, it follows the course of most other herbivore pathogens which normally are not pathogenic in animals that prey on their hosts.

Experimental approaches to risk assessment

Accurate recovery and enumeration of *B. t.* from field samples is of prime importance in assessment of its fate and effects. The principal challenge is to develop sensitive methods of detecting introduced strains of *B. t.* against the background of wild-type strains commonly present in the environment. Laboratory microcosm studies are a useful way to develop techniques and models, but care must be taken in extrapolating results due to the increased number of variables and interactions in the field.

B. t. strains may be marked for detection in environmental samples using antibiotic or heavy metal resistance markers (as recommended for other bacteria by Lindow and Panopoulos, 1983) introduced into the chromosome by mutation and selection (Jarrett and Stephenson, 1990), or possibly by the insertion of a gene using recombinant technology. The use of such chromosomally located markers may be preferable in studies designed to assess the fate of *B. t.* itself in the environment as such markers are probably less likely to be transferred to other organisms than if they were located on plasmids.

A number of classes of *B. t.* mutants can be used as potential donor and recipient strains in the study of gene transfer. Such strains may lack δ-endotoxin-encoding plasmids and, hence, are acrystalliferous, or lack the ability to produce spores. Mutant strains have been used to study the transfer of *B. t.* plasmids within insects (Jarrett and Stephenson, 1990)—an interesting example of the use of a microcosm relevant to *B. t.*—and such a study could be extended to investigate plasmid transfer in soil. For example, small quantities of prepared soil could be inoculated with two or more strains of *B. t.* in the laboratory. The donor strain might be crystalliferous and carry a plasmid-encoded antibiotic resistance marker and the acrystalliferous recipient could carry a chromosomally-encoded antibiotic resistance marker. Detection of a previously acrystalliferous strain producing crystals on an appropriate selective medium would indicate plasmid transfer. Toxicity of the recipient strain to an indicator insect species and plasmid analysis would confirm the transfer and expression of introduced δ-endotoxin genes. Growth of the bacteria in soil could be stimulated by autoclaving the soil or amending it with nutrients (West *et al.*, 1985a). This protocol could then be extended to study possible plasmid transfer between *B. t.* and other soil bacteria. Initial experiments for the study of gene transfer could be carried out with sterile soil in enclosed microcosms in the laboratory to avoid dealing with a complex medium containing a diverse population of microorganisms. The study could then proceed to non-sterile soil in enclosed systems, more complex microcosms based on the intended environment for release and finally to controlled releases in the environment.

Although *B. t.* has demonstrated no environmental risk even when applied at high concentrations as an insecticide, in some cases it may be desirable to detect *B. t.* at very low concentrations, for example, if *B. t.* is used as a model bacterium for further development of protocols for the study of release of genetically modified bacteria. Because of the difficulties of recovering and isolating rare microorganisms from complex microbial communities using the usual methods of selective culturing, the recently developed technique of DNA amplification by polymerase chain reaction (PCR) (Saiki *et al.*, 1985) could be applied to studies with *B. t.* A PCR can increase the number of copies of a target sequence of DNA without having to culture the organism. This improves the sensitivity of trying to detect DNA sequences present in small amounts in samples containing DNA from mixed populations, and reduces the problem of detecting possibly viable, but not necessarily culturable, organisms. Specific, unique regions of DNA from the *B. t.* strain under study could be used as target sequences which could then be detected in field samples. Steffen and Atlas (1988) have used such a technique to increase the sensitivity of detection of a herbicide-degrading bacterium, *Pseudomonas cepacia*, by dot-blot analysis. They reported detection of one bacterium per gram of soil sediment against a background of 10^9 non-target organisms per gram. Such sensitivity of detection would be more than adequate in the case of *B. t.*, as it must be present at much higher concentrations to be of environmental significance for insect control.

PCR may also be developed for detection of *B. t.* in water courses, in particular strains such as *B. t. israelensis*, applied for mosquito control. Amici *et al.* (1992) used the rapid amplification by PCR of a synthetic target sequence of DNA to monitor a genetically modified bacterium, *Pseudomonas putida*, in a freshwater environment.

CONCLUDING REMARKS

Across the whole field of microbiology, exploitation of natural biological diversity and use of recombinant DNA technology bears great promise for pest control, but only if tempered with thorough research and informed discussion on possible environmental risks. Because of its exceptionally good safety record and prognosis, *B. t.* can play a major role in the application of biotechnology in two ways. Firstly, the scope for application of *B. t.*, both on its own, and included in Integrated Pest Management programmes will be increased by the exploitation of the diversity of wild-type isolates, allied with creation of recombinant strains of *B. t.* and transgenic plants. Secondly, *B. t.* is a good model microorganism, requiring only the minimum of containment restrictions, for the study of the possible risks entailed in environmental release of genetically modified bacteria.

Although there is a good basis of knowledge of the behaviour of *B. t.* in the environment, its ecological role is still debatable. The somewhat speculative nature of some of the sections in this chapter highlights the need for additional research to provide information on its ecology, physiology and population genetics. Progress in many areas of molecular biology, microbial technology and ecology provides a great opportunity for a multidisciplinary approach to the study of this important microorganism.

ACKNOWLEDGEMENTS

The author thanks Paul Jarrett and Denis Burges for reviewing the manuscript and for many helpful and stimulating discussions on the topics covered in this chapter. MPM acknowledges funding from Ciba-Geigy AG,

Basel, Switzerland, in cooperation with the Agricultural Genetics Company, Cambridge, UK.

Certain passages of this chapter have been reproduced by permission from a chapter by Martin P. Meadows (1992) entitled 'Environmental release of *Bacillus thuringiensis*'. In *Environmental Release of Genetically Engineered and Other Microorganisms* (Eds. J. C. Fry & M. J. Day), Cambridge University Press, Cambridge, 120–136.

REFERENCES

Adang, M. J., Staver, M. J., Rocheleau, T. A., Leighton, J., Barker, R. F. and Thompson, D. V. (1985). 'Characterized full-length and truncated plasmid clones of the crystal protein of *Bacillus thuringiensis* subsp. *kurstaki* HD-73 and their toxicity to *Manduca sexta*'. *Gene*, **36**, 289–300.

Ahmad, W. and Ellar, D. J. (1990). 'Directed mutagenesis of selected regions of a *Bacillus thuringiensis* entomocidal protein'. *FEMS Microbiol. Lett.*, **68**, 97–104.

Aizawa, K. and Fujiyoshi, N. (1964). 'The growth of *Bacillus thuringiensis* in dead larvae of the silkworm, *Bombyx mori*'. *J. Sericulture Sci. Japan*, **33**, 399–402 (in Japanese).

Akiba, Y. (1986). 'Microbial ecology of *Bacillus thuringiensis* VI. Germination of *Bacillus thuringiensis* spores in the soil'. *Appl. Entomol. Zool.*, **21**, 76–80.

Akiba, Y., Sekijima, Y., Aizawa, K. and Fujiyoshi, N. (1980). 'Microbial ecological studies on *Bacillus thuringiensis*. IV. The growth of *B. thuringiensis* in soils of mulberry plantations'. *Jap. J. Appl. Entomol. Zool.*, **24**, 13–17.

Amici, A., Bazzicalupo, M., Gallori, E. and Rollo, F. (1992). 'Monitoring a genetically engineered bacterium in a freshwater environment by rapid enzymatic amplification of a synthetic DNA "number-plate"'. *Appl. Microbiol. Biotechnol.*, **36** 222–227.

Armstrong, J. L., Rohrmann, G. F., and Beaudreau, G. S. (1985). 'Delta-endotoxin of *Bacillus thuringiensis* subspecies *israelensis*'. *J. Bacteriol.*, **161**, 39–46.

Aronson, A. I., Beckman, W. and Dunn, P. (1986). 'Bacillus thuringiensis and related insect pathogens'. *Microb. Rev.*, **50**, 1–24.

Barton, K. A., Whiteley, H. R. and Ning-Sun Yang (1987). '*Bacillus thuringiensis* δ-endotoxin expressed in transgenic *Nicotiana tabacum* provides resistance to lepidopteran'. *Plant Physiol.*, **85**, 1103–1109.

Battisti, L., Green, B. D. and Thorne, C. B. (1985). 'Mating system for transfer of plasmids among *Bacillus anthracis*, *Bacillus cereus*, and *Bacillus thuringiensis*'. *J. Bacteriol.*, **162**, 543–550.

Baum, J. A., Coyle, D. M., Pearce-Gilbert, M., Jany, C. S. and Gawron-Burke, C. (1990). 'Novel cloning vectors for *Bacillus thuringiensis*'. *Appl. Env. Microbiol.*, **56**, 3420–3428.

Baumann, L., Okamoto, K., Unterman, B. M., Lynch, M. J. and Baumann, P. (1984). 'Phenotypic characterisation of *Bacillus thuringiensis* and *Bacillus cereus*'. *J. Invert. Pathol.*, **44**, 329–341.

Bone, E. J. and Ellar, D. J. (1989). 'Transformation of *Bacillus thuringiensis* by electroporation'. *FEMS Microbiol. Lett.*, **58**, 171–178.

Burges, H. D. (1973). 'Enzootic diseases of insects'. In *Regulation of Insect Populations by Microorganisms* (Ed. L. A. Bulla Jr), *Annals of the New York Academy of Sciences*, **217**, 31–49.

Burges, H. D. (1981). 'Safety, safety testing and quality control of microbial pesticides'. In *Microbial Control of Pests and Plant Diseases 1970–1980* (Ed. H. D. Burges), pp. 737–767, Academic Press, London.

Burges, H. D. and Daoust, R. A. (1986). 'Current status of the use of bacteria as biocontrol agents'. In *Fundamental and Applied Aspects of Invertebrate Pathology* (Eds. R. A. Samson, J. M. Vlak and D. Peters), pp. 514–517, Agricultural University, Wageningen.

Burges, H. D. and Hurst, J. A. (1977). 'Ecology of *Bacillus thuringiensis* in storage moths'. *J. Invert. Pathol.*, **30**, 131–139.

Burges, H. D. and Jarrett, P. (1990). 'Preparation of strains of *Bacillus thuringiensis* having an improved activity against certain Lepidopterous pests and novel strains produced thereby'. US patent no. 4,935,353.

Casida, L. E. (1988). 'Response in soil of *Cupriavidus necator* and other copper-resistant bacterial predators of bacteria to the addition of water-soluble nutrients, various bacterial species, or *Bacillus thuringiensis* spores and crystals'. *Appl. Env. Microbiol.*, **54**, 2161–2166.

Caulcott, C. A., Dunn, A., Robertson, H. A., Cooper, N. S., Brown, M. E. and Malcolm Rhodes, P. (1987). 'Investigation of the effect of growth environment on the stability of low-copy-number plasmids in *Escherichia coli*'. *J. Gen. Microbiol.*, **133**, 1881–1889.

Claus, D. and Berkeley, R. C. W. (1986). 'Genus *Bacillus* Cohn 1872'. In *Bergey's Manual of Systematic Bacteriology*, Vol. 2 (Eds. P. H. A. Sneath, F. G. Priest, M. Goodfellow and C. Todd), pp. 1105–1140, Williams and Wilkins, Baltimore.

Cunningham, J. C. (1990). 'Use of microbials for control of defoliating pests of conifers'. *Proceedings Vth International Colloquium on Invertebrate Pathology and Microbial Control*, Adelaide, Australia, 20–24 August 1990, pp. 164–168, Society for Invertebrate Pathology.

de Barjac, H. (1978). 'A new subspecies of *Bacillus thuringiensis* very toxic for mosquitoes *Bacillus thuringiensis* var. *israelensis* sero-type 14'. *C. R. Acad. Sci. Paris, ser D.*, **286**, 797–800.

de Barjac, H. (1981). 'Identification of H-serotypes of *Bacillus thuringiensis*'. In *Microbial Control of Pests and Plant Diseases 1970–1980* (Ed. H. D. Burges), pp. 35–45, Academic Press, London.

de Barjac, H. and Lajudie, J. (1974). 'Mise en evidence de facteurs antagonistes du type des bacteriocins chez *Bacillus thuringiensis*'. *Ann. Microbiol.*, **125B**, 529–537.

DeLucca, A. J., Simonson, J. G. and Larson, A. D. (1981). '*Bacillus thuringiensis* distribution in soils of the United States'. *Can. J. Microbiol.*, **27**, 865–870.

DeLucca, A. J., Palmgren, M. S. and Ciegler, A. (1982). '*Bacillus thuringiensis* in grain elevator dusts'. *Can. J. Microbiol.*, **28**, 452–456.

Dulmage, H. T. (1970). 'Insecticidal activity of HD-1, a new isolate of *Bacillus thuringiensis* variety *alesti*'. *J. Invert. Pathol.*, **15**, 232–239.

Dulmage, H. T. and Cooperators (1981). 'Insecticidal activity of isolates of *Bacillus thuringiensis* and their potential for pest control'. In *Microbial Control of Pests and Plant Diseases 1970–1980* (Ed. H. D. Burges), pp. 193–222, Academic Press, London.

Edlund, T., Siden, I. and Boman, H. G. (1976). 'Evidence for two immune inhibitors from *Bacillus thuringiensis* interfering with the humoral defense system of Saturniid pupae'. *Infect. Immun.*, **14**, 934–941.

Ellar, D. J. (1990). 'Pathogenic determinants of entomopathogenic bacteria'. *Proceedings Vth International Colloquium on Invertebrate Pathology and*

Microbial Control. Adelaide, Australia, 20–24 August 1990, pp. 298–302. Society for Invertebrate Pathology.

Ficksell, J. R. and Covello, V. T. (1986). 'The suitability and applicability of risk assessment methods for environmental applications of biotechnology'. In *Biotechnology Risk Assessment* (Eds. J. F. Ficksell and V. T. Covello), p. 1, Pergamon Press, New York.

Fischoff, D. A., Bowdish, K. S., Perlack, F. J., Marrone, P. G., McCormick, S. M., Niedermeyer, J. G., Dean, D. A., Kusano-Kretzmer, K., Mayer, E. J., Rochester, D. E., Rogers, S. G. and Fraley, R. T. (1987). 'Insect tolerant transgenic tomato plants'. *Bio/Technology*, **5**, 807–813.

Fuxa, J. R. (1989). 'Environmental risks of genetically engineered entomopathogens'. In *Safety of Microbial Insecticides* (Eds. M. Laird, L. Lacey and E. W. Davidson), pp. 203–207, CRC Press, Florida.

Garduno, F., Thorne, L., Walfield, A. M. and Pollock, T. J. (1988). 'Structural relatedness between mosquitocidal endotoxins of *Bacillus thuringiensis* subsp. *israelensis*'. *Appl. Environ. Microbiol.*, **54**, 277–279.

Gilbert, R. J. (1979). '*Bacillus cereus* gastroenteritis'. In *Food-borne Infections and Intoxications* (Eds. H. Reimann and F. L. Bryan), 2nd edn, pp. 495–518, Academic Press, New York.

Godwin, D. and Slater, J. H. (1979). 'The influence of the growth environment on the stability of a drug resistance plasmid in *Escherichia coli* K12'. *J. Gen. Microbiol.*, **111**, 201–210.

Goldberg, L. H. and Margalit, J. (1977). 'A bacterial spore demonstrating rapid larvacidal activity against *Anopheles sergentii*, *Uranotaenia unguiculata*, *Culex univittatus*, *Aedes aegypti* and *Culex pipiens*'. *Mosquito News*, **637**, 355–358.

Gonzalez, J. M., Jr., Brown, B. J. and Carlton, B. C. (1982). 'Transfer of *Bacillus thuringiensis* plasmids coding for δ-endotoxin among strains of *Bacillus thuringiensis* and *Bacillus cereus*'. *Proc. Natl Acad. Sci. USA*, **79**, 6951–6955.

Gordon, R. E. (1977). 'Some taxonomic observations on the genus *Bacillus*'. In *Biological Regulation of Vectors: the Saprophytic and Aerobic Bacteria and Fungi*, pp. 67–73, US Department of Health Education and Welfare.

Graham, J. B. and Istock, C. A. (1978). 'Gene exchange in *Bacillus subtilis* in soil'. *Mol. Gen. Genet.*, **166**, 287–290.

Graham, J. B. and Istock, C. A. (1979). 'Gene exchange and natural selection cause *Bacillus subtilis* to evolve in soil culture'. *Science* (Washington, DC), **204**, 637–639.

Guillet, P. (1984). 'Development and field evaluation of *Bacillus thuringiensis* H-14 against blackflies' (Review of research activities within the special programme). WHO TDR/BCV-SWG-7/84.

Helling, R. B., Kinney, T. and Adams, J. (1981). 'The maintenance of plasmid-containing organisms in populations of *Escherichia coli*'. *J. Gen. Microbiol.*, **123**, 129–141.

Hofmann, C., Vanderbruggen, H., Höfte, H., Van Rie, J., Jansens, S. and Van Mallaert H. (1988). 'Specificity of *Bacillus thuringiensis* δ-endotoxins is correlated with the presence of high affinity binding sites in the brush border membrane at target insect midguts'. *Proc. Natl Acad. Sci. USA*, **85**, 7844–7848.

Höfte, H. and Whiteley, H. R. (1989).'Insecticidal crystal proteins of *Bacillus thuringiensis*'. *Microb. Rev.*, **53**, 242–255.

Honée, G., Vriezen, W. and Visser, B. (1990). 'A translational fusion product of two different insecticidal crystal protein genes of *Bacills thuringiensis* exhibits an enlarged insecticidal spectrum'. *Appl. Env. Microbiol.*, **56**, 823–825.

Jarrett, P. (1980). 'Persistence of *Bacillus thuringiensis* on tomatoes'. Annual Report, Glasshouse Crops Research Institute, 1980, 117 pp.

Jarrett, P. (1985). 'Potency factors in the delta-endotoxin of *Bacillus thuringiensis* var. *aizawi* and the significance of plasmids in their control'. *J. Appl. Bacteriol.*, **58**, 437–448.

Jarrett, P. and Stephenson, M. (1990). 'Plasmid transfer between strains of *Bacillus thuringiensis* infecting *Galleria mellonella* and *Spodoptera littoralis*'. *App. Env. Microbiol.*, **56**, 1608–1614.

Johnson, D. E., Brookhart, G. L., Kramer, K. J., Barnett, B. D. and McGaughey, W. H. (1990). 'Resistance to *Bacillus thuringiensis* by the Indian Meal Moth, *Plodia interpunctella*: comparison of midgut proteinases from susceptible and resistant larvae'. *J. Invert. Pathol.*, **55**, 235–244.

Johnson, K. M. (1984). '*Bacillus cereus* foodborne illness—an update'. *J. Food Protocol.*, **47**, 145–153.

Khawaled, K., Ben-Dov, E., Zaritsky, A. and Barak, Z. (1990). 'The fate of *Bacillus thuringiensis* var. *israelensis* in *B. thuringiensis* var. *israelensis*-killed pupae of *Aedes aegypti*'. *J. Invert. Pathol.*, **56**, 312–316.

King, S.R. (1985). 'Economic impacts of biotechnology'. In *Biotechnology and the Environment* (Eds. A. H. Teich, M. A. Levin and J. A. Pace) p. 179, American Society for the Advancement of Science, Washington, DC.

Kinghorn, J. M., Fisher, R. A., Angus, T. A. and Heimpel, A. M. (1961). 'Aerial spray trials against the blackheaded budworm in British Columbia'. *Department of Forestry Bimonthly Report.*, **17**, 3–4.

Klier, A., Bourgouin, C. and Rapoport, G. (1983). 'Mating between *Bacillus cereus* and *Bacillus thuringiensis* and transfer of cloned crystal genes'. *Mol. Gen. Genet.*, **191**, 257–262.

Kostka, S. J. (1991). 'The design and execution of successive field releases of genetically engineered microorganisms'. In *Biotechnology Field Test Results* (Ed. D. R. MacKenzie.) Academic Press, New York (in press).

Krieg, A. (1969). 'In vitro determination of *Bacillus thuringiensis*, *Bacillus cereus*, and related Bacilli'. *J. Invert. Pathol.*, **15**, 313–320.

Krieg, A. (1970). 'Thuricin, a bacteriocin produced by *Bacillus thuringiensis*'. *J. Invert. Pathol.*, **15**, 291.

Krieg, A. and Langenbruch, G. A. (1981). 'Susceptibility of arthropod species to *Bacillus thuringiensis*'. In *Microbial Control of Pests and Plant Diseases 1970–1980* (Ed. H. D. Burges), pp. 737–767, Academic Press, London.

Kumari, S. M. and Neelgund, Y. F. (1985). 'Preliminary infectivity tests using six bacterial formulations against the red flour beetle, *Tribolium castaneum*'. *J. Invert. Pathol.*, **46**, 198–199.

Lacey, L. A., Escaffre, H., Philippon, B., Seketeli, A. and Guillet, P. (1982). 'Large river treatments with *Bacillus thuringiensis* (H-14) for the control of *Simulium damnosum* s.1. in the Onchocerciasis Control Programme'. *Tropenmedizin und Parasitologie*, **33**, 97–101.

Lemille, F., de Barjac, H. and Bonnefoi, A. (1969). 'Essai sur la classification biochemique de 97 bacillus du groupe 1 appartenant a 9 especes differentes'. *Extrait Annals Institute Pasteur*, Juin 1969, **116**, 808–819.

Lereclus, D., Arantes, O., Chaufaux, J. and Lecadet, M.-M. (1989). 'Transformation and expression of a cloned δ-endotoxin gene in *Bacillus thuringiensis*'. *FEMS Microbiol. Lett.*, **60**, 211–218.

Levin, M. A. (1982). 'Review of environmental risk assessment studies sponsored by EPA'. *Recombinant DNA Technology Bulletins*, **5**, 177.

Lindow, S. E. and Panopoulos, N. J. (1983). Request for permission to test

Pseudomonas syringae pv. *syringae* and *Erwinia herbicola* carrying specific deletions in ice nucleation genes under field conditions as biocontrol agents of frost injury to plants. Revised protocol for Recombinant DNA Committee, National Institute of Health, Washington, D.C.

Lüthy, P. (1989). 'Large-scale use of *Bacillus thuringiensis* H-14 in a mosquito-infected area in a southern region of Switzerland'. *Proceedings and Abstracts Society for Invertebrate Pathology XXIInd Annual Meeting*, p. 82, University of Maryland, USA, 20–24 August 1989.

Lysenko, O. (1983). '*Bacillus thuringiensis*: evolution of a taxonomic conception'. *J. Invertebrate Pathol.*, **42**, 295–298.

Margalit, J. and Dean, D. (1985). 'The story of *Bacillus thuringiensis* var. *israelensis* (*B.t.i.*)'. *Journal of the American Mosquito Control Association*, **1**, 1–7.

Martens, J. W. M., Honée, G., Zuidema, D., van Lent, J. W. M., Visser, B. and Vlak, J. M. (1990). 'Insecticidal activity of a bacterial crystal protein expressed by a recombinant baculovirus in insect cells'. *Appl. Env. Microbiol.*, **56**, 2764–2770.

Martin, P. A. W. and Travers, R. S. (1989). 'Worldwide abundance and distribution of *Bacillus thuringiensis* isolates'. *Appl. Env. Microbiol.*, **55**, 2437–2442.

Martin, W. F. and Reichelderfer, C. F. (1989). '*Bacillus thuringiensis*: persistence and movement in field crops'. *Proceedings and Abstracts Society for Invertebrate Pathology XXIInd Annual Meeting*, p. 25, University of Maryland, USA, 20–24 August, 1989.

McGaughey, W. H. and Johnson, D. E. (1987). 'Toxicity of different serotypes and toxins of *Bacillus thuringiensis* to resistant and susceptible Indianmeal Moths (Lepidoptera: Pyralidae)'. *J. Econ. Entomol.*, **80**, 1122–1126.

Meadows, M. P. (1992). 'Environmental release of *Bacillus thuringiensis*'. In *Environmental Release of Genetically Engineered and Other Microorganisms* (J. C. Fry and M. J. Day), pp. 120–136, Cambridge University Press, UK.

Meadows, M. P., Ellis, D. J., Butt, J., Jarrett, P. and Burges, H. D. (1992). 'Distribution, frequency and diversity of *Bacillus thuringiensis* in an animal feed mill'. *Appl. Environ. Microbiol.*, **58**, 1344–1350.

Meadows, M. P., Ellis, D. J., Pethybridge, N. J., Bernhard, K., Burges, H. D. and Jarrett, P. (1990). 'Activity of new isolates of *Bacillus thuringiensis* against five insect pests'. *Proceedings Vth International Colloquium on Invertebrate Pathology and Microbial Control*, p. 479, Adelaide, Australia, August 20–24 1990.

Miller, R. V. (1988). 'Potential of transfer and establishment of engineered genetic sequences'. In *Planned Release of Genetically Engineered Organisms* (Trends in Biotechnology/Trends in Ecology and Evolution Special Publication) (Eds. J. Hodgson and A. M. Sugden), pp. 23–27, Elsevier, Cambridge.

Morris, O. N. (1982). 'Bacteria as pesticides: forest applications'. In *Microbial and Viral Pesticides* (Ed. E. Kurstak), p. 239, Marcel Dekker, New York.

Mott, D. G., Angus, T. A., Heimpel, A. M. and Fisher, R. A. (1961). 'Aerial application of Thuricide against spruce budworm in New Brunswick'. *Department of Forestry Bimonthly Progress Report*, **17**, 2.

National Academy of Sciences (1987). *Introduction of Recombinant DNA Engineered Organisms into the Environment: Key Issues.* National Academy Press, Washington, DC.

Obukowicz, M. G., Perlack, F. J., Kusano-Kretzmer, K., Mayer, E. J., Bolton, S. W. and Watrud, L. S. (1986). 'Tn5-mediated integration of the delta-endotoxin gene of *Bacillus thuringiensis* into the chromosome of root-colonizing pseudomonads'. *J. Bacteriol.*, **168**, 982–989.

Ohba, M. and Aizawa, K. (1986a); 'Insect toxicity of *Bacillus thuringiensis* isolated from soils in Japan'. *J. Invert. Pathol.*, **47**, 12–20.

Ohba, M. and Aizawa, K. (1986b). 'Distribution of *Bacillus thuringiensis* in soils of Japan'. *J. Invertebrate Pathol.*, **47**, 277–282.

Ohba, M. and Aizawa, K. (1986c). 'Frequency of acrystalliferous spore-forming bacteria possessing flagellar antigens of *Bacillus thuringiensis*'. *J. Basic Microbiol.*, **26**, 185–188.

Padua, L. E., Gabriel, B. P., Aizawa, K. and Ohba, M. (1982). '*Bacillus thuringiensis* isolated from the Philippines'. *The Philippine Entomologist*, **5**, 199–208.

Parry, J. M., Turnbull, P. C. B. and Gibson, J. R. (1983). *A Colour Atlas of Bacillus Species*, Wolfe Medical Publications, London.

Pendleton, I. R. (1969). 'Ecological significance of antibiotics in some varieties of *Bacillus thuringiensis*'. *J. Invert. Pathol.*, **13**, 235–240.

Petras, S. T. and Casida, L. E. (1985). 'Survival of *Bacillus thuringiensis* spores in soil'. *Appl. Env. Microbiol.*, **50**, 1496–1501.

Pinnock, D. E., Brand, R. J. and Milstead, J. E. (1971). 'The field persistence of *Bacillus thuringiensis* spores'. *J. Invert. Pathol.*, **18**, 405–411.

Prasertphon, S., Areekul, P. and Tanada, Y. (1973). 'Sporulation of *Bacillus thuringiensis* in host cadavers'. *J. Invert. Pathol.*, **21**, 205–207.

Priest, F. G., Goodfellow, M. and Todd, C. (1988). 'A numerical classification of the genus *Bacillus*'. *J. Gen. Microbiol.*, **134**, 1847–1882.

Pruett, C. J. H., Burges, H. D. and Wyborn, C. H. (1980). 'Effect of exposure to soil on potency and spore viability of *Bacillus thuringiensis*'. *J. Invert. Pathol.*, **35**, 168–174.

Rahmet-Alla, M. and Rowley, A. (1989). 'Studies of the pathogenicity of different strains of *Bacillus cereus* for the cockroach, *Leucophaea maderae*'. *J. Invert. Pathol.*, **53**, 190–196.

Reanney, D. C., Roberts, W. P. and Kelly, W. J. (1982). 'Genetic interactions among microbial communities'. In *Microbial Interactions and Communities* (Eds. A. T. Bull and J. H. Slater), Vol. 1, Academic Press, New York.

Reddy, A., Battisti, L. and Thorne, C. B. (1987). 'Identification of self transmissible plasmids in four *Bacillus thuringiensis* subspecies'. *J. Bacteriol.*, **169**, 5263–5270.

Rigby, S. (1991). *Bt in Crop Protection.* PJB Publications, Richmond, UK.

Saiki, R. K., Scharf, S., Fallona, F., Mullis, K. B., Horn, G. T., Erlich, H. A. and Arnheim, N. (1985). 'Enzymatic amplification of beta-globin genomin sequences and restriction site analysis for diagnosis of sickle cell anaemia'. *Science*, **230**, 1350–1354.

Salama, H. S., Foda, M., Zaki, F. and Ragaei (1986). 'On the distribution of *Bacillus thuringiensis* and closely related *Bacillus cereus* in Egyptian soils and their activity against cotton insects'. *Angewandte Zoologie*, **3**, 257–265.

Schurter, W., Geiser, M. and Mathe, D. (1989). 'Efficient transformation of *Bacillus thuringiensis* and *Bacillus cereus* via electroporation: transformation of acrystalliferous strains with a cloned δ-endotoxin gene'. *Mol. Gen. Genet.*, **218**, 177–181.

Sebesta, K., Farkas, J., Horska, K. and Vankova, J. (1981). 'Thuringiensin, the beta-exotoxin of *Bacillus thuringiensis*'. In *Microbial Control of Pests and Plant Diseases 1970–1980* (Ed. H. D. Burges), pp. 193–222, Academic Press, London.

Shivakumar, A. G., Gundling, G. J., Benson, T. A., Casuto, D., Miller, M. F. and Spear, B. B. (1986). 'Vegetative expression of the delta-endotoxin gene of *Bacillus thuringiensis* subsp. *kurstaki* in *Bacillus subtilis*'. *J. Bacteriol.*, **166**, 194–204.

Siegel, J. P., Shadduck, J. A. and Szabo, J. (1987). 'Safety of the entomopathogen *Bacillus thuringiensis* var. *israelensis* for mammals'. *J. Econ. Entomol.*, **80**, 717–723.

Smirnoff, W. A. and Macleod, C. F. (1961). 'Study of the survival of *Bacillus thuringiensis* var. *thuringiensis* Berliner in the digestive tracts and in feces of a small mammal and birds'. *J. Insect Pathol.*, **3**, 266–270.

Smith, R. A. and Couche, G. A. (1991). 'The phylloplane as a source of *Bacillus thuringiensis* variants'. *Appl. Env. Microbiol.*, **57**, 311–331.

Snarksi, V. M. (1990). 'Interactions between *Bacillus thuringiensis* subspecies *israelensis* on Fathead Minnows, *Pinnephales promelas* Rafinesue, under laboratory conditions'. *Appl. Env. Microbiol.*, **56**, 2618–2622.

Steffen, R. J. and Atlas, R. M. (1988). 'DNA amplification to enhance detection of genetically engineered bacteria in environmental samples'. *Appl. Env. Microbiol.*, **54**, 2185–2191.

Stone, T. B., Sims, S. R. and Marrone, P. G. (1989). 'Selection of Tobacco Budworm for resistance to a genetically engineered *Pseudomonas fluorescens* containing the δ-endotoxin of *Bacillus thuringiensis* subsp. *kurstaki*'. *J. Invert. Pathol.*, **53**, 228–234.

Tabashnik, B. E., Cushing, N. L., Finson, N. and Johnson, M. W. (1990). 'Field development of resistance to *Bacillus thuringiensis* in Diamondback moth (Lepidoptera: Plutellidae)'. *J. Econ. Entomol.*, **83**, 1671–1676.

Taylor, A. J. and Gilbert, R. J. (1975). '*Bacillus cereus* food poisoning: a provisional serotyping scheme'. *J. Med. Microbiol.*, **8**, 543–550.

Thomas, W. E. and Ellar, D. J. (1983). '*Bacillus thuringiensis* variety *israelensis* crystal δ-endotoxin: effect on insect and mammalian cells *in vitro* and *in vivo*'. *J. Cell Sci.*, **60**, 181–197.

Travers, R. S., Martin, P. A. W. and Reichelderfer, C. F. (1987). 'Selective process for the efficient isolation of soil *Bacillus* species'. *Appl. Env. Microbiol.*, **53**, 1263–1266.

Trevors, J. T., Barkey, T. and Bourquin, A. W. (1986). 'Gene transfer among bacteria in soil and aquatic environments: a review'. *Can. J. Microbiol.*, **33**, 191–198.

Undeen, A., Takaoka, H. and Hansen, K. (1981). 'A test of *Bacillus thuringiensis* var. *israelensis* de Barjac as a larvicide for *Simulium ochraceum*, the Central American vector of onchocerciasis'. *Mosquito News*, **41**, 37–40.

Vaeck, M., Raynaerts, A., Höfte, H., Jansens, S., De Beukeleer, M., Zabeau, M., Van Montagu, M. and Leemans, J. (1987). 'Transgenic plants protected from insect attack'. *Nature (London)*, **328**, 33–37.

Vaeck, M., Raynaerts, A. and Höfte, H. (1989). 'Protein engineering in plants: expression of insecticidal protein genes'. *Cell Culture and Somatic Cell Genetics of Plants*, **6**, 425–439.

Van Elsas, J. D., Govaert, J. M. and Van Veen, J. A. (1987). 'Transfer of plasmid pFT30 between Bacilli in soil as influenced by bacterial population dynamics and soil conditions'. *Soil Biol. Biochem.*, **19**, 639–647.

Van Rie, J., McGaughey, W. H., Johnson, D. E., Barnett, B. D. and Van Mallaert, H. (1990). 'Mechanism of insect resistance to the microbial insecticide *Bacillus thuringiensis*'. *Science*, **247**, 72–74.

West, A. W. and Burges, H. D. (1985). 'Persistence of *Bacillus thuringiensis* and *Bacillus cereus* in soil supplemented with grass or manure'. *Plant Soil*, **83**, 389–398.

West, A. W., Crook, N. E. and Burges, H. D. (1984). 'Detection of *Bacillus thuringiensis* in soil by immunofluorescence'. *J. Invert. Pathol.*, **43**, 150–155.

West, A. W., Burges, H. D., Dixon, T. J. and Wyborn, C. H. (1985a). 'Survival of *Bacillus thuringiensis* and *Bacillus cereus* spore inocula in soil: effects of pH, moisture, nutrient availability and indigenous microorganisms'. *Soil Biol. Biochem.*, **17**, 657–665.

West, A. W., Burges, H. D., Dixon, T. J. and Wyborn, C. H. (1985b). 'Effect of incubation in non-sterilised and autoclaved arable soil on survival of *Bacillus thuringiensis* and *Bacillus cereus* spore inocula'. *NZ J. Agricultural Res.*, **28**, 559–566.

Yamvrias, C. (1962). 'Contribution a l'étude du mode d'action de *Bacillus thuringiensis* Berliner vis-à-vis de la teigne de la farine *Anagasta (Ephestia) Kuhniella* Zeller (Lepidoptere)'. *Entomophaga*, **7**, 101–159.

Yong-Man Yu, Ohba, M. and Gill, S. S. (1991). 'Characterization of mosquiticidal activity of *Bacillus thuringiensis thuringiensis* subsp. *fukuokaensis* crystal proteins'. *Appl. Env. Microbiol.*, **57**, 1075–1081.

Zahner, V., Momen, H., Salles, C. A. and Rabinovitch, L. (1989). 'A comparative study of enzyme variation in *Bacillus cereus* and *Bacillus thuringiensis*'. *J. Appl. Bacteriol.*, **67**, 275–282.

10 Resistance to *Bacillus thuringiensis* and Resistance Management

PAMELA G. MARRONE AND SUSAN C. MACINTOSH
Entotech, Inc. (A Novo Nordisk Subsidiary), 1497 Drew Avenue, Davis, CA 95616-4880, USA

INTRODUCTION

Insects have shown a remarkable ability to develop resistance to insecticidal chemicals. Over 500 species of insects are now resistant to all known classes of insecticidal chemistry (Georghiou and Lagunes, 1988). The development of resistance is one of the most critical issues facing growers and pest managers today. The large development costs, estimated to be $20–40 million, for registration of a new chemical pesticide highlights the critical need for new and existing pesticides to be managed wisely.

Products based on *Bacillus thuringiensis* (*B. t.*) are the most successful microbiological pesticides used in agriculture, forestry and public health. Despite their success, usage in comparison to chemical pesticides has been small, accounting for only about 1% of the world insecticide market. Several developments have led to rapidly expanding usage of *B. t.* products. These are increased environmental concerns, use of *B. t.* as a resistance management tool, withdrawal of chemical insecticides from the market, prices competitive with chemicals and improved product efficacy.

During the past 30 years of *B. t.* use as a microbial insecticide, there have been very few documented cases of *B. t.* resistance development in actual field use. In these situations, Indian meal moth, *Plodia interpunctella*, in stored grain and diamondback moth, *Plutella xylostella*, in certain vegetables, the insects were under continuous, intensive *B. t.* use. A large number of companies and universities have undertaken research programmes to manipulate *B. t.* toxin genes with the aim of producing vastly improved insect control products. Some microbial products under development are plant endophytes engineered to contain certain *B. t.*

Bacillus thuringiensis, An Environmental Biopesticide: Theory and Practice. Edited by P. F. Entwistle, J. S. Cory, M. J. Bailey and S. Higgs
© 1993 John Wiley & Sons Ltd

protein genes (Crop Genetics International) and *B. t.* protein genes engineered into *Pseudomonas* species reportedly with longer residual activity than conventional *B. t.* products (Mycogen) (see Gelernter and Schwab, Chapter 4). Several companies have transferred *B. t.* genes into crop plants such as cotton, potato and tomato (Monsanto, Plant Genetics Systems, Agracetus, Calgene, Lubrizol) (see Ely, Chapter 5). *B. t.* products based on recombinant DNA technology have the potential to improve *B. t.* efficacy and persistence, which will lead to wider use of *B. t.*-based biotechnology products. Resistance development is one of *the* critical issues in implementation of these new biotechnology *B. t.*-based products.

This chapter provides an overview of reported incidences of *B. t.* resistance, laboratory research, cross resistance, genetic and biochemical mechanisms of resistance, and possible management strategies to maximize *B. t.* efficacy.

FIELD RESISTANCE DEVELOPMENT TO *B. THURINGIENSIS*

Only a few years ago, there were no reports of field resistance development to *B. t.*, prompting many to conclude that *B. t.* resistance was unlikely. The Indian meal moth, *P. interpunctella*, a pest in grain bins, was the first insect with documented resistance to *B. t.*. Kinsinger and McGaughey (1979) reported up to a 42-fold difference in susceptibility to Dipel® (*B. t. kurstaki*) between laboratory and field strains. In a subsequent study (McGaughey, 1985a), populations surveyed over a five state area showed a moderate reduction in susceptibility to Dipel®. Laboratory selection of a colony derived from a field population resulted in a 27-fold decrease in susceptibility after two generations and a 97-fold decrease after 15 generations. It was concluded that the decrease in susceptibility was true resistance development, as evidenced by the similarities in the slopes of the dose–mortality lines (McGaughey, 1985b).

Until very recently, the development of resistance by the Indian meal moth was considered unique due to the prolonged, intimate association between the pest and the *B. t.* product. Tabashnik *et al.* (1990) documented the first case of *B. t.* resistance development by a field crop pest, the diamondback moth, *P. xylostella*. On one particular farm in Hawaii, diamondback moths on watercress were treated with *B. t.* Dipel® 50–100 times from 1978 to 1982. Because resistance was suspected, this site was then not treated until 1989–90, when Javelin® (*B. t. kurstaki*) was applied 15 times. Resistance ratios of insects from this and another heavily sprayed farm ranged from 20 to 33. Field mortality was only 34% in treated populations and 90–100% in susceptible laboratory populations. Subsequent laboratory selection of insects from one of these watercress farms resulted in 150–190-fold resistance after five generations and 430–820-fold resistance after nine generations of selection. However, five repeated field applications to the other field did not increase resistance (Tabashnik *et al.*, 1991). Field selected resistance declined slowly in the absence of *B. t.* treatments.

Shelton and Wyman (1990) surveyed 11 populations of diamondback moths from six states of the USA and from Indonesia. Cooperators collected 50–100 larvae and pupae from heavily sprayed commercial cabbage fields, especially in areas where growers experienced control difficulties. High levels of resistance to Dipel® and Javelin®, the two commercial *B. t. kurstaki* preparations tested with leaf dip bioassays, revealed extremely high resistance ratios respectively (240 and 470) in Florida and New York populations. Some of the resistance in New York insects could be traced to diamondback moths on cabbage transplants from Florida. Insects resistant to *B. t. kurstaki* (resistance ratio 112.5) had only a 3.3 resistance ratio to *B. t. aizawai*, marketed as Florbac®.

The variation in susceptibility of a forest pest, the gypsy moth, *Lymantria dispar*, to *B. t.* (HD-1 strain) was studied to determine the potential for resistance development (Rossiter *et al.*, 1990). There were significant differences in the susceptibility of four field populations and also families within each population. Based on oviposition sequence, 42% of the total variation was due to qualitative differences among siblings and 16% was due to familial differences. The authors concluded that the variation in susceptibility to *B. t.* is based on vigour differences in growth and developmental capability. In addition, they concluded that resistance genotypes will be favoured if these genotypes are also more vigorous.

LABORATORY SELECTION FOR RESISTANCE

Several researchers have attempted selection of resistance to *B. t.* in laboratory experiments with varying results. Unsuccessful attempts to develop resistance to *B. t.* in the laboratory have been summarized by Georghiou (1988) and Briese (1981). Since those reviews, there have been a number of successful selection experiments. As discussed earlier, both Tabashnik *et al.* (1991) and McGaughey (1985b) increased the resistance through laboratory selections in colonies derived from field populations. McGaughey and Beeman (1988) reported that five Indian meal moth colonies derived from populations in midwestern grain bins developed resistance at different rates. The highest resistance was 250-fold at the F36 generation, and the lowest 15-fold at the F39. A different stored grain species, *Ephestia cautella*, the almond moth, showed only a 7.5-fold decrease in susceptibility after 21 generations of selection. However selection was carried out on a laboratory strain with little genetic hetero-geneity (McGaughey and Beeman, 1988). Brewer (1991) exposed sunflower moths, *Homeosoma electellum*, to *B. t.* for 12 generations and found approximately a two-fold difference in larval and pupal mortality between selected and non-selected immatures.

Stone *et al.* (1989b) reported the first incidence of laboratory selection to a field crop pest. *Heliothis virescens*, tobacco budworm, was exposed to increasing doses of a recombinant *Pseudomonas fluorescens*, which expressed the *crylA(b)* gene protein product (Watrud *et al.*, 1985). Three-fold resistance developed after three generations of selection at LC_{60-80}. At the F14, resistance was approximately 24-fold. In subsequent studies, from the F14–18, the strain was subcultured on Dipel®, and from the F18–F22 was selected on the recombinant *Pseudomonas*. The strain had approximately 75-fold resistance to the *Pseudomonas*, 57-fold resistance to Dipel® (which contains CryIA(a), CryIA(b), CryIA(c), and CryIIA pro-teins), 71-fold resistance to the CryIA(b) purified protein, and 16-fold resistance to CryIA(c) (MacIntosh *et al.*, 1991).

Another field crop insect that has since been shown to develop resistance under laboratory selection is the Colorado potato beetle, *Leptinotarsa decemlineata*. The selected colony, which originated from field populations sprayed with M-One®, was selected in the laboratory for 13 generations on M-One®-treated foliage. Bioassays showed 67-fold resistance at the F10 generation (Miller *et al.*, 1990).

Several studies have been carried out to determine the potential for mosquitoes to develop resistance to *B. t. israelensis*. Only a two-fold decrease in susceptibility was reported by Goldman *et al.* (1986) and Gharib and Szalay-Marzso (1986) in response to selection of *Aedes aegypti* to *B. t. israelensis* for 14 and 25 generations, respectively. Georghiou and Vasquez-Garcia (1982) and Georghiou (1984) selected field-derived labora-tory populations of *Culex quinquefasciatus* with *B. t. israelensis* for 11–36 generations, producing a 6–16.5-fold decrease in susceptibility. After relaxation of selection for three generations, the susceptibility increased by 50%.

It has been hypothesized that the reason for limited *B. t. israelensis* resistance development in mosquitoes is the presence of multiple protein toxin genes. Dai and Gill (1989) selected *C. quinquefasciatus* for 20 genera-tions on a preparation containing *B. t. israelensis* spores and endotoxin, and then subcultured the mosquitoes on purified 72 kDa protein endo-toxin. The mosquitoes developed a 70-fold resistance to purified 72 kDa protein, but only a three-fold resistance to the spore/endotoxin preparation.

CROSS RESISTANCE OF *B. THURINGIENSIS* TO CHEMICALS AND OTHER *B. THURINGIENSIS* STRAINS

B. t. has a unique mode of action and therefore it is not surprising that to date there have been no reported cases of cross resistance to insecticidal chemicals. However, there have been reports of cross resistance among *B. t.* strains or proteins. Knowledge on cross resistance among *B. t.* strains and proteins will be critical when devising resistance management strategies.

The cross resistance to Dipel® of 57 strains active against *P. interpunctella* was determined. Twenty-one strains were equally effective on Dipel® resistant and Dipel® sensitive insects. The insects were also resistant to *Escherichia coli*-cloned CryIA(a–c) proteins. The researchers concluded that resistance in *P. interpunctella* was limited to the HD-1 type strains with CryIA(a–c) proteins (McGaughey and Johnson, 1987; Han *et al.*, 1988).

H. virescens, selected for resistance to a *Pseudomonas* species containing *cryIA(b)* protein gene and Dipel® had varying levels of cross resistance (2.4–57-fold) to all of the approximately 20 different *B. t.* strains from subspecies *aizawai*, *colmeri*, *darmstadiensis*, *entomocidus*, *kurstaki*, *thuringiensis* and others (Stone *et al.*, 1989a).

GENETICS OF RESISTANCE

The tactics used to manage resistance will depend in part on the genetic basis of the resistance. Therefore, genetic studies are a fundamental part of any resistance research programme. Roush and McKenzie (1987) concluded that laboratory strains of insects, initiated with a small sample of the total species genetic variability, usually develop polygenic resistance mechanisms. If studies are carried out on laboratory colonies, caution must be taken because they may or may not reflect field situations.

A laboratory strain of *H. virescens* resistant to a *B. t.* CryIA(b) expressing *Pseudomonas* was found to be autosomally inherited, incompletely dominant, and controlled by several genetic factors (Sims and Stone, 1991). Resistance of 67-fold in one selected line declined to 13-fold by the fifth generation without selection. A follow-up study (Sims and Stone, personal communication) investigated the inheritance in *H. virescens* of resistance to the 63 kDa activated fragment of the δ-endotoxin protein. Resistance (139-fold) was controlled by several genetic factors, was partially recessive, and control was predominantly autosomal, with some sex-linked factors involved. An interesting result was that the stability of resistance depended on the endotoxin used for selection. Resistance to

Dipel® and a recombinant *Pseudomonas* expressing CryIA(b) protein was unstable in the absence of selection pressure. In contrast, resistance to HD-73, which contains a different protein CryIA(c), was stable over seven generations without selection.

McGaughey (1985a) and McGaughey and Beeman (1988) studied the genetics of *B. t.* (Dipel®) resistance in *P. interpunctella* with a field history of prior *B. t.* exposure. They concluded that resistance is variably recessive and suggested that it is controlled by a single major factor. After three generations without selection, resistance in one strain declined from 63–70-fold to 2–4-fold, suggesting that the strain still contained susceptible genotypes and had not reached a resistance plateau.

MECHANISM OF RESISTANCE

The mechanism by which an insect evolves resistance to a particular toxin is unavoidably tied to the toxin's mode of action. The majority of published data examines the mechanism of resistance of chemical insecticides. Chemical insecticide resistance is so widespread, it allows a multitude of study systems (Wilkinson, 1983; Knipple et al., 1988). Genetic, biochemical, physiological, and ecological elements are a few of the many interdependent factors influencing the onset of resistance.

Since development of resistance to *B. t.* has only recently been documented in the field, reports on the mechanism of resistance have focused entirely on laboratory-selected insect colonies. One study has been carried out on a field population of diamondback moth (Ferré et al., 1991). This section will review the results from laboratory-selected insect colonies and one field population with supporting data on the mode of action of *B. t. kurstaki*.

B. t. has a distinct yet complex mode of action; once the *B. t.* protoxin is ingested, it is first proteolysed to an activated toxin protein (Fast, 1981, 1983; Huber and Lüthy, 1981) that diffuses through the peritrophic membrane and binds to high affinity receptors present on the midgut epithelium. The gut becomes paralysed and the larvae cease feeding (Dulmage et al., 1978; Salama and Sharaby, 1985). Once the activated toxin binds (Hofmann et al., 1988a,b; Van Rie et al., 1989, 1990a; Wolfersberger, 1990; MacIntosh et al., 1991), the punitive receptor/toxin interaction becomes irreversible and the toxin is believed to insert into the membrane causing a lesion or pore to form (Knowles and Ellar, 1986; Ellar, 1989; Hendrick et al., 1990). The pore formation disrupts the inward K^+ gradient (Sacchi et al., 1986) leading to the swelling of microvilli and destruction (Lüthy et al., 1982; Percy and Fast, 1983; deLello et al., 1984). If enough midgut cells are destroyed the larvae will die.

McGaughey's (1985a,b) studies of *B. t.* susceptibility in wild populations of Indian meal moth led to the laboratory selection of a colony collected from stored grain bins. The resultant resistant colony had a resistance ratio of >200-fold when compared to a susceptible laboratory strain (Van Rie *et al.*, 1990b). Two possible resistance mechanisms were investigated, proteolytic processing of *B. t.* proteins and specific binding of *B. t.* toxin proteins. Post-binding events, such as membrane insertion and pore formation, were not evaluated.

The proteolytic digestion of the full-length *B. t.* protoxins (130–140 kDa) to the trypsin stable proteins (55–70 kDa) is necessary for insecticidal activity. A mode of resistance could involve the decrease in pH or proteolytic properties of the insect midgut. Since the peritrophic membrane acts as a molecular sieve, the intact crystal proteins would be prevented from interacting with the midgut epithelium and be excreted. In the case of Indian meal moth, no differences in midgut pH or midgut proteinase patterns were found (Johnson *et al.*, 1990). Proteolytic processing of crystal proteins from either *B. t. kurstaki* or *B. t. aizawai* strains produced indistinguishable patterns when combined with midgut homogenate of susceptible and resistant insects.

Additional studies focused on the specific binding of CryIA(b) and CryIC toxin proteins to isolated brush border membrane vesicles from the resistant and susceptible Indian meal moth strains (Van Rie *et al.*, 1990b). Resistance ratios of sensitivity to the CryIA(b) toxin was > 300-fold while the sensitivity to CryIC was actually enhanced in the resistant insects by 3.6-fold. The binding parameters correlated with the observed differences in toxicity. The binding affinity of CryIA(b) was significantly reduced, 50-fold, in the resistant Indian meal moth strain whereas the binding site concentration was virtually unchanged. Although the binding affinities of CryIC were identical between the resistant and susceptible insect strains, the binding site concentration was increased by a factor of three for resistant Indian meal moth. Therefore, the resistance development in this laboratory-selected Indian meal moth was accompanied by two distinct molecular changes in the specific binding of two different toxin proteins, CryIA(b) and CryIC.

The tobacco budworm, the first field insect to be selected for resistance to *B. t.* (Stone *et al.*, 1989b), displayed a mechanism of resistance also related to specific binding of the toxin proteins. The proteolytic composition of the gut juice and proteolytic processing of the toxin proteins, CryIA(b) and CryIA(c), were unchanged in the resistant insect strain (MacIntosh S. C., personal communication). The resistance ratio of sensitivity of CryIA(c) and CryIA(b) was 16- and 70-fold, respectively, towards the resistant insects but only a four- and two-fold reduction in receptor affinity was observed (MacIntosh *et al.*, 1991). Surprisingly, this loss of affinity was accompanied by an increase in binding site concentration for both toxins, four-fold for CryIA(c) and six-fold for CryIA(b). Competition experiments revealed differential receptor affinities in the resistant tobacco budworm. CryIA(b) was much less efficient in competing with CryIA(c) in the resistant than in the susceptible strain.

In the first report of the mechanism of resistance in field populations, Ferré *et al.* (1991) determined that 200-fold resistance to Dipel® by diamondback moths from the Philippines could be explained by a complete lack of binding of CryIA(b) protein in the midgut. However, binding to CryIB and CryIC proteins, which are not present in Dipel®, was similar in resistant and susceptible insects. Heterologous competition experiments indicate that diamondback moth has three distinct high-affinity binding sites (CryIA(b), CryIB and CryIC).

The mechanism of resistance of all insects studied to date was influenced by alterations in specific toxin binding and not tied to the proteolytic processing of the protoxin proteins. The type of proteolytic cleavage that occurs in the insect gut may determine insect specificity as with *B. t. colmeri*, which has dual activities against lepidopteran and dipteran cell lines (Ellar *et al.*, 1986; Haider *et al.*, 1986). A similar resistant mechanism is feasible but has not been found to date. Instead, modulations in the specific binding of the toxin proteins to brush border membrane vesicles (BBMV) have played an important role in the mechanism of resistance of the three reported cases (Van Rie *et al.*, 1990b; MacIntosh *et al.*, 1991). Receptor binding of a variety of *B. t.* proteins has been analysed from BBMVs of a number of lepidopteran species (Hoffman *et al.*, 1988a,b; Van Rie *et al.*, 1989, 1990a; Wolfersberger, 1990; Ferré *et al.*, 1991; MacIntosh *et al.*, 1991). In most cases, insect sensitivity correlated with the presence of high affinity binding sites. Two studies found an inverse correlation of binding with toxicity, CryIC in tobacco budworm (Van Rie *et al.*, 1990a) and CryIA(b) and CryIA(c) in gypsy moth (Wolfersberger, 1990).

The complexity of the resistance mechanism and, of course, the mode of action is evident from studies of the laboratory selected resistant insects. In Indian meal moth and tobacco budworm at least two different molecular changes occurred, affecting binding affinity and binding site concentration. Other aspects of the possible mechanisms of resistance remain to be explored, including post-binding events such as membrane insertion and pore formation and behaviour studies. These types of studies are difficult to interpret and in the case of pore formation, can only be evaluated indirectly by analysing amino acid transport (Sacchi *et al.*, 1986; Wolfersberger *et al.*, 1987).

Since the properties and characterization of the putative *B. t.* receptor are still elusive (Ellar, 1989; Muthukumar and Nickerson, 1987), the physiological function of the receptor also remains known. Therefore, it is not surprising to find that the resistant insects have developed multiple modifications to overcome the effects of *B. t.*

STRATEGIES FOR DELAYING RESISTANCE DEVELOPMENT

The successful use of new *B. t.*-based biotechnology products depends on development and implementation of sound management strategies. In the past, resistance management to chemical insecticides has occurred after resistance reached crisis proportions. With current *B. t.* products, resistance has developed in fields under continuous, intensive use of *B. t.* This provides a lesson for future management and use of new *B. t.*-based biotechnology products. At present, there is an opportunity to develop strategies in advance of marketing products such as first generation, single gene transgenic plants, which are thought by many experts to have a high risk for resistance development (Gould, 1988; Raffa, 1989; Tabashnik *et al.*, 1991).

The *B. t.* Management Working Group, a consortium of biotechnology and agrochemical companies, was formed in 1988 to address the issue of *B. t.* resistance development (Marrone, 1989). The goals of this group are to determine the potential for resistance development in key pests, determine the mechanism of resistance, develop resistance management strategies, and serve as a resource on *B. t.* resistance. Over a 2 year period, the group provided funds to five research groups. The purpose of the grants is to develop a base of information necessary to develop strategies for maximizing efficacy of *B. t.* products.

Roush (1989) states that without question the most effective way to manage resistance is to avoid pesticide applications. In addition, Tabashnik *et al.* (1991) conclude that the best way to manage resistance to *B. t.* is to prevent its occurrence. One obvious approach is rotation (alternation) across time of *B. t.* products with other pesticides of different modes of action. The basic principle behind rotation is that fitness disadvantages will cause the decline of resistance in the absence of the selecting pesticide (Roush, 1989; Georghiou, 1983). Rotation across generations is thought to be considerably more effective than a programmed rotation on a weekly or 5-day schedule. This thinking was the basis behind the Australian pyrethroid resistance strategy, where the cotton growing season was divided into three periods for different pesticide classes. Pyrethroids could only be used during a 6 week period (Daly *et al.*, 1988). For *B. t.* products, a rotation strategy across generations would only be practical with microbial products, rather than transgenic plants expressing *B. t.* protein the entire crop season.

Another possible strategy for both microbial and transgenic plants is mixtures. Currently, microbial *B. t.* products are primarily applied in mixtures with chemical insecticides. However, there are no studies that demonstrate that this will delay resistance. In fact, it is possible that it may actually enhance resistance development.

Roush (1989) provides some advice on the choice between mixtures and rotation. In order for mixtures to be very effective for delaying resistance, the initial frequencies of resistance should be low, the fraction escaping treatment should be high relative to dominance and linkage, the pesticides must be nearly 100% effective against treated susceptible heterozygotes, and equal in persistence. If all of these conditions are met, then mixtures are the choice over rotation. Rotation is the choice if any one condition is not met. In reality, it is very difficult to meet these conditions and rotation may be the wisest choice.

Another possible strategy is the use of mosaics. The number of B. t. expressing or B. t.-treated plants in a field would be diluted with non-expressing or untreated plants. Gould (1986a,b) indicates that a pest will evolve much more slowly to a mixture of plants comprising 80% with resistance factors and 20% with no resistance factors. He cites the example of the hessian fly, Mayetiola destructor, where an 80%/20% mixture will take the flies five times longer to adapt than to a widely used cultivar with two resistance factors. He is sceptical that this strategy would be practical to those developing B. t.-expressing transgenic plants, because it is easier to develop genetically homogeneous cultivars than mixtures, and farmers will buy from the company offering the most toxic plants (Gould, 1988).

Roush (1989) states that alternation is sometimes much better and certainly never worse than mosaics of more than one pesticide. His model assumes completely random mating for the mosaic. Where mating is not completely random, mosaics will be as durable as rotation. When resistance is completely recessive to both pesticides and there are no fitness disadvantages, rotation and mosaics are equal in delaying resistance.

An obvious strategy for transgenic plants is the incorporation of more than one gene into the crop plant. This strategy would only be effective if the two genes had unique modes of action. Unfortunately, insecticidal genes available for incorporation into plants are currently limited to B. t. and protease inhibitors (Sanchez-Serano et al., 1987; Hilder et al., 1987; MacIntosh et al., 1990).

A number of companies developing transgenic plants view the incorporation of two or more B. t. genes into the crop plant as the most promising strategy for delaying resistance. Plant Genetics Systems filed a patent on this approach for managing resistance to B. t. The idea behind the strategy is that a particular B. t. protein is active at one midgut binding site, while a different protein is active at another binding site. The combination of the two in the plant should delay resistance when compared to a plant expressing only one protein.

Although a seemingly simple solution, the situation is more complicated than it appears. The model assumes that protein activity in the gut and development of resistance is primarily due to binding in the midgut. Wolfersberger (1990) found that B. t. activity in gypsy moth was not always explained by binding. Likewise, MacIntosh *et al.* (1991) concluded that while binding was extremely important in explaining *B. t.* resistance in *H. virescens*, post-binding events may be even more important. Resistance could develop to both proteins if post-binding events or some aspects of the *B. t.* 'receptor' share a common physiological function. In fact, we now know that field insects can develop resistance to products with multiple genes, such as Dipel®. Tabashnik *et al.* (1991) state that the use of mixtures of toxin genes will not necessarily halt the development of resistance. Clearly more knowledge of the resistance mechanism and *B. t.* mode of action is required.

A long-term approach for *B. t.* expressing transgenic plants is tissue and temporal specific expression of the *B. t.* protein. The protein will be expressed only when and where the insect feeds, thus limiting selection pressure. Gould (1988) suggests that if a *B. t.* protein were expressed only in the cotton boll tissues, selection would be discontinuous because only one of three to four generations of bollworms, *Heliothis* spp., feeds on the fruiting structures. Scientists have succeeded in transferring a proteinase inhibitor gene from potato into tobacco. The protein was only expressed in response to wounding (Sanchez-Serano *et al.*, 1987).

CONCLUSIONS

Due to environmental concerns, the use of *B. t.* and other microbial products will continue to increase. Biotechnology can provide exciting new tools in the quest for new, environmentally friendly pest control methods. As new biotechnology products based on *B. t.* are developed, the challenge to use these products in a sound manner that maximizes their field life becomes critical. Pest control specialists view resistance as the single most important issue in the development of single gene *B. t.*-based products, such as first generation, single gene transgenic plants. With chemical insecticides, resistance management strategies have been implemented only after widespread field control failures. We now have the opportunity to demonstrate the lessons of past use of chemical insecticides by developing sound strategies for using *B. t.*-based biotechnology products in pest management systems in advance of their actual commercialization and marketing.

REFERENCES

Brewer, G. J. (1991). 'Resistance to *B. t.* in the sunflower moth'. *Env. Entomol.*, **20**, 316–322.

Briese, D. T. (1981). 'Resistance of insect species to microbial pathogens'. In

Pathogenesis of Invertebrate Microbial Diseases (Ed. E. Davidson), pp. 511–545, Allenheld, Osmun, Totowa, N.J.

Dai, S.-M. and Gill, S. (1989). 'Development of resistance to the 72 kDa toxin of *Bacillus thuringiensis* in *Culex quinquefasciatus*'. In *Proceedings of International Symposium of Molecular Insect Science*, October 22–27, Tucson, AZ.

Daly, J., Fisk, J. H. and Forrester, N. W. (1988). 'Selective mortality in field trials between strains of *Heliothis armigera* (Lepidoptera: Noctuidae) resistant and susceptible to pyrethroids: functional dominance of resistance and age class'. *J. Econ. Entomol.*, **81**, 1000–1007.

deLello, E., Hanton, W. K., Bishoff, S. T. and Misch, D. W. (1984). 'Histopathological effects of *Bacillus thuringiensis* on the midgut of tobacco hornworm larvae (*Manduca sexta*): low doses compared with fasting'. *J. Invert. Pathol.*, **43**, 169–181.

Dulmage, H. T., Graham, H. M. and Martinez, E. (1978). 'Interactions between the tobacco budworm, *Heliothis virescens*, and the delta-endotoxin produced by the HD-1 isolate of *Bacillus thuringiensis* var. *kurstaki*: relationship between length of exposure to the toxin and survival'. *J. Invert. Pathol.*, **32**, 40–50.

Ellar, D. J. (1989). 'Investigation of the molecular basis of *Bacillus thuringiensis* delta-endotoxin specificity and toxicity'. Abstracts of the Soc. Invert. Path. Meeting, College Park, MD, 20–24 August.

Ellar, D. J., Knowles, B. H., Haider, M. Z. and Drobniewski, F. A. (1986). 'Investigation of the specificity, cytotoxic mechanisms and relatedness of *Bacillus thuringiensis* insecticidal δ-endotoxins from different pathotypes'. In *Bacterial Protein Toxins* (Eds. P. Falmagne, F. J. Fehrenbach, J. Jeljaszewics and M. Thelestam), p. 41, Gustav Fischer, New York,

Fast, P. G. (1981). 'Bacteria: the crystal toxin of *Bacillus thuringiensis*'. In *Microbial Control of Pests and Plant Diseases, 1970–1980* (Ed. H. D. Burges), pp. 223–248, Academic Press, London.

Fast, P. G. (1983). '*Bacillus thuringiensis* parasporal toxin: aspects of chemistry and mode-of-action'. *Toxicon*, Suppl. 3, 123–125.

Ferré, J., Real, M. D., Van Rie, J., Jansens, S. and Peferoen, M. (1991). 'Resistance to *Bacillus thuringiensis* bioinsecticides in a field population of *Plutella xylostella* is free to change in midgut membrane receptor'. *Proc. Natl Acad. Sci. USA*, **88**, 5119–5123.

Georghiou, G. P. (1983). 'Management of resistance in arthropods'. In *Pest Resistance to Pesticides* (Eds. G. P. Georghiou and T. Saito), pp. 769–792, Plenum, New York.

Georghiou, G. P. (1984). 'Insecticide resistance in mosquitoes: research on new chemicals and techniques for management'. *Ann. Rept. Mosquito Control Research*, University of California, pp. 97–99.

Georghiou, G.P. (1988). 'Implications of potential resistance to biopesticides'. In *Biotechnology, Biological Pesticides, and Novel Plant Pest Resistance for Insect Pest Management* (Eds. D. W. Roberts and R. R. Granados), pp. 137–146, Boyce Thompson Institute, Ithaca, NY.

Georghiou, G. P. and Lagunes, A. (1988). 'The occurrence of resistance to pesticides: cases of resistance reported worldwide through 1988'. *F.A.O.*, Rome, p. 325.

Georghiou, G. P. and Vazquez-Garcia, M. (1982). 'Assessing the potential for development of resistance to *Bacillus thuringiensis* subsp. *israelensis* toxin (*Bti*) by mosquitoes'. *Ann. Rept, Mosquito Control Research*, University of California, pp. 80–81.

Gharib, A. H. and Szalay-Marzso, L. (1986). 'Selection for resistance to *Bacillus thuringiensis* serotype H-14 in a laboratory strain of *Aedes aegypti* L'. In *Fundamental and Applied Aspects of Invertebrate Pathology* (Eds. R. A. Samson, J. M. Vlak and D. Peters), p. 37, Found. Fourth Int. Colloq. Invert. Pathol., Wageningen, Netherlands.

Goldman, I. F., Arnold, J. and Carlton, B. C. (1986). 'Selection of resistance to *Bacillus thuringiensis* subsp. *israelensis*, in field and laboratory populations of *Aedes aegypti*'. *J. Invert. Pathol.*, **47**, 317–324.

Gould, F. (1986a). 'Simulation models for predicting durability of insect resistant germplasm: a deterministic diploid, two-locus model'. *Env. Entomol.*, **15**, 1–10.

Gould, F. (1986b). 'Simulation models for predicting durability of insect resistant germplasm: hessian fly (Diptera: Cecidomyiidae)-resistant winter wheat'. *Env. Entomol.*, **15** 11–23.

Gould, F. (1988). 'Evolutionary biology and genetically engineered crops'. *BioScience*, **38**, 26–32.

Haider, M. Z., Knowles, B. H. and Ellar, D. J. (1986). 'Specificity of *Bacillus thuringiensis* var. *colmeri* insecticidal d-endotoxin is determined by differential proteolytic processing of the protoxin by larval gut proteases'. *Eur. J. Biochem.*, **156**, 531–540.

Han, E.-S., McGaughey, W. H., Johnson, D. E., Dunn, P. and Aronson, A. I. (1988). 'Characterization of *Bacillus thuringiensis* isolates effective on resistant Indian meal moth'. In *Genetics and Biotechnology of the Bacilli* (Eds. A. T. Ganesan and J. A. Hoc) Vol. 2, pp. 233–238, Academic Press, New York.

Hendrick, K., De Loof, A. and Van Mellaert, H. (1990). 'Effects of *Bacillus thuringiensis* delta-endotoxin on the permeability of brush border membrane vesicles from tobacco budworm (*Manduca sexta*) midgut'. *Comp. Biochem. Physiol.*, **95C**, 241–245.

Hilder, V. A., Gatehouse, A. M. R., Sheerman, S. E., Barker, R. F. and Boulter, D. (1987). 'A novel mechanism of insect resistance engineered into tobacco'. *Nature (London)*, **330**, 160–163.

Hofmann, C., Lüthy, P., Hütter, R. and Pliska, V. (1988a). 'Binding of the delta-endotoxin from *Bacillus thuringiensis* to brush-border membrane vesicles of the cabbage butterfly (*Pieris brassicae*)'. *Eur. J. Biochem.*, **173**, 85–91.

Hofmann, C., Vanderbruggen, H., Höfte, H., Van Rie, J., Jansens, S. and Van Mellaert, H. (1988b). 'Specificity of *Bacillus thuringiensis* δ-endotoxins is correlated with the presence of high-affinity binding sites in the brush border membrane of target insect midguts'. *Proc. Natl Acad. Sci. USA*, **85**, 7844–7848.

Huber, H. E. and Lüthy, P. (1981). '*Bacillus thuringiensis* delta-endotoxin composition and activation'. In *Pathogenesis of Invertebrate Microbial Diseases* (Ed. E. W. Davidson), pp. 209–234, Allenheld, Osmun., Totowa, NJ.

Johnson, D. E., Brookhart, G. L., Kramer, K. J., Barnett, B. D. and McGaughey, W. H. (1990). 'Resistance to *Bacillus thuringiensis* by the Indian mealmoth, *Plodia interpunctella*: comparison of midgut proteinases from susceptible and resistant larvae'. *J. Invert. Pathol.*, **55**, 235–244.

Kinsinger, R. A. and McGaughey, W. H. (1979). 'Susceptibility of populations of Indian meal moth and almond moth to *Bacillus thuringiensis*'. *J. Econ. Entomol.*, **72**, 346–349.

Knipple, D. C., Bloomquist, J. R. and Soderlund, D. M. (1988). 'Molecular genetic approach to the study of target-site resistance to pyrethroids and DDT in insects'. In *Biotechnology for Crop Protection* (Eds. P. A. Hedin, J. J. Menn and R. M. Hollingsworth), pp. 199–214, Amer. Chem. Soc., Washington, DC.

Knowles, B. H., and Ellar, D. J. (1986). 'Characterization and partial purification of a plasma membrane receptor for *Bacillus thuringiensis* var. *kurstaki* lepidopteran-specific delta-endotoxin'. *J. Cell Sci.*, **83**, 89–101.

Lüthy, P., Jaquet, F., Huber-Lukac, H. E. and Huber-Lukac, M. (1982). 'Physiology of the delta-endotoxin of *Bacillus thuringiensis* including the ultrastructure and histopathological studies'. In *Basic Biology of Microbial Larvicides of Vectors of Human Diseases* (Ed. F. Michal), pp. 29–36, UNDP/World Bank/WHO, Geneva, Switzerland.

MacIntosh, S. C., Kishore, G. M., Perlak, F. J., Marrone, P. G., Stone, T. B., Sims, S. R. and Fuchs, R. L. (1990). 'Potentiation of *Bacillus thuringiensis* insecticidal activity by serine protease inhibitors'. *J. Agric. Food Chem.*, **38**, 1145–1151.

MacIntosh, S. C., Stone, T. B., Jokerst, R. S. and Fuchs, R. L. (1991). 'Binding of *Bacillus thuringiensis* proteins to a laboratory selected line of *Heliothis virescens*'. *Proc. Natl Acad. Sci. USA.*, **88**, 8930–8933.

Marrone, P. G. (1989). '*B. t.* Working Group'. *Pest Resistance Management, Western Regional Coordinating Committee (WRCC) Newsletter.* 1, p. 13.

McGaughey, W. H. (1985a). 'Evaluation of *Bacillus thuringiensis* for controlling Indian meal moths (Lepidoptera: Pyralidae) in farm grain bins and elevator silos'. *J. Econ. Entomol.*, **78**, 1089–1094.

McGaughey, W. H. (1985b). 'Insect resistance to the biological insecticide *Bacillus thuringiensis*'. *Science*, **229**, 193–195.

McGaughey, W. H. and Beeman, R. W. (1988). 'Resistance to *Bacillus thuringiensis* in colonies of the Indian meal moth and almond moth (Lep.: Pyralidae)'. *J. Econ. Entomol.*, **81**, 28–33.

McGaughey, W. H. and Johnson, D. E. (1987). 'Toxicity of different serotypes of *Bacillus thuringiensis* to resistant and susceptible Indian meal moths (Lepidoptera: Pyralidae)'. *J. Econ. Entomol.*, **80**, 1122–1126.

Miller, D. L., Rahardja, U. and Whalon, M. E. (1990). 'Development of a strain of Colorado potato beetle resistant to the delta-endotoxin of *B. t.*'. *Pest Resistance Management, WRCC Newsletter* 2, p. 25.

Muthukumar, G. and Nickerson, K. W. (1987). 'The glycoprotein toxin of *Bacillus thuringiensis* subsp. *israelensis* indicates a lectin like receptor in the larval mosquito gut'. *Appl. Env. Microbiol.*, **53**, 2650–2655.

Percy, J. and Fast, P. G. (1983). '*Bacillus thuringiensis* crystal toxin: ultrastructural studies of its effect on silkworm midgut cells', *J. Invert. Pathol.*, **41**, 86–98.

Raffa, K. F. (1989). 'Genetic engineering of trees to enhance resistance to insects'. *BioScience*, **39**, 524–533.

Rossiter, M., Yendol, W.G. and Dubois, N. R. (1990). 'Resistance to *Bacillus thuringiensis* in gypsy moth (Lepidoptera: Lymantriidae): genetic and environmental causes'. *J. Econ. Entomol.*, **83**, 2211–2218.

Roush, R (1989). 'Designing resistance management programs: how can you choose?' *Pestic. Sci.*, **26**, 423–441.

Roush, R. T. and McKenzie, J. A. (1987). 'Ecological genetics of insecticide and acaracide resistance'. *Ann. Rev. Entomol.*, **32**, 361–380.

Sacchi, V. F., Parenti, P., Hanozet, G. M., Giordana, B., Lüthy, P. and Wolfersberger, M. G. (1986). '*Bacillus thuringiensis* toxin inhibits K^+-gradient-dependent amino acid transport across the brush border membrane of *Pieris brassicae* midgut cells'. *Federation European Biochemical Societies (FEBS) Letter*, **204**, 213–218.

Salama, H. S., and Sharaby, A. (1985). 'Histopathological changes in *Heliothis armigera* infected with *Bacillus thuringiensis* as detected by electron microscopy'. *Insect Sci. Applic.*, **6**, 503–511.

Sanchez-Serrano, J. J., Keil, M., O'Connor, A., Schell, J. and Willmitzer, L. (1987).

'Wound induced expression of a potato proteinase inhibitor II gene in transgenic tobacco plants'. *EMBO J.*, **6**, 303–306.

Shelton, A. M. and Wyman, J. A. (1990). 'Insecticide resistance of diamondback moth (Lepidoptera: Plutellidae) in North America'. Abstracts of the International Diamondback Workshop, Tainan, Taiwan, 10–14 December, 1990.

Sims, S. R. and Stone, T. B. (1991). 'Genetic basis of tobacco budworm resistance to an engineered *Pseudomonas fluorescens* expressing the delta endotoxin of *Bacillus thuringiensis kurstaki*'. *J. Invert. Pathol.*, **57**, 206–210.

Stone, T. B., Sims, S. R., MacIntosh, S. C., Marrone, P. G., Armbruster, B. A. and Fuchs, R. L. (1989a). 'Insect resistance to *Bacillus thuringiensis* delta-endotoxins'. Presented at International Symposium of Molecular Insect Science, 22–27 October, Tucson, AZ.

Stone, T. B., Sims, S. R. and Marrone, P. G. (1989b). 'Selection of tobacco budworm to a genetically engineered *Pseudomonas fluorescens* containing the delta-endotoxin of *Bacillus thuringiensis* subsp. *kurstaki*'. *J. Invert. Pathol.*, **53**, 228–234.

Tabashnik, B. E., Cushing, N. L., Finson, N. and Johnson, M. (1990). 'Field development of resistance to *Bacillus thuringiensis* in diamondback moth (Lepidoptera: Plutellidae)'. *J. Econ. Entomol.*, **83**, 1671–1676.

Tabashnik, B. E., Finson, N. and Johnson, M. (1991). 'Managing resistance to *Bacillus thuringiensis*: lessons from the diamondback moth (Lepidoptera: Plutellidae)'. *J. Econ. Entomol.*, **84**, 49–55.

Van Rie, J., Jansens, S., Höfte, H., Degheele, D. and Van Mellaert, H. (1989). 'Specificity of *Bacillus thuringiensis* delta-endotoxins: Importance of specific receptors on the brush border membrane of the midgut of target insects'. *Eur. J. Biochem.*, **186**, 239–247.

Van Rie, J., Jansens, S., Höfte, H., Degheele, D. and Van Mellaert, H. (1990a). 'Receptors on the brush border membrane of the insect midgut as determinants of the specificity of *Bacillus thuringiensis* delta-endotoxins'. *Appl. Env. Microbiol.*, **56**, 1378–1385.

Van Rie, J., McGaughey, W. H., Johnson, D. E., Barnett, B. D. and Van Mellaert, H. (1990b). 'Mechanism of insect resistance to the microbial insecticide *Bacillus thuringiensis*'. *Science*, **247**, 72–74.

Watrud, L. S., Perlak, F. J., Tran, M.-T., Kusano, K., Mayer, E. J., Miller-Wideman, M. A., Obukowicz, M. G., Nelson, D. R., Kreitenger, J. P. and Kaufman, R. J. (1985). 'Cloning of the *Bacillus thuringiensis* subsp. *kurstaki* delta-endotoxin gene into *Pseudomonas fluorescens*: Molecular biology and ecology of an engineered microbial pesticide'. In *Engineered Organisms in the Environment* (Eds. H. O. Halverson, D. Pramer, and M. Rogul), pp. 40–46, Amer. Soc. Microbiol., Washington.

Wilkinson, C. F. (1983). 'Role of mixed-function oxidases in insecticide resistance'. In *Pest Resistance to Pesticides* (Eds. G. P. Georghiou and T. Saito), p. 175, Plenum Press, New York.

Wolfersberger, M. G. (1990). 'The toxicity of two *Bacillus thuringiensis* delta-endotoxins to gypsy moth larvae is inversely related to the affinity of binding sites on midgut brush border membranes for the toxins'. *Experientia*, **46**, 475–477.

Wolfersberger, M. G., Lüthy, P., Maurer, A., Parenti, P., Sacchi, F. V., Giordana, B. and Hanozet, G. M. (1987). 'Preparation and partial characterization of amino acid transporting brush border membrane vesicles from the larval midgut of the cabbage butterfly (*Pieris brassicae*)'. *Comp. Biochem. Physiol.*, **86A**, 301–308.

11 The Use of *Bacillus thuringiensis* in Developing Countries

H. S. SALAMA[1] **AND O. N. MORRIS**[2]
[1]*National Research Centre, Tahrir Street, Dokki, Cairo, Egypt and*
[2]*Agriculture Canada Research Station, 195 Dafoe Road, Winnipeg, Manitoba R3T 2M9, Canada*

HISTORY AND USAGE OF *B. THURINGIENSIS* IN DEVELOPING NATIONS

Research on the microbial control agent *Bacillus thuringiensis* (*B. t.*) has only recently been initiated in developing countries where its use is limited in comparison with that of chemical insecticides. Several factors make the local production and use of *B. t.* highly appropriate for pest control in developing nations. In particular, *B. t.* can be cheaply and easily produced on a wide variety of low value, organic substrates and local production results in considerable savings in hard currency which otherwise would be spent on importation of chemical insecticides.

Although the use of *B. t.* in most developing countries depends on imported commercial preparations, several countries have been looking at the development and production of their own *B. t.* and China has been a pioneer in this. Xie Tianjian *et al.* (1990) reported that Wuhan has been the main site for *B. t.* production in the People's Republic of China since 1964. *B. t.* subspecies *galleriae*, *B. t. wuhanensis*, *B. t. dendrolimus* and *B. t. kurstaki* HD-1 were used in the 1970s, while *B. t. kurstaki* HD-1 was again the main strain used in 1980.

In Egypt, the leading role in research and development of *B. t.* has been taken by the National Research Centre (NRC). The efforts of the NRC team led to the establishment of scientific cooperation with the US Department of Agriculture (1977–1985) and Agriculture Canada (1985 onwards), to combat the key lepidopterous pests of field, oilseed and vegetable crops. *B. t.* has not yet been registered in Egypt and its use is so far limited to large field trials.

Bacillus thuringiensis, An Environmental Biopesticide: Theory and Practice. Edited by P. F. Entwistle, J. S. Cory, M. J. Bailey and S. Higgs
© 1993 John Wiley & Sons Ltd

STRAIN SURVEY AND SELECTION

A considerable amount of survey work has been conducted in developing countries to isolate novel B. t. strains. This type of activity led, for example, to the discovery of B. t. israelensis with its considerable capacity for the control of mosquitoes and blackflies (Goldberg and Margalit, 1977). A survey of the distribution of B. t. throughout China has been coordinated since 1956 by the Institute of Zoology in Peking (Hussey and Tinsley, 1981). Efforts led to the isolation of B. t. hubeinsis (Wang Ying et al., 1988), B. t. pectinophorae (Chen Tao Liang et al., 1988), B. t. neoleonensis (Luis et al., 1988) and B. t. kumamotoensis (Roldan et al., 1988). In addition new isolates active against mosquitoes have also been discovered (see Becker and Margalit, Chapter 7).

In Egypt, a survey was made of the distribution of B. t. and Bacillus cereus in the soil of 15 Egyptian Governorates (Salama et al., 1986). High bacillus counts were detected in most fertile soils, while sandy soils always had low counts. Some isolates were found to be active against the beet armyworm, Spodoptera exigua, and the African bollworm, Heliothis armigera, but none of the isolates exhibited significant activity against cotton leafworm, Spodoptera littoralis. B. t. strains were also isolated from insects such as the Indian meal moth, Plodia interpunctella, and the pink bollworm, Pectinophora gossypiella (Abdel-Rahman, 1966, Abul-Nasr et al., 1979).

In Pakistan most isolates were shown to belong to B. t. subspecies sotto (129 out of 150 isolates), while the remainder were primarily serotypes of dendrolimus (3), kurstaki (4), kenyae (1) and pakistani (10) (Shaikh et al., 1987). During a study in South India, samples of soil, water and larvae were collected from various mosquito breeding habitats. One hundred and one strains of B. t. HD-14 were obtained and assessed. From the 11 non-HD-14 strains, one was identified as a new subgroup of serotype H–20, designated B. t. pondicherriensis (Manomani et al., 1987). Padua (1988) isolated the B. t. subspecies designated PG-14 from a soil sample in Philippines, which was considered to be the most effective mosquito larvicide developed in that country.

PRODUCTION LEVELS AND COSTS

The cost involved in obtaining registration and approval for microbial pesticides is relatively low compared to chemical insecticides. As the development and commercialization costs are much cheaper, the minimum size of the target market can also be smaller. Traditionally, less than 1% of the cost of developing chemical pesticides has been spent on insect pathogens (Dulmage, 1971). Several factors may significantly increase the use of *B. t.* in future, in particular, the increase in social costs associated with the usage of chemical pesticides and the rapid escalation of direct costs of development and production of petrochemicals.

Production in China

Large-scale production on Chinese communes is either by solid fermentation or liquid fermentation in tanks (Hussey and Tinsley, 1981). Wheat bran, corn meal, soybean, defatted cottonseed cake and peanut bran are the main media components used in *B. t.* production. Wuhan has been the main site for *B. t.* production for many years (Xie Tianjian *et al.*, 1988, 1990). In the pilot plant at Hubei Academy of Agricultural Sciences, production has grown from 26 tonnes in 1983 to 90 in 1984, 160 in 1985, 260 in 1986, 360 in 1987, 472 in 1988, 732 in 1989 to 900 tonnes in 1990. *B. t.* is now widely used in 30 provinces for the control of various insect pests of agriculture and forests as well as vectors of human diseases. Alternative estimates put the production in 1990 at 1500 tonnes (Ziniu Yu and Xixia Luo, personal communication). A portion of the *B. t.* local product is exported to Thailand and South East Asia.

Production in China between 1960 and 1970 started with powder formulations but, by 1980, a flowable formulation had been developed which now represents 70–80% of total *B. t.* production. *B. t. israelensis* mosquito larvicide is produced using 500–7000 litre fermenters. The Chinese claim to have had no problems in practice and, because they have mass produced *B. t.* with elegantly simple methods suitable for farm workers, several techniques have become widespread (Hussey and Tinsley, 1981). More than eight million hectares of farmland are reported to be protected using insect microbial pathogens.

Production in Egypt

In Egypt, increased production of *B. t.* has been a major goal during the last 2 years. Pilot-scale production of promising *B. t.* subspecies, primarily *galleriae* HD-234 and *kurstaki* HD-341, has been carried out in a mobile fermentation unit (5 m^3 capacity) located in a sugar and distillery company in Hawamdia, Giza. This unit represents a prototype fermentation line that can be adjusted and modified (Figure 11.1). The sugar and distillery company produces fodder yeast and molasses which can be used as cheap local by-products for production media. The medium used for the *B. t.* production (g/l) consists of dry fodder yeast (40); molasses (15); K_2HPO_4 (1), $MgSO_4$ (7); and water (0.2). The local products have been used effectively in field trials against *S. littoralis*, greasy cutworm, *Agrotis ipsilon*, *P. gossypiella* and spiny bollworm, *Earias insulana*.

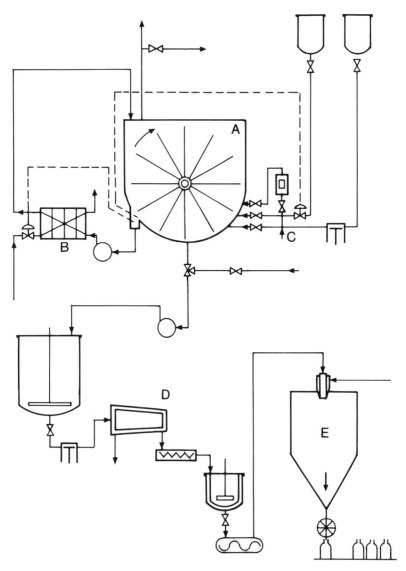

Figure 11.1. Diagrammatic sketch of the fermentation unit for *B. t.* production in the Sugar and Distillery Company, Hawamdia, Egypt. A = paddle-wheel fermenter; B = fedder water chiller; C = compressed air set; D = decanter centrifuge; e = spray dryer

Production in the former USSR

In the former USSR approximately 110 million hectares are under cultivation and biological control methods currently make up about 17% of crop protection techniques. Entomophagous insects are used on about 14 million hectares and bacterial insecticides on about 3 million hectares. In 1987, the Soviet microbiological industry produced some 6100 tonnes of microbial pesticides (*B. t.*, viruses and fungi). The main subspecies used for production were *B. t. caucasicus, thuringiensis, dendrolimus* and *galleriae*. Among problems which inhibit the development of microbial pesticides are the resistance of agricultural workers to new methods, the effects of residual material upon consumers, the high price of these preparations compared to chemical pesticides, poor quality packaging and their lack of adhesiveness after spraying. One of the key problems is quality control.

The republics of the former USSR are seeking advice and adopting Western technology to help resolve problems associated with scaling-up production of microbial pesticides. The USSR Academy of Sciences has recently put the construction of a pilot production plant out to tender to Western companies. Production processes developed at the plant can then be transferred to factories belonging to the microbiological industry, allowing large-scale output to begin (Rimmington, 1989).

Table 11.1. Fermentation media components used for *Bacillus thuringiensis* production in some developing countries, other than Egypt

Country	Media components	Authors
Mexico	Molasses, soybean flour, corn steep liquor, $CaCO_3$ + water	Roldan *et al.* (1988)
South Korea	Fish meal, soybean, red bran, sesame dregs, rice, bran	Yoon *et al.* (1987)
China	Wheat bran, rice husk, lime powder, defatted soybean cake or peanut cake, cottonseed cake or wheat bran and corn meal	Hussey and Tinsley (1981); Wang Tao (1988)
Nigeria	Fermented cassava, ground and whole maize, whole cowpeas	Ejiofar and Okafer (1989);
Brazil	Residues of paper and cellulose industry plus soluble starch	Moscardi (1988)
India	Powder of leguminous seeds or soybean plus soluble starch or molasses	Mummigatti and Raghunathan (1990)

INEXPENSIVE AND LOCAL FERMENTATION MEDIA

The wider use of *B. t.* in developing countries has been restricted for economic reasons whilst the feasibility of local production depends to a large degree on the cost at which it can be produced with high potential insecticidal activity. Different fermentation media including agricultural and industrial by-products have been used for *B. t.* production in several developing countries such as Mexico, South Korea, Nigeria, Brazil and India (Table 11.1). In Egypt studies were initiated to explore the feasibility of producing *B. t.* using fermentation media with cheap, locally available components, including agroindustrial by-products, leguminous seeds, fodder yeast, extracts of potato tubers and citrus peel (Table 11.2) (Salama *et al.*, 1983a,c,d; Foda and Salama, 1986). Most of the preparations derived

Table 11.2. Activity of *Bacillus thuringiensis* subspecies grown on media with inexpensive components (Salama *et al.*, 1983a,d)

		Larval mortality (%)	
B. t. subspecies	Media	*Spodoptera littoralis*	*Heliothis armigera*
kurstaki HD-1	*RSH	55	40
	Beef blood	70	100
	Fodder yeast	65	90
	Peanuts	30	40
	Lima beans	86	40
	Horse beans	75	60
	Soybeans	72	90
kurstaki HD-73	*RSH	45	100
	Beef blood	20	100
	Fodder yeast	35	80
	Fish meal	15	70
	Cottonseed meal	50	60
	Chick peas	81	60
entomocidus HD-635	Kidney beans	90	0
	Chick peas	90	0
	Peanuts	90	0
	Lima beans	70	0
	Horse beans	80	20
	Soybeans	92	20
	Fodder yeast	84	—
kurstaki HD-251	Sweet potato extract	100	—
	Potato extract	100	—
	Fodder yeast (FY)	100	—
	FY + date seeds (2%)	—	100
	FY + minced citrus peel (10%)	—	50
	FY + wheat bran (6%)	—	70

* RSH = residue of slaughterhouses.

from fermentations using leguminous seeds as the sole sources of proteins have a high spore yield associated with high activity against *S. littoralis* and *H. armigera*. Extracts of potato tubers, sweet potato roots, minced citrus peel, ground seeds of dates and wheat bran, can also be successfully used in combination with fodder yeast as media for producing endotoxins active against *H. armigera*. Different protein sources produce widely differing spore yields and activities but there is no simple relationship between spore yield and insecticidal activity. Oxygen availability during fermentation also has a profound effect on viable spore yield and δ-endotoxin activity. A significant interaction of aeration and medium buffering has also been demonstrated (Foda *et al.*, 1985).

LOW TECHNOLOGY FERMENTATION PROCEDURES

In China solid media for the production of *B. t.* is inoculated with a pure *B. t.* culture. After 24 h growth in test tubes the medium reaches the 'solid white state' and is then incorporated in a larger volume of media in litre bottles. Spore count is assessed after 24 h. This inoculum is then mixed with piles of peanut bran and soybean meal, dampened and put in shallow trays. It is ready for use in 30 h. Alternatively, on some communes, this procedure is achieved in 2.5–3 days in trays or soil pits. Fermentation tanks are also used with electric pumps providing filtered air. The medium in this case is 4% peanut bran, 0.5% yeast and 0.5% silkworm pupal-powder suspension in water. The second stage fermentation takes place within 300 litre tanks and is completed in 8 h. This is then used to seed the final fermentation (in 1500 litre tanks), which takes place at room temperature and is complete within 16 h. The final product is forced under pressure through silk filters which are squeezed in a giant press to remove the water. The solids are scraped from the screens and then dried under shelters in the sun. They are then powdered in an electric mill, diluted with chalk and packed into plastic bags for distribution (Hussey and Tinsley, 1981).

In Brazil, the use of *B. t.* has been mainly based on commercially available formulations applied against Lepidoptera on various crops. *B. t. israelensis* is also being used on a very small scale against larvae of Culicidae and Simuliidae. Two Brazilian industries have undertaken small-scale production of experimental formulations. Culture media mainly consist of agro-industrial by-products or residues (Moscardi, 1988). Small-scale industrial production by fermentation of *B. t.* had begun by 1990. The potential for the use of *B. t.* in Brazil is large because of the extensive cultivated area, diversity of crops and high numbers of insect pests which could be controlled.

Other countries have also successfully undertaken pilot production of *B. t.* For example, in Thailand, several formulations (liquid acetone dried, air and spray dried) have been produced in 100 litre batches (Amaret Bhumivatana, 1988). In India, pilot-scale production of *B. t. israelensis* was also successfully accomplished in 100 litre fermentation chambers. Twelve batches of a water-dispersible powder formulation of the bacterial spores and crystals were produced with an average yield of 800–1150 g/batch per 100 litre of medium. The potency of batches in the field was high (Balaraman *et al.*, 1986).

STABILITY AND FORMULATION

Stability in storage

Angus (1956) reported that pH values of more than 12 or lower than 3.3 destroyed the activity of the protein crystal of *B. t.* Nishiitsutsuji-Uwo and Ohsawa (1975), however, found that the activity of *B. t.* was unaffected by exposure to acetic acid (low pH) but it decreased after exposure to sodium or potassium hydroxide at pH 11–12. The effect of pH and temperature on the activity of *B. t. kurstaki* HD-1 has been assessed against *H. armigera* (Ragaei, 1985). Storing a *B. t.* formulation for 2–4 days at pH values between 3 and 13 caused an obvious decrease in potency against this species.

Temperature may also have some effect on *B. t.* The storage of Bactospeine® for 4 years under room conditions ranging between 11 and 37 °C at 41–47% relative humidity caused some reduction in the viability of spores (Afify, 1969). The optimum temperature for the effectiveness of the pathogen was found to be 30 °C in *S. littoralis* and 35 °C in *S. exigua* and *A. ipsilon* (Afify *et al.*, 1969). There was a decrease in the potency of *B. t.* exposed to 55 °C at pH 7–9, but the decrease was more obvious at pH 11 (Ragaei, 1985). Nishiitsutsuji-Uwo *et al.* (1977) reported that no degradation of the activity of *B. t.* occurred up to 75 °C, at pH 6–9 and up to 55 °C at pH 11.

Field stability

The low field persistence of *B. t.* is a major problem in its effective use for pest control. If physical loss of *B. t.* after field application is excluded, solar irradiation would seem to be the major factor affecting *B. t.* persistence on treated plants. In Egypt, studies were made to determine the persistence of different formulations of *B. t.* spores after spray application on cotton. The results showed that the viability of *B. t.* spores declined following application (Salama *et. al.*, 1983b). For example, the half-life of *B. t. entomocidus* was 109 h in May and this decreased to 89 h in June, when the number of hours of sunlight reached a maximum. In the last week of July, the spore half-life decreased further to 61 h at temperatures ranging between 33 °C and 45 °C on exposed cotton leaves. The results indicate that the half-life of the spores is not consistently correlated with the temperatures attained on cotton leaves (Salama *et al.*, 1983e). Different formulations of *B. t.* vary in the degree to which they afford radiation protection to spores. Relative humidity fluctuation showed no obvious correlation with the spore viability. Addition of phagostimulants or adjuvants, such as extracts of cotton plant leaves and Jews' mallow, molasses and Coax®, had no effect on protecting *B. t.* from ultraviolet (UV) irradiation (Salama *et al.*, 1983e). It has been reported, however, that Coax® and Gustal® act as phagostimulants and as protectants when used against *S. littoralis* (Sneh and Gross, 1983). The concentration of the formulations tested had no consistent correlation with their persistence.

Enhancement of field stability

Additives are required to achieve the desired physical properties of agricultural formulations of *B. t.* including ability to flow, wet, disperse and suspend with little foam and also to have stable, physical storage. Dilution of the commercial formulation into a spray for use on a target insect usually requires water and also additives to increase persistence. Typical additives are wetting agents, which affect the physicochemical interaction of the formulations, and plant surface stickers, which improve weathering by forming a film to retard wash-off by rain. Thickeners can also be added to reduce evaporation, and humectants to retard desiccation during spraying, as well as phagostimulants and UV protectants. Some substances can serve more than one function.

In order to overcome the short persistence of *B. t.* caused by solar UV, combination with substances with a high degree of UV absorbance is valuable. Some of these substances provide good protection and maintain spore viability in the laboratory and field. Evaluation of 30 potentially protective materials revealed that 6% sodium lignosulphonate (Orzan®) was strongly protective. Molasses in combination with Orzan® also proved to be a good UV protectant. Congo red, starch-encapsulated *B. t.*, charcoal, chitin and oxybenzone, peptonized milk (5%), brewer's yeast (5%) and egg albumen (5%) singly or combined with brewer's yeast, gave a high degree of protection against UV (Ragaei, 1985).

FIELD APPLICATION

Methodologies

Pneumatic knapsack sprayers are the most widely used method of ground application. The spray is delivered via a lance and nozzle to the crop.

Application volumes are usually 100 litres/ha or greater. The lance can direct the spray onto the crop, but underleaf cover may be poor; in such cases a drop lance with some upwardly directed nozzles is useful. Other relatively high volume sprayers include the motorized back sprayers. These deliver 200–600 litres/ha. At these volumes, plants are sprayed to run off so that much of the active ingredient drips off on to the soil. Farmers in most developing countries will continue to use knapsack sprayers since they are readily available and can be used on several different crops. Using machines producing smaller droplet-sizes can result in improved crop coverage and underleaf cover, as well as reduced spray volumes. The optimal choice of a sprayer will depend on crop type, whilst a study of pest biology will indicate where spray coverage is most needed.

Case histories

In the People's Republic of China the farmers select their own methods of production and application of *B. t.* aided by regional centres that supply inocula and advice. The first large field trials with *B. t.* were organized by the Institute of Zoology in Peking against *Dendrolimus punctatus* on pine (*Pinus* sp.) in 1959. *B. t. wuhanensis* has been used since 1974 to control more than 40 pests in various provinces (Hussey and Tinsley, 1981). *B. t.* is used in more than 30 provinces in south-central China and on more than 16 million mu (1 mu = 0.066 ha) against a wide range of pests including rice caseworm (*Nymphula depunctalis*), yellow rice borer (*Tryporyza incertulus*), rice stem borer, cabbage butterfly (*Pieris rapae*), diamondback moth (*Plutella xylostella*), pine caterpillar, oriental tobacco budworm (*Heliothis assulta*), *Plusia agrata*, *Spodoptera litura* and *S. exigua* (Hussey and Tinsley, 1981; Wu *et al.*, 1987). Luo *et al.* (1986) recommend the use of commercial formulations of *B. t. wuhanensis*, *morrisoni* and *kurstaki*, singly or in combination with Sumicidin®, for the control of *P. gossypiella* on cotton. The application of *B. t. israelensis* is widespread in China for the control of various mosquito species (see Becker and Margalit, Chapter 7).

Large-scale field experiments to evaluate the potency of *B. t.* formulations in comparison with conventional chemical insecticides (pyrethroids or carbamates) and in combinations, for the control of *S. littoralis* and *S. exigua* on soybeans, have also been carried out in Egypt. One spray with Dipel 2X® or fenvalerate gave an obvious reduction in larval counts of *S. exigua* associated with a significant increase in crop yield. Treatments with combinations of *B. t.* and fenvalerate (62 g and 50 ml per 0.4 ha) showed high potential and caused a 2.8-fold increase in crop yield. Dipel 2X® (250 g/0.4 ha) was as effective as Lannate® (methomyl) (300 g/0.4 ha) in controlling *S. littoralis* on soybeans. When the treatment was applied once, the larval population was reduced and the yield was 1.1 times the control. When the treatment was applied three times at 4 to 5 week intervals, the yield per 0.4 ha was higher, showing a 2.0-, 2.1- and 2.1-fold increase after treatment with Dipel 2X® (250 g), its combination with molasses 20% or Lannate® (300 g), respectively (Salama et. al., 1990b,d).

The effectiveness of wheat-bran baits based on B. t. kurstaki HD-1 compared with the organophosphorous insecticide Hostathion® used against A. ipsilon on horse beans (Vicia fabae) was also examined in field experiments (Salama et al., 1990a). Baits based on Dipel 2X® (375 or 625 g/ha) combined with or without K_2CO_3 (375 g/ha) proved to be as efficient as baits based on Hostathion® (Table 11.3). The incorporation of chemical additives such as calcium carbonate, calcium oxide, calcium sulphate and zinc sulphate (125–375 g/ha) in baits also potentiated the effectiveness of Dipel 2X® against A. ipsilon and led to significant yield increases in some vegetable crops including lentils, peas, egg plants, okra, chillies and potatoes (Salama et al., 1990c).

A large-scale field programme, implemented in Egypt in 1990, assessed the potential of several B. t. formulations against pests on cotton plants. The efficacy levels of baits based on B. t. against A. ipsilon, and spray application against S. littoralis, P. gossypiella and E. insulana, were highly encouraging. The yield of cotton after a spray application with B. t. was almost equal to that obtained with conventional chemical insecticides (672.8 kg/ha compared with 697.4 kg/ha). However, the cost of treatment

Table 11.3. Assessment of yield of horse beans following combinations of Bacillus thuringiensis and chemical insecticides in baits against Agrotis ipsilon (Salama et al., 1990a)

Treatments and dose/hectare	Mean yield kg/0.01 ha ($\pm$ s.e)	Increase in yield
Control bait (A):		
Molasses (7.5 litres) + wheatbran (62.5 kg)	13.40 $\pm$ 1.569[a]	—
Chemo-bait:		
A + Hostathion[c] (3.75 litres)	20.44 $\pm$ 1.736[b]	1.56
Bio-bait:		
(A + 625 g Dipel 2X®) = B	19.38 $\pm$ 1.504[b]	1.45
(A + 375 g Dipel 2X®) = C	21.24 $\pm$ 2.022[b]	1.59
Bio-bait + K_2CO_3:		
B + 375 g	20.60 $\pm$ 1.927[b]	1.54
C + 375 g	21.40 $\pm$ 2.083[b]	1.60

[a,b] Means in a column followed by the same letter are not significantly different ($p < 0.05$) using Duncan's multiple range test.
[c] Triazophos.

with commercial *B. t.* products was higher than that for chemical insecticides (Salama *et al.*, unpublished data). This favours the concept of cheap local production of *B. t.*

In Israel *B. t.* has been used to control *Boarmia* (*Ascotis*) *selenaria* caterpillars in avocado groves since 1977 (Wysoki *et al.*, 1981; Wysoki and Izhar, 1986). In another set of trials, nine chemical insecticide applications (seven sprays plus two baits) were required to control *S. littoralis* on alfalfa, whereas five applications of *B. t. entomocidus* (three spray plus two baits) produced the same effect (Broza *et al.*, 1986).

The application of *B. t.* has also been reported from Indonesia and Malaysia (Iman *et al.*, 1986; Brunner and Stevens, 1986). In Pakistan, India and West Africa, tests with *B. t.* against many pests are also underway. Larvae of the potato tuber moth, *Phthorimaea operculella*, were controlled effectively by *B. t. kenyae*, which was shown to retain its potency against this pest for up to 60 days in India (Amonkar *et al.*, 1979). *B. t.* has also been used on coffee in Kenya to control *B. selenaria* (Waikwa and Mathenge, 1977).

B. t. israelensis has been found to be very useful as a larvicide for some Diptera (de Barjac, 1987), particularly for the control of mosquito and blackfly larvae. Tons of commercial *B. t. israelensis* products are applied each year to control *Culex*, *Anopheles*, *Aedes* and *Simulium* larvae thus helping to prevent the transmission of such tropical diseases as filariasis, malaria, dengue, various types of encephalitis, yellow fever and onchocerciasis (see Becker and Margalit, Chapter 7).

FURTHER DEVELOPMENT OF *B. THURINGIENSIS*

Research on the genetic manipulation of *B. t.* in developing countries is limited but there have been some attempts to broaden its host range. In Egypt, for example, Salama *et al.* (1991, 1992) were able to obtain *Azotobacter chroococcum* strains that acquired the *B. t.*. δ-endotoxin gene using phage DNA isolated from *B. t.* Some of the transformants obtained were highly potent against *S. littoralis*. These strains were also able to fix nitrogen and so could possibly be used to benefit both pest control and crop nutrition. Salama *et al.* (1984a,b, 1991) also attempted to develop high temperature and UV-resistant *B. t.* strains and to determine the effect of this resistance on their potency against insects. The UV-resistant strains showed high toxicity.

In China, Li (1986) studied the pattern and transfer of plasmids carrying the crystal protein gene of 19 strains of *B. t.*. Wu (1986) found that out of 865 strains, isolated from mutants of *B. t.* 7216 and treated with UV, 512 had a glossy, milky white and lustre surface and few or no crystals were produced, while 353 strains had a pale dull and rough surface and produced crystals in rhombic, square or irregular shapes. The virulence of the glossy crystal strains was significantly lower than that of the parent strains, as well as the standard *kurstaki* HD-1, against *H. armigera* and *Spodoptera* (*Prodenia*) *litura* and *P. gossypiella*, whereas that of the rough strains was higher.

Several collaborative research programmes have been run between Latin American countries and elsewhere (Ibarra *et al.*, 1988). One of the successful examples of this collaboration is the UN supported project on the development of crop plants made resistant to insect attack by incorporation of genes encoding *B. t.* δ-endotoxins. Argentina is providing local *B. t.* strains and assaying their insecticidal efficacy, Cuba is developing immunoassays and constructing synthetic probes, and Mexico is cloning the toxin genes, developing the methodology to transform maize and sugar cane and constructing vectors for expression in plants. The reported success of different *B. t.* toxins expressed in transgenic plants (Vaeck *et al.*, 1987), and the close relationship with Plant Genetics Systems, Belgium, encouraged researchers in Latin America to follow similar approaches. Reasonable success has been reported in cloning a *B. t.* K45 KB-type gene using *Agrobacterium tumefaciens* transformation. In Mexico plants have been successfully transformed and have been shown to kill the larvae of tobacco hornworm, *Manduca sexta*. Further ongoing work is directed towards food crops such as tomato and potato.

FUTURE ROLE FOR *B. THURINGIENSIS* IN DEVELOPING COUNTRIES

B. t. is a desirable agent for insect pest control and ideal for use in developing countries because of its low production cost and lack of toxicity. The use of *B. t.* in developing countries is still, however, very limited, although the increasing production levels in some countries may eventually replace chemical insecticides The possibility of increasing regional production using local, inexpensive material is a particularly attractive option and more effort is needed to improve yield, efficacy, application and assessment technology. Genetic modification of *B. t.* is also expected to play an increasingly important role, particularly in the expansion of host range.

ACKNOWLEDGEMENTS

The authors express sincere thanks to the International Development Research Centre (IDRC) Canada for sponsoring some of the research work carried out in Egypt and included in this Chapter. The technology for *B. t.* production in China has been abstracted from Hussey and Tinsley (1981), to whom the authors are indebted.

REFERENCES

Abdel-Rahman, H. A. (1966). 'Study of the pathogenicity of crystalline inclusion of *Bacillus thuringiensis*'. *Ain-Shams Sci. Bull.*, **10**, 89–95.

Abul-Nasr, S., Amar, E., Merdan, A. and Farrag, S. (1979). 'Infectivity tests on *Bacillus thuringiensis* and *B. cereus* isolated from resting larvae of *Pectinophora gossypiella*'. *Z. Ang. Entomol.*, **88**, 60–69.

Afify, A. M. (1969). 'Effect of storage of Bactospeine under varying room conditions on the viability and virulence of the spores'. *Entomophaga*, **14**, 215–223.

Afify, A. M., Abdel-Rahman, H. A. and Atwa, W. (1969). 'Combined effect of temperature and humidity on the response of *Pieris rapae* to a bacterial insecticide'. *Ain-Shams Sci. Bull.*, **13**, 271–278.

Amaret Bhumivatana (1988). 'Local production of *Bacillus thuringiensis* for controlling insect pests and vectors'. *International Symposium on insecticide of Bacillus thuringiensis, Hubei Academy of Agricultural Sciences, Wuhan*. October 1988.

Amonkar, S., Pol, A. K., Vijayalakshimi, I. and Rao, A. S. (1979). 'Microbial control of potato tuber moth *Phthorimaea operculella* Zall'. *Indian J. Exp. Biol.*, **17** (10), 1127–1133.

Angus, T. A. (1956). The biochemistry and mode of action of *Bacillus thuringiensis berliner* and its varieties. Entomophaga, **2**, 165–173.

Balaraman, K., Bhatia, M. C., Hoti, S. I. and Trpathin, S. C. (1986). 'Pilot scale production and evaluation of *Bacillus thuringiensis* H-14'. *Indian J. Med. Res.*, **83**, 462–465.

Broza, M., Sneh, B. and Levi, M. (1986). 'Evaluation of the effectiveness of *Bacillus thuringiensis* var. *entomocidus* as a pest control agent to replace chemical pesticides in alfalfa fields in Israel'. *Anz. Schadlingskde Pflanzenschutz, Umweltschutz*, **9**(18), 152–156.

Brunner, E. and Stevens, P. F. (1986). 'The control of diamondback moth with Thuricide'. *Proc. 1st International Workshop*, Taiwan, 1985, pp. 213–217.

Chen Tao Liang, Shiping, Chen Aongsheng, Luo Chebn, Yuac Li, Lei Zasheng and Xie Baiging (1988). 'A new subspecies of *Bacillus thuringiensis* isolated in China; *Bacillus thuringiensis* subsp. *pectinophorae* and its effect on pea pests'. *International Symposium on Insecticide of Bacillus thuringiensis, Hubei Academy of Agricultural Sciences, Wuhan*, October 1988.

de Barjac, H. (1987). 'Operational bacterial insecticides and their potential for future improvement'. In *Biotechnology in Invertebrate Pathology and Cell Culture* (Ed. K. Maramorosch), pp. 63–73, Academic Press, London.

Dulmage, H. T. (1971). 'Economics of microbial control'. In *Microbial Control of Insects and Mites* (Eds. H. D. Burges and N. W. Hussey), p. 581, Academic Press, London.

Ejiofar, A. O. and Okafer, N. (1989). 'Production of mosquito larvicidal *Bacillus thuringiensis* serotype HD-14 on raw material media from Nigeria'. *J. Appl. Bacteriol.*, **67** (1), 5–10.

Foda, M. S. and Salama, H. S. (1986). 'New fermentation technologies for the production of bacterial insecticides'. *Zbl. Microbiol.*, **141**, 151–157.

Foda, M. S., Salama, H. S. and Selim, A. (1985). 'Factors affecting growth physiology of *Bacillus thuringiensis*'. *Eur. J. Microbiol. and Biotechnology*, **22**, 50–52.

Goldberg, L. J. and Margalit, J. (1977). 'A bacterial spore demonstrating rapid larvicidal activity against *Anopheles sergentii*, *Uranotaenia unguiculata*, *Culex univitatus*, *Aedes aegypti* and *Culex pipiens*'. *Mosquito News*, **37**, 353–358.

Hussey, N. W. and Tinsley, T. W. (1981). 'Impressions of insect pathology in the People's Republic of China'. In *Microbial Control of Pests and Plant Diseases (1970–1980)* (Ed. H. D. Burges), pp. 785–795, Academic Press, London.

Ibarra, J., Alba, J., Xconostle, B. *et al.* (1988). 'Transgenic plants with novel *Bacillus thuringiensis* toxin genes for insect pest control in Latin America'. *Biotech. Biological Pesticides and Novel Plant-Pest Resistance for Insect Pest Management*, Proceedings of a conference (Eds. D. W. Roberts and R. R. Granados), organized by Insect Pathology Resource Center, Boyce Thompson Institute for Plant Research, at Cornell University, Ithaca, NY 14853, USA, July 18–20, 1988.

Iman, M., Scekarna, D., Situmorang, J., Adiputra, I. M. and Manti, I. (1986). Effect of insecticides on various field strains of diamond moth and its parasitoid in Indonesia. Diamond Back moth management. *Proc. 1st International Workshop*, Taiwan, March 1985.

Li, R. S. (1986). 'The plasmids carrying protein gene of *Bacillus thuringiensis* and their transfer between various strains'. *Acta Microbiol. Sinica*, **26**(2), 143–150.

Luis, J., Galan, C. R., Padilla, H., de Barjac, R. S., Tamez-Guerra, L. R. T., Maxino, R. C., Dulmage, H. and Hiram, R. (1988). 'Production and toxicity of strains of *Bacillus thuringiensis* against plague insects of agricultural and medical importance'. *Symposium on Insecticide of Bacillus thuringiensis, Hubei Academy of Agricultural Sciences, Wuhan*.

Luo, S. B., Van, H., Chai, C. J., Liang, S., Zhang, Y.M ., Zhang, Y. and Le G. (1986). Control of pink bollworm, *Pectinophora gossypiella* with *Bacillus thuringiensis* in cotton fields'. *Chinese J. Biol. Control.*, **2**(4), 167–169.

Manomani, A. M., Hoti, S. L. and Balarman, K. (1987). 'Isolation of mosquito pathogenic *Bacillus thuringiensis* strains from mosquito breeding habitats in Tamil Nadu India'. *Indian J. Med. Res.*, **86**, 462–468.

Moscardi, F. (1988). 'Production and use of entomopathogens in Brazil'. *Biotech. Biological Pesticides and Novel Plant Pest-Resistance for Insect Pest Management*, Proceedings of a conference (Eds. D. W. Roberts and R. R. Granados), organized by Insect Pathology Resource Center, Boyce Thompson Institute for Plant Research, at Cornell University, Ithaca, NY14853, USA, July 18–20, 1988.

Mummigatti, S. G. and Raghunathan, A. M. (1990). 'Influence of media composition on the production of *delta* endotoxin by *Bacillus thuringiensis* var. *thuringiensis*'. *J. Invert. Pathol.*, **55**(2), 147–151.

Nishiitsutsuji-Uwo, J. and Ohsawa, A. (1975). 'Effect of alkaline and acid solutions on insecticidal activity of *Bacillus thuringiensis*'. *Batvu-Kagaka*, **40**, 96–102.

Nishiitsutsuji-Uwo, J., Ohsawa, A. and Nishimura, M. S. (1977). 'Factors affecting the insecticidal activity of δ-endotoxin of *Bacillus thuringiensis*'. *J. Invert. Pathol.*, **29**, 162–169.

Padua, L. E. (1988). '*Bacillus thuringiensis* isolated from the Philippines with vector control potential'. *International Symposium on Insecticide Bacillus thuringiensis, Hubei Academy of Agricultural Sciences, Wuhan*, October 1988.

Ragaei, M. (1985). 'Studies on the effect of some environmental and chemical factors on the potency of *Bacillus thuringiensis* against some cotton pests'. M.Sc. Thesis. Faculty of Science, Cairo University, Egypt.

Rimmington, A. (1989). 'The production and use of microbial pesticides in the USSR'. *Int. Ind. Biotechnol.*, **9**(5), 10–14.

Roldan, M., Solis, I., Perez, O., Casillas, A., Vega, N., Cordova, V., Peraza, L., Garcia, G., Padilla, R. D., Galan-Wong, L. and Tamez Guerra, T. (1988). 'Production of a bioinsecticide from a native *Bacillus thuringiensis* var. *kumamotoensis* and its application against insect pests of maize'. *International Symposium on*

Insecticide Bacillus thuringiensis, Hubei Academy of Agricultural Sciences, Wuhan, October 1988.

Salama, H. S., Foda, M. S., Dulmage, H. T. and El-Sharaby, A. (1983a). 'Novel fermentation media for production of δ-endotoxin for *Bacillus thuringiensis*'. *J. Invert. Pathol.*, **41**, 8–19.

Salama, H. S., Foda, M. S. and El-Sharaby, A. (1983b). 'Biological activity of mixtures of *Bacillus thuringiensis* against some cotton pests'. *Z. Ang. Entomol.*, **95**, 69–71.

Salama, H. S., Foda, M. S. and Selim, M. (1983c). 'A novel approach for whey recycling in production of bacterial insecticides'. *Entomophaga*, **23**(2), 151–160.

Salama, H. S., Foda, M. S. and Selim, M. (1983d). 'Utilization of fodder yeast and agro-industrial byproducts in production of spores and biologically active endotoxins from *Bacillus thuringiensis*'. *Zbl. Mikrobiol.*, **138**, 553–563.

Salama, H. S., Foda, M. S., Zaki, F. N. and Khalafallah, M. (1983e). 'Persistence of *Bacillus thuringiensis* Berliner spores in cotton cultivation'. *Z. Ang. Entomol.*, **95**, 321–326.

Salama, H. S., Foda, M. S. and Selim, M. (1984a). 'Mutation in relation to sporulation and potency of *Bacillus thuringiensis* vs. cotton pests'. *Z. Ang. Entomol.*, **97**, 29–35.

Salama, H. S., Foda, M. S. and Selim, M. (1984b). 'Isolation of *Bacillus thuringiensis* mutants resistant to physical and chemical factors'. *Z. Ang. Entomol.*, **97**, 139–145.

Salama, H. S., Foda, M. S., Zaki, F. N. and Ragaei, M. (1986). 'On the distribution of *Bacillus thuringiensis* and closely related *Bacillus cereus* in Egyptian soils and their activity against lepidopterous cotton pests'. *Z. Ang. Zool.*, **73**, 257–265.

Salama, H. S., Moawed, S., Salah, R. and Ragaei, M. (1990a). 'Field tests on the efficacy of baits based on *Bacillus thuringiensis* and chemical insecticides against the greasy cutworm *Agrotis ypsilon* Rdf. in Egypt'. *Anz. Schadlingskde Pflanzenschutz Pflanzeuschutz Umweltschutz*, **63**, 33–36.

Salama, H. S., Saleh, M. S., Moawed, S. and Shams El-Din, A. (1990b). 'Evaluation of microbial and chemical insecticides for the control of *Spodoptera exigua* on soybean plants'. *Anz. Schadlingskde Pflanzenschutz Umweltschutz*, **63**, 100–102.

Salama, H. S., Salem, S., Zaki, F. N. and Matter, M. (1990c). 'Control of *Agrotis ypsilon* (Hofn.) on some vegetable crops in Egypt using the microbial agent *Bacillus thuringiensis*'. *Anz. Schadlingskde Pflanzenschutz Umweltschutz.*, **63**, 147–151.

Salama, H. S., Zaki, F. N., Salem, S. A. and Shams El-Din, A. (1990d). 'Comparative effectiveness of *Bacillus thuringiensis* and lannate against *Spodoptera littoralis*'. *J. Islam Acad. Sci.*, **3**, 325–329.

Salama, H. S., Ali, A. and Sharaby, A. (1991). '*Bacillus thuringiensis* resistant to high temperature and ultraviolet radiation'. *Z. Ang. Entomol.*, **112**(5), 520–524.

Salama, H. S., Ali, A. and Sharaby, A. (1992). 'The cloning of *Bacillus thuringiensis* endotoxin gene in genetically stable *Azotobacter chroococcum* transformants'. *Discovery and Innovation* (in press).

Skaikh, M. R., Nagui, B. S., Shailh, D. and Khan, A. F. (1987). 'Distribution of *Bacillus thuringiensis* serotypes in Pakistan'. *J. Sci. Ind. Res.*, **29**(4), 295–296.

Sneh, B. and Gross, S. (1983). 'Biological control of the Egyptian cotton leaf-worm, *Spodoptera littoralis* Boisd. (Lep., Noctuidae) in cotton and alfalfa field using a preparation of *Bacillus thuringiensis* supplemented with adjuvants'. *Z. Ang. Entomol.*, **95**, 418–424.

Tianjian, A. and Wang, B. (1988). 'Production and application of *B. t.* insecticide in China'. International Symposium of Insecticide *Bacillus thurgiensis*, Hubei Academy of Agricultural Sciences, Wuhan, October 1988.

Tianjian, A., Bingao, H. and Wu, G. (1990). 'Commercial production and application of *B. t.* insecticide in China'. *Proceedings Vth International Colloquium Invertebrate Pathology, Adelaide, Australia*, p. 16.

Vaeck, M., Reynaerts, A., Höfte, H., Jansens, S., Beuckleleer, C., Zabeau, M., Montagu, V. and Leeman, J. (1987). 'Transgenic plants protected from insect attack'. *Nature (London)*, **328** (6125): 33–37.

Waikwa, J. W. and Mathenge, W. M. (1977). 'Field studies on the effect of *Bacillus thuringiensis* on the larvae of the giant coffee looper *Ascotis selenaria reciprocaria* and its side effects on the larval parasites of the leaf miner (*Leucoptera* spp.)'. *Kenya Coffee*, **42**, 95–101.

Wang Tao (1988). 'A simple and economic semi-solid fermentation process for *Bacillus thuringiensis*'. *International Symposium on Insecticide of Bacillus thuringiensis, Hubei Academy of Agricultural Sciences, Wuhan*, October 1988.

Wang Ying, Wen Jie and Wang Yuzhen (1988). 'Presence of flagellar antigenic subfactors in serotype 1 of *Bacillus thuringiensis* with the description for new serotype *B. t.* var. *hubeinsis*'. *International Symposium on Insecticide of Bacillus thuringiensis, Hubei Academy of Agricultural Sciences, Wuhan*, October 1988.

Wu, H. F., Yang, X. M. and Chen, Y. P. (1987). 'The control of vegetable pests with *Bacillus thuringiensis* emulsion'. *Chinese J. Biol. Control*, **3**, 141.

Wu, J. A. (1986). 'Selection in *Bacillus thuringiensis* (7215) by induced mutation'. *Natural Enemies of Insects*, **8**(1), 14–17.

Wysoki, M. and Izhar, Y. (1986). 'Fluctuation of the male population of *Boarmia selenaria* estimated by virgin female baited traps'. *Acta Oecol. Appl.*, **7**, 251–259.

Wysoki, M., Swirski, E. and Izhar, Y. (1981). 'Biological control of Avocado pests in Israel'. *Prot. Ecol.*, **3**, 25–28.

Xie Tianjian, J., Wang, B., Zhong, L. and Wu, G. (1988). 'Production and application of *B. t.* insecticide in China'. *International Symposium on Insecticide Bacillus thuringiensis, Hubei Academy of Agricultural Sciences, Wuhan*. October 1988.

Xie Tianjian, J., Jung, B. G., Zhong, L. S. and Wu, G. X. (1990). 'Commerical production and application of *B. t.* insecticide in China'. In *Proceedings Vth International Colloq. Invert. Pathol. Adelaide, Australia*, p. 16.

Yoon, K. H., Han, M. D. and Yu, H. S. (1987). 'Production of the spore crystal complex *Bacillus thuringiensis* var. *israelensis* with some natural media'. *Korean J. Entomol.*, **17**(4), 237–243.

12 Production of *Bacillus thuringiensis* Insecticides for Experimental and Commercial Uses

K. BERNHARD AND R. UTZ

Ciba-Geigy AG, Plant Protection Division, CH-4002 Basel, Switzerland

INTRODUCTION

The potential of *Bacillus thuringiensis* (*B. t.*) for biological pest control was first recognized by Berliner (1915). Since then it has received considerable attention in the research community. To date it is the most successful bioinsecticide and estimated to account for 80–90% of all biological pest control agents sold worldwide. Nevertheless, *B. t.* insecticides have not yet made much impact and only account for a 1–2% share of the total insecticide market. Although entry of *B. t.* insecticides into the market has been slow, trends in recent years appear to be working in their favour: (a) In some crop/pest situations *B. t.* products are as efficacious as chemical standards. (b) The work of Dulmage and collaborators (1981) showed that *B. t.* strains active against lepidopteran larvae differ considerably in potency and insecticidal spectra. This triggered studies to identify more potent strains than those currently used in commercially available products. (c) Discovery of strains active against Diptera (Goldberg and Margalit, 1977) and Coleoptera (Krieg *et al.*, 1983) demonstrated that the spectra of potential uses is wider than previously thought. (d) Increasing public concern about residues of pesticides in food and in the environment, as well as shifts in agricultural practice towards integrated pest management, are leading to a more favourable evaluation of the merits of *B. t.* insecticides in comparison with chemical insecticides. (e) Industry is faced with increasing requirements for registration of new synthetic insecticides. Due to their impeccable safety record, registration requirements for *B. t.* insecticides are currently much lower. This makes development of *B. t.* insecticides economically feasible, although their selectivity makes

Bacillus thuringiensis, An Environmental Biopesticide: Theory and Practice. Edited by P. F. Entwistle, J. S. Cory, M. J. Bailey and S. Higgs
© 1993 John Wiley & Sons Ltd

their market sizes generally much smaller than those for chemical insecticides. In future, the market share of B. t. insecticides is expected to reach 5–10% of the total insecticide market by the year 2000.

Current production outside the former USSR and China is estimated at 2000–3000 t/year. If the predicted increase in market share materializes production of B. t. insecticides is expected to approach the same level as classical biotechnological products such as enzymes or antibiotics. In striking contrast to these forecasts is the lack of information on mass production of B. t. available in the published literature. A search of a computer database showed that out of more than 1800 publications on B. t. published between 1982 and 1990 only 32 papers dealt with aspects of mass production. The majority of them were by authors affiliated with research institutions in developing nations or the former USSR. There are, however, several explanations for this apparent lack of interest. Details of a production process are either not patentable at all, or cannot be sufficiently protected by patents. In a competitive situation it is therefore preferable for manufacturers to guard them as trade secrets, rather than reveal them to the public. Researchers at academic institutions tend to conduct larger field trials with commercially available preparations or experimental preparations obtained from industrial collaborators. For laboratory trials or small-scale field trials, standard laboratory media and equipment such as shaker flasks or 10-litre laboratory fermenters are usually sufficient to provide the required amounts of material.

OPTIMIZATION OF PRODUCTION METHODS

B. t. grows in many standard laboratory media. None of them will, however, give satisfactory results with every strain. Most allow cell densities of only 10^8 cells/ml to 10^9 cells/ml at reasonable sporulation rates. With industrial mass production, on the other hand, cell densities of more than 5×10^9 cells/ml at sporulation rates of more than 90% are achieved. These improvements are only possible because the culture conditions are optimized according to the specific needs of the strain to be produced. Accordingly, production parameters optimized for efficient production of one strain may not give satisfactory results when applied to production of other strains. Therefore the optimization process has to be carried out for each strain being produced on a large scale, which is laborious and time consuming. Once mass production of a strain has been decided upon, the cost of optimizing production, harvesting, formulation and packaging are then justified. For experimental uses, however, where only small quantities are needed, it is very often impossible to devote much time and labour to optimization of production. To overcome this conflict, a strain should be tested early in development against only a limited number of different standard media for the selection of the most suitable one. If not enough material for the intended trials is obtained in a single batch, it is usually easier to increase the number of batches produced, or to use a larger fermenter, rather than carrying out further media optimization. Within reasonable limits, δ-endotoxin concentration is not critical in preparations for experimental uses since rates for field trials may be adjusted to correct suboptimal δ-endotoxin concentrations if required.

Extensive media optimization is only carried out with strains selected for mass production. The decision to start optimizing production of a strain should, however, not be delayed too long. In many countries safety testing for registration must be done with 'typical batches', produced with the same procedure and media components, although not the same scale, as commercial production. Therefore media optimization should be completed before registration trials are initiated. Later changes in the media composition may require further safety tests, causing delays in development and additional costs.

Growth requirements

Although strains of *B. t.* differ, they all share some common properties. It is therefore possible to define some basic requirements for production of *B. t.* Production of δ-endotoxin occurs only during sporulation, so optimal sporulation is essential. Under anaerobic conditions, growth of *B. t.* is slow and sporulation may be inhibited; good oxygen supply is therefore an essential requirement. All strains are good producers of amylases and proteinases, which allows them to utilize a wide range of raw materials as substrates. Other exoenzymes, e.g. lecitinases and haemolytic enzymes, are also produced. Growth occurs between 15 °C and 45 °C, with the optimum between 26 °C and 30 °C. High frequency plasmid loss was induced in *B. t.* subspecies *israelensis* by cultivating it at 43 °C (Gonzalez and Carlton, 1984). Cultivation at higher temperatures should therefore be avoided. Typically, temperature for all productions is kept at 28 °C. *B. t.* is not particularly sensitive to pH and growth will occur between pH 5.5 and pH 8.5 with the optimum between pH 6.5 and pH 7.5.

Glucose is catabolized by several pathways with the major intermediates being acetate, lactate, pyruvate and acetoin (Nickerson *et al.*, 1974; Benoit *et al.*, 1990). With some strains large amounts of acids are excreted during early logarithmic growth which may require pH control during fermentation. Some strains also utilize galactose and mannose. Granules of poly-β-hydroxybutyrate, stimulated by an excess of sugars in the media, are produced by most strains during late-log and early stationary phases. The granules are virtually completely consumed during sporulation, but may be mistaken for the parasporal crystals of irregular spherical shape which are produced by some strains.

Parameters to optimize production

An important criterion for the efficiency of a production process is the amount of δ-endotoxin produced per batch, which is dependent on cell density, sporulation rate and the quantity of δ-endotoxin produced per sporulating cell. Judged by the size of the parasporal inclusion bodies, for most strains, production of δ-endotoxin is much less sensitive to changes in environmental conditions than cell density and sporulation rate. Cell density and sporulation are easily measured by counting vegetative and sporulating cells in a counting chamber under a phase contrast microscope, which allows rapid examination of many samples. This method is only slightly hampered by the presence of other particles in the culture media, e.g. non-degradable media components. Sporulation rates should be counted before lysis of the sporangia. Due to their hydrophobic surfaces, liberated spores tend to attach to foam and glass surfaces, resulting in reduced spore counts. δ-Endotoxin may be directly quantitated by bioassays or biochemically by ion exchange chromatography (Bernhard, 1992), electrophoresis (Brussock and Currier, 1990) or immunological methods (Groat et al., 1990). Bioassays take 1–6 days, biochemical assay methods several hours and microscopic investigation only a few minutes. Therefore use of microscopy is preferable when large numbers of samples need to be analysed or rapid analysis is required.

PRODUCTION METHODS FOR B. THURINGIENSIS

Since δ-endotoxins are synthesized during sporulation, which occurs in the stationary growth phase, batch production is the obvious method of choice. Continuous culture methods have not been used for mass production. In mass production of B. t. insecticides two approaches have been used: (a) decentralized production in small facilities located close to the users, as implemented in some developing countries; and (b) centralized production at large facilities, which is the predominant practice in industrialized countries. The concept of having farmers produce their own B. t. insecticides with locally available raw materials is attractive for countries which otherwise depend upon imports from developed countries for their insecticide supply, particularly where traditional fermentations, e.g. in food processing are done. As B. t. has similar growth requirements to other bacilli, e.g. B. cereus, mass production only succeeds under sterile conditions, which requires skilled labour supervised by competent microbiologists. If infections occur, it is important to recognize them early enough to prevent release of inactive material and to localize the source of infection quickly, in order to minimize interruptions in the production process. Since decentralized production needs more facilities, more skilled personnel are needed. Equipment costs and operating expenses are also likely to be higher. The price of equipment needed for production does not increase equally with size. Since a 100-litre fermenter is only three to four times more expensive than a 10-litre fermenter, it is more economical to produce one batch in a large facility, rather than producing several small batches in smaller facilities. Additionally, the more batches produced, the greater the amount of testing required for quality control. Some advantages may, however, result from decentralized production: products can be made upon request, which reduces requirements for storage stability, which in turn may reduce or eliminate requirements for formulation altogether. An extensive infrastructure for distribution is not required either.

A current example of decentralized production of *B. t.* exists in Cuba where large, agricultural cooperatives operate laboratories for the production of *B. t.* and other biological control agents. *B. t.* is cultivated in 1 litre batches in sterilized liquid media in static culture. Presumably because of the low oxygen supply, it takes more than 1 week to complete sporulation. The seed cultures are supplied by a central laboratory, which also maintains stock cultures. The material is stored in sterile flasks and applied to the field without further formulation (J. Brassel, personal communication; also see Salama and Morris, Chapter 11).

In biotechnology two types of production methods are used: (a) production by surface cultures on semisolid media and (b) production by liquid media in submerged culture. With semisolid media, nutrients are contained in a coarse porous matrix with a large surface area. Humidity has to be maintained within narrow limits to allow growth of the micro-organism, without reducing surface area through clumping induced by excessive moisture. In larger cultures, oxygen transfer, humidity and temperature are maintained by either placing the media on shallow trays, or keeping them in aerated drums. Semisolid production of *B. t.* has been reviewed by Dulmage (1983). Mass production of *B. t.* on semisolid media in the People's Republic of China was reported by Hussey and Tinsley (1981). Production on semisolid media allows cultivation of micro-organisms which either do not grow at all or fail to produce the desired product in submerged culture. Semisolid media are, however, difficult to sterilize and to keep sterile during production. Maintaining defined, uniform conditions during the production process as well as adjusting parameters like pH, is also difficult. Submerged culture is therefore the preferred method for industrial mass production.

Production in submerged culture

Small quantities of liquid culture are grown in shaker flasks incubated on rotary shakers at constant temperature. Cell densities up to

1×10^{10} cells/ml may be achieved in 500 ml fluted flasks containing not more than 50 ml culture medium. For production of larger amounts of material as well as for further optimization of production methods, fermenters measuring 15–200 litres in total volume are used. Media volume does usually not exceed two-thirds of the total volume. For *B. t.* production there is only one specific requirement regarding fermenters: the equipment should be sterilizable under constant agitation of the culture media. Media for industrial mass production contain large quantities of insoluble particles, which sediment quickly without agitation. In sediments, complete heat inactivation of *Bacillus* spores may not be attainable, even if sterilization time is increased. It is therefore important that these media are agitated during sterilization to prevent sedimentation. Even if heating is provided by electricity, steam should also be available for sterilization of sample valves during a fermentation run. It is advisable to use stainless steel rather than glass vessels as the latter may develop hidden cracks and break, with potentially disastrous consequences if it happens during sterilization.

The fermenter will be heavily contaminated by endospores after fermentation. This may not be a serious problem if several batches of the same strain are produced in sequence. If, however, different strains are to be produced in succession in the same fermenter, care has to be taken to avoid cross contamination. Recommended procedures involve mechanical cleaning and rinsing with tap water followed by a full sterilization cycle. If time permits. the fermenter should be allowed to cool slowly overnight. It is then drained and prepared for the next fermentation cycle.

Most fermenters are agitated by paddle blades on a vertical rotating shaft, which, together with static baffles on the inner wall of the vessel, cause turbulence in the culture medium. Fermenters agitated by impeller and draught tubes or airlift fermenters may also be used. With most types of fermenters, aeration occurs through perforated pipes which introduce filter sterilized air below the paddle blades. The pipes should be designed to allow easy access and cleaning. Incoming air must be sterilized by filtration. Although air sterilization by membrane filters or ceramic filters gives satisfactory results, *B. t.* fermentations are short and air filters have to be frequently sterilized. As membrane filters deteriorate after 10–15 sterilization cycles the more expensive ceramic filters are more economical, because they last much longer. During *B. t.* fermentations foaming is often a problem. It occurs during sterilization and after inoculation throughout the early log-phase when the medium still contains large amounts of undigested protein. Towards the end of sporulation, foaming may again become a problem because liberated spores attach themselves to air bubbles and stabilize the foam. Silicon based antifoam emulsion and polypropylene based antifoam liquid are efficient in suppressing foaming. Although they may temporarily cause slightly reduced rates of oxygen transfer, they have no impact on growth. Antifoam chemicals may form aggregates which cause problems during harvesting and spray drying and their use should be minimized. Antifoam agents are usually added to the fermenter vessel before sterilization. Foam build-up during the fermentation is monitored by a probe and sterile antifoam emulsion pumped automatically into the fermenter vessel as required.

Oxygen concentration and pH can be monitored online using probes, which provides additional indicators to monitor progress of a fermentation process. If agitation and aeration rates remain unchanged during fermentation, oxygen concentration in the medium gradually declines during logarithmic growth. Entry into the stationary phase is characterized by a gradual increase in oxygen concentration and pH. As long as oxygen consumption is low, agitation may be kept low, which reduces foaming. If oxygen concentration falls below 20% saturation, agitation is increased. When oxygen demand decreases in the stationary phase, and oxygen concentration increases above 75% saturation, agitation may be reduced again.

A typical *B. t.* fermentation is started by inoculating a flask of culture medium with a loopful of bacteria from an agar plate. The flask is then incubated at 28 °C on a shaker. At this stage spores germinate and the vegetative cells adapt to the culture medium. After 10–16 hours, the cells are in the mid-log phase. The culture is then used to inoculate a 20 litre fermenter containing the same culture medium. Growth in the fermenter is monitored by counting cells microscopically. After 6–8 hours, the culture is in the late log-phase and is transferred into the 200 litre fermenter. If the transfer is not delayed too long, the cells will continue to grow exponentially. Growth in the production fermenter is again monitored microscopically until the sporangia start to lyse.

Selection of media components

Conditions for the culture of *B. t.* are optimized to achieve two objectives: high cell densities and high sporulation rates. In media which contain, for example, a high ratio of sugar, high cell densities may be achieved with little or no sporulation. For optimal sporulation, a careful balance of substrates and synchronized growth must be achieved.

Most strains will not grow in mineral glucose media unless they are supplemented by some source of amino acids, e.g. peptones or yeast extract. Since *B. t.* produces exoproteases, amino acids may be substituted by proteins or peptides. Apart from protein sources most media also contain sugar, mainly as energy sources. With most strains, growth is stimulated by small amounts of yeast extract. Sporulation is stimulated by inorganic ions, particularly Ca^{2+} and Mn^{2+}. Phosphate salts are usually added in larger quantities than required for growth because of their buffering capacity. Adjusting concentrations of Mg^{2+}, Cu^{2+}, Fe^{3+}, Co^+ and Zn^+ ions may also improve growth and sporulation, even if complex substrates are used.

Media optimization is always accompanied by a search for inexpensive components. Most protein sources are found among agricultural byproducts used in the production of animal feedstuffs. Typical substrates are soya flour, cotton seed flour, bean powders, corn steep liquor and molasses. Apart from supporting growth of *B. t.* several other factors, such as availability and price, are also important for selecting substrates for mass production. Reproducibility of physical properties is also important. Since the effects on growth usually cannot be predicted from available quality parameters, testing of batches in shaker flasks or even fermenter cultures may be necessary. All substrates should be finely ground flours, as large particles can cause considerable difficulties during production due to clogging of pipes and nozzles.

Since *B. t.* produces amylases, glucose may be substituted by starch. Starch is less susceptible to thermal breakdown during sterilization than glucose and is osmotically inert, which allows higher initial concentrations in the media. Starch also seems to reduce acid excretion into the medium during early logarithmic growth.

Many different qualities of yeast extracts are available which differ in price, yeast species composition and production method. Extracts from baker's yeast are the most suitable, but brewer's yeast is not, as it contains compounds which inhibit the growth of *B. t.*

Formulation

Although sporulated cultures may be used directly in pest control, *B. t.* preparations are normally further processed to make their physical properties more suitable for field application. In some instances, e.g. control of the European corn borer, *Ostrinia nubilalis* and cutworms, *Agrotis* spp., granular or bait formulations are applied directly, without further dilution, to the plant or soil. In most instances, however, insecticides are supplied as concentrates, diluted with water and sprayed on to the plants. With commercial *B. t.* products, application rates range from 0.5 kg to 2 kg per hectare. Depending upon crop and mode of application, e.g. ground or aerial application, spray volumes vary between 10 and 1000 litres per hectare. Stickers, spreaders or even other chemical pesticide concentrates are added, which should not have negative effects upon each other.

Concentrated *B. t.* preparations are either wettable powders, wettable granules (also called 'dry flowables'), or liquid, flowable suspension concentrates which must mix readily with water to give a lump free, homogeneous suspension. The suspension should be stable for some time without aggregation of particles or sedimentation and be fine enough not to cause any blockage of spray nozzles.

Sedimentation behaviour and distribution on plant surfaces are strongly influenced by particle size. Particle sizes of commercial and experimental formulations of *B. t.* were found to be between 3 μm and 25 μm. Sedimentation behaviour of *B. t.* preparations is generally inferior to wettable powder formulations of chemical insecticides with 10–40% sedimentation observed after 30 min. High concentration of δ-endotoxin in the technical product allow improvement of physical properties by mixing with additives without too much dilution of the active ingredient. δ-Endotoxin concentrations in commercial and experimental formulations were determined by ion exchange chromatography and were found to contain between 0.3% and 1.7% δ-endotoxin (R. Utz and K. Bernhard, unpublished data).

Storage stability is also important. For chemical insecticides 2 years' storage stability is a standard requirement. In terms of biological activity, dry preparations of *B. t.* are normally stable for years at ambient temperature. Short exposure to temperatures above 50 °C, which may occur in warehouses where insecticides are stored, are also tolerated. In aqueous suspensions, however, stability is more critical because of possible microbial growth in the product.

In the first formulation step, *B. t.* cultures are concentrated to reduce bulk. Occasionally sedimentation overnight at 4 °C and careful removal of the supernatant may be sufficient. In most instances, however, centrifugation is needed for complete recovery of solids. With some strains reduction of pH in the culture may improve recovery of product. Volumes of up to 2–3 litres are conveniently processed in laboratory centrifuges fitted with fixed angle rotors. For larger volumes, continuous flow centrifugation is needed. During centrifugation, the culture is continuously pumped into the rotor and the unwanted clear supernatant removed. For processing large amounts of culture, a centrifuge system is needed which also allows removal of the collected solids during operation. With such a system, solids are periodically ejected from the rotor as a sludge. Concentration of the sludge depends upon ejection time and intervals between ejections. This equipment is not suited for harvesting small quantities of culture, because even small size centrifuges need at least 10–20 litres to operate efficiently.

The collected sludge still contains nutrients which support growth of microorganisms other than *B. t.* and should therefore not be left for more than 24 hours at room temperature. It needs to be stabilized, which may be achieved in two different ways; in the first, flowable suspension concentrates are made by mixing with additives to suppress microbial growth and maintain desirable physical characteristics. They are easy to manufacture but difficult to develop because extensive stability testing is required.

They are also bulky, because they contain 80–90% water. With this approach, stabilization and formulation are achieved in one step.

The second technique is spray drying. This yields a fine stable powder which can be stored and shipped before further formulation. In the spray drying process, the sludge is pumped as a fine spray into a large vessel through which hot air is blown by a fan. The water evaporates quickly and the droplets are converted into dust particles. Particle size depends mainly upon the initial droplet size. Spraying occurs either by compressed air in a nozzle or over a disc spinning at high speed. The energy requirement of evaporation cools the air in the spray chamber and prevents thermal denaturation of δ-endotoxin. In laboratory spray dryers, air inlet temperature in the spray chamber typically is 180–200 °C and the outlet temperature 80–90 °C. The dust is separated from the air by a cyclone and collected in a container. Since milling is expensive and may also cause thermal denaturation of δ-endotoxin, generation of a fine spray is desirable.

QUALITY CONTROL

Quality control aims to prevent the manufacture and release of bad batches which would fail to control insect pests in the field. Descriptions of the production process and quality control procedures are amongst the requirements for registration of biological pesticides. With experimental formulations some form of quality control is also needed to make adjustments of the field trial protocols and for interpretation of results. Many strains of *B. t.* produce several δ-endotoxins; it is therefore important that the cultures used to start a fermentation process are not only microbiologically pure, but also genetically unaltered. This is the objective of the first level of quality control. Only small amounts of cells are needed to start a fermentation process. Therefore, a small culture which does not need much storage space may be sufficient to start fermentations for several years. This reduces the amount of testing required, but suitable procedures for culture maintenance must be available. Cultures on agar plates or slants are unsuitable since they dry out and tend to go mouldy after some months' storage. Endospores of bacilli are known to survive in dry soil for many years. Dulmage (1983) reported a method to store dried spores of *B. t.* in clay. In our laboratories, permanent cultures are prepared by adding aliquots of sporulated *B. t.* cultures to either sterilized quartz sand or pieces of sterilized filter paper contained in brown screwcap glass vials. They are thoroughly dried under a laminar flow hood and stored at room temperature in the dark. Another option is storage of vegetative cells in 30% glycerol. Cell numbers rapidly declined when stored at −20 °C (Bernhard, unpublished observations) whereas better results were achieved by storage at $-70\,^{\circ}$C (P. Jarrett, personal communication). Freeze drying and sealing in glass vials is laborious, but probably the safest way for long-term storage of *B. t.* stock cultures. Testing of stock cultures depends upon the information available on a particular strain. Early in development, when little information is available on a strain, studies of insecticidal spectra may be needed to detect loss of δ-endotoxin genes. Studies of plasmid profiles may indicate genetic instability as well. Once more information is available on the δ-endotoxins produced, and suitable probes for their identification and differentiation are developed, biochemical tests may reduce the need for bioassays.

The objective of the second level of quality control is to monitor the production process and performance of the final product. Much more testing is needed at this level. It is therefore worthwhile to streamline testing protocols to save labour. 'In process' controls, which are necessary during the production process anyway, can be useful tools for quality control. For production of experimental batches, optimization procedures must be used initially. Once more experience has been gained with a particular process, control procedures may be reduced and simplified. Determinations of cell counts may be replaced by monitoring pH and oxygen concentration. For quality control of the final product, the intended use has to be considered. With products for experimental use, made from strains which are not yet well characterized, extensive bioassay work may be the only available approach. Extensive bioassays are, however, time consuming and may not allow deadlines to be met. It is therefore helpful to consider which data are required before the intended field trial in order, for example, to modify application rates, and which data are only needed for the interpretation of field trial results.

With commercial products, field application rates are specified on the label. Therefore product specifications need to be defined, which take into account normal batch-to-batch variation as well as confidence limits of the proposed quality control tests. Use of biochemical tests may limit the need for bioassays. Regulating agencies, particularly in the USA and in Canada, are increasingly concerned about the presence of human pathogens in *B. t.* products, and request additional hygiene tests, particularly with products intended for use on food crops.

CONCLUSION

B. t. is the only type of bioinsecticide manufactured on an industrial scale and available on the market at prices which the farmer can afford. Nevertheless, *B. t.* insecticides are still expensive in comparison to synthetic chemical insecticides. Further improvements in the production technology may help to reduce production costs. Furthermore, the physical properties of *B. t.* formulations are generally inferior to the standard achieved with synthetic chemical insecticides, and this will also need improvement. Basic research provides new methods, which can be made into tools for improved quality control procedures. Due to their selectivity, more *B. t.* insecticides based on different strains will be developed and marketed. This will require more work on fermentation optimization as well as development of formulations and quality control procedures. Streamlining the development process in order to minimize cost and time will be a major challenge. Recombinant DNA technology is a potentially powerful tool for improving the production of *B. t.* insecticides. Since, however, acceptance of recombinant *B. t.* insecticides by the public is still an unresolved issue, predictions about the impact of recombinant DNA technology are premature.

REFERENCES

Benoit, T. G., Wilson, G. R. and Baugh, C. L., (1990). 'Fermentation during growth and sporulation of *Bacillus thuringiensis* HD-1'. *Lett. Appl. Microbiol.*, **10.1**, 15–18.
Berliner, E. (1915). 'Uber die Schlaffsucht der Mehlmottenraupe *Ephestia kühniella* Zell. und ihren Erreger *Bacillus thuringiensis* n. sp'. *Z. Ang. Entomol.*, **2**, 29–56.
Bernhard, K. (1992). 'Quantitative determination of δ-endotoxin contents in spray-dried preparations of *Bacillus thuringiensis* strain GC-91'. *World J. Microbiol. Biotechnol.*, **8**, 24–29.
Brussock, S. M. and Currier, T. C. (1990). 'Use of sodium dodecyl sulfate-polyacrylamide gel electrophoresis to quantify *Bacillus thuringiensis* δ-endotoxins'. In *Analytical Chemistry of Bacillus thuringiensis* (Eds. L. A. Hickle and W. L. Fitch), pp. 78–87. American Chemical Society, Washington, D.C.
Dulmage, H. T. and Collaborators (1981). 'Insecticidal activity of isolates of *Bacillus thuringiensis* and their potential for pest control'. In *Microbial Control of Pests and Plant Diseases 1970–1980* (Ed. H. D. Burges), pp. 193–222.
Dulmage, H. T. (1983). 'Guidelines for production of Bacillus thuringiensis H-14'. In *Proceedings of Consultation Geneva, Switzerland 25–28 Oct. 1982* (Eds. M. Vandekar and H. T. Dulmage). UNDP/World Bank/WHO Special programme for Research and Training in Tropical Diseases, 124 pp.
Groat, R. G., Mattison, J. W. and French, E. J. (1990). 'Quantitative immunoassay of insecticidal proteins in *Bacillus thuringiensis* products'. In *Analytical Chemistry of Bacillus thuringiensis* (Eds. L. A. Hickle and W. L. Fitch), pp. 78–87.
Goldberg, L. J. and Margalit, J. (1977). 'A bacterial spore demonstrating rapid larvicidal activity against *Anopheles sergentii*, *Uranotaenia unguicultata*, *Culex univittatus*, *Aedes aegypti* and *Culex pipiens*'. *Mosquito News*, **37**, 246–251.
Gonzalez, J. M. Jr and Carlton, B. C. (1984). 'A large transmissable plasmid is required for crystal toxin production in *Bacillus thuringiensis* variety *israelensis*'. *Plasmid*, **11**, 28–38.
Hussey, N. W. and Tinsley, T. W. (1981). 'Impressions of insect pathology in the People's Republic of China'. In *Microbial Control of Pests and Plant Diseases 1970–1980* (Ed. H. D. Burges), pp. 193–222.
Krieg, A., Huger, A. M., Langenbruch, G. A. and Schnetter, W. (1983). '*Bacillus*

thuringiensis var. *tenebrionis*: ein neuer, gegenüber Larven von Coleopteren wirksamer Pathotyp'. *Z. Ang. Entomol.*, **96**, 500–508.

Nickerson, K. W., St. Julian, G. and Bulla, L. A. Jr (1974). 'Physiology of spore-forming bacteria associated with insects: Radiorespirometric survey of carbohydrate metabolism in 12 serotypes of *Bacillus thuringiensis*'. *Appl. Microbiol.*, **28.1**, 129–132.

Appendix: Classification of Organisms Referred to in this Book

The taxonomic order of the organisms mentioned in this book is listed below together with common names, where available. Some common alternative generic and specific names are also provided. All species are listed in the general index against appropriate page numbers. The names of animals employed in safety tests have not been included here but are listed in Tables 7.3 and 8.2.

Abbreviations: (Pr) = predator; (Pa) = parasite or pathogen.

VIRUSES

Baculoviridae

Nuclear polyhedrosis virus (NPV)

Caulimovirus

Cauliflower mosaic virus (Pa)

PROKARYOTA

Bacteria

Azotobacteraceae

Azotobacter chroococcum

Bacillaceae

Bacillus amyloliquifaciens
Bacillus anthracis
Bacillus cereus
Bacillus megaterium
Bacillus sphaericus
Bacillus subtilis

Bacillus thuringiensis (B. t.)
- *B. t. aizawai*
- *B. t. alesti*
- *B. t. berliner*
- *B. t. canadensis*
- *B. t. causcasicus*
- *B. t. colmeri*
- *B. t. darmstadiensis*
- *B. t. dendrolimus*
- *B. t. donegani*
- *B. t. entomocidus*
- *B. t. finitimus*
- *B. t. galleriae*
- *B. t. hubeinsis*
- *B. t. israelensis*
- *B. t. kenyae*
- *B. t. kumamotoensis*
- *B. t. kurstaki* HD-1
- *B. t. kurstaki* HD-73
- *B. t. kurstaki* HD-2510
- *B. t. kyushuensis*
- *B. t. medellin*
- *B. t. morrisoni*
- *B. t. neoleonensis*
- *B. t. ostriniae*
- *B. t. pakistani*
- *B. t. pectinophorae*
- *B. t. pondicherriensis*
- *B. t. san diego*
- *B. t. shandongiensis*
- *B. t. sotto*
- *B. t. subtoxicus*
- *B. t. tenebrionis*
- *B. t. thompsoni*
- *B. t. thuringiensis*
- *B. t. tohokuensis*
- *B. t. tolworthi*
- *B. t. wuhanensis*
- *B. t. yunnanensis*

Enterobacteriaceae

Escherichia coli

Pseudomonadaceae

Pseudomonas cepacia
Pseudomonas fluorescens
Pseudomonas putida

Rhizobiaceae

Agrobacterium tumefaciens
Bradyrhizobium sp.
Rhizobium leguminosarum

Streptococcaceae

Streptococcus faecalis

Irregular, non-sporing, Gram-positive rods:

Clavibacter (Corynebacterium) xyli
Clavibacter xyli cynodontis

Cyanophyta

Cyanobacteria (blue-green bacteria/algae)

Agmenellum quadriplicatum
Synechocystus PCC6803

EUKARYOTA

Protozoa

Coccidia

Plasmodium vivax (Pa)   A malarial parasite

Hymenostomatida

Tetrahymena pyriformis

METAZOA

Nematoda

Filariidae

Brugia malayi (Pa)

Brugia timori (Pa)
Onchocerca volvulus (Pa)　　　　　River blindness agent
Wuchereria bancrofti (Pa)　　　　Elephantiasis agent

Neoplectanidae

Steineronema feltia (Pa)

Tylenchidae

Meloidogyne incognita

ARTHROPODA

Arachnida

Tetranychidae

Metaseiulus occidentalis (Pr)　　　A predatory mite
Tetranychus urticae　　　　　　　A spider mite

Insecta

HEMIPTERA

Pentatomidae

Perillus bioculatus

NEUROPTERA

Chrysopidae

Chrysoperla (see Chrysopa)
Chrysopa carnea (Pr)　　　　　　Green lacewing, lacewing

LEPIDOPTERA

Cossidae

Zeuzera pyrina　　　　　　　　　Leopard moth

Glyphipterigidae

Simaethis pariana　　　　　　　　Apple leaf skeletonizer

Yponomeutidae

Plutella xylostella   Diamondback moth

Gelechiidae

Keiferia lycopersicella   Tomato pinworm
Pectinophora gossypiella   Pink bollworm
Phthorimaea opercullela   Potato tuber moth
Sitotraga cerealella   Angoumis grain moth

Tortricidae

Acleris gloverana   Western black-headed budworm
Archips argyrospilus
Choristoneura fumiferana   Spruce budworm
Choristoneura occidentalis   Western spruce budworm
Choristoneura pinus   Jack-pine budworm
Cydia pomonella   Codling moth
Platynota stultana   Omnivorous leaf roller
Tortrix viridana   Green oak tortrix
Zeiraphera diniana   Larch budmoth

Cochylidae

Clysiana (Eupoecilia) ambiguella   Grape berry moth, or vine moth

Pyralidae

Chilo partellus   Spotted stalk borer
Cryptoblades gnidiella
Desmia funeralis   Grape leafroller
Diaphania hyalinata   Melonworm
Ephestia cautella   Tropical warehouse moth
Ephestia elutella
Ephestia (Anagasta) kuehniella   Flour moth or Mediterranean flour moth

Galleria mellonella   Greater wax moth
Homoeosoma electellum   Sunflower moth
Loxostege sticticalis   Webworm
Nymphula depunctalis   Rice caseworm
Ostrinia nubilalis   European cornborer
Plodia interpunctella   Indian meal moth
Tryporyza incertulus   Yellow rice borer

Papilionidae

Papilio cresphontes   Orange dog butterfly

Pieridae

Ascia monuste   Gulf white cabbage worm
Pieris brassicae   Large white cabbage butterfly
Pieris (Artogeia) rapae   Small white cabbage butterfly,
                                           imported cabbageworm

Lycaenidae

Lycaeidus melissa semnelis   Karner blue butterfly

Bombycidae

Bombyx mori   Silk moth

Lasiocampidae

Dendrolimus punctatus
Dendrolimus sibiricus   Siberian silk worm
Malacosoma spp.   Tent caterpillar
Malacosoma disstria

Saturniidae

Automeris io   Io moth
Hylesia nigricans   Tent caterpillar

Geometridae

Alsophila pometaria   Cankerworm
Boarmia (Ascotis) selenaria
Lambdina fiscellaria fiscellaria   Eastern hemlock looper
Operophtera brumata   Winter moth
Palaecrita vernata   Cankerworm

Sphingidae

Manduca sexta   Tobacco hornworm

Notodontidae

Datana integerrima	Walnut caterpillar
Schizura concinna	Red humped caterpillar

Thaumetopeoidae

Thaumetopoea pityocampa	Pine processionary moth

Lymantriidae

Lymantria dispar	Gypsy moth
Orgyia leucostigma	

Arctiidae

Estigmene acraea	Saltmarsh caterpillar
Hyphantria cunea	Fall webworm

Noctuidae

Actebia (Ochropleura) fennica	
Agrotis ipsilon	Black cutworm, greasy cutworm
Anomis texana	Cotton worm
Anticarsia gemmmatalis	Velvet bean caterpillar
Autographa california	
Busseola fusca	Maize stalk borer
Earias insulana	Spiny bollworm
Helicoverpa (Heliothis) zea	Corn earworm
Heliothis armigera	Old World cotton bollworm
Heliothis asulta	Oriental tobacco budworm
Heliothis virescens	Tobacco budworm
Mamestra brassicae	Cabbage moth
Mamestra configurata	Bertha armyworm
Panolis flammea	Pine beauty moth
Plathypena scabra	Green cloverworm
Plusia spp.	
Plusia agrata	
Prodenia	see *Spodoptera*
Sesamia spp.	
Spodoptera exigua	Beet armyworm
Spodoptera frugiperda	Fall armyworm
Spodoptera littoralis	Egyptian cotton leafworm, cotton leafworm

Spodoptera litura	Cotton leafworm
Trichoplusia ni	Cabbage looper

COLEOPTERA

Carabidae

Callosoma frigidum (Pr)
Zabrus tenebriodes (Pr)

Elateridae

Agriotes lineatus

Dermestidae

Attagenus unicolor
  (= megatoma, piceus)
Trogoderma granarium

Nitidulidae

Meligethes aeneus

Coccinelidae

Coccinella septempunctata (Pr)	Seven-spotted ladybird
Epilachna varivestris	Melon beetle, Mexican bean beetle

Ptinidae

Gibbium psylloides

Tenebrionidae

Tenebrio molitor	Yellow meal worm
Tribolium casteaneum	Red flour beetle
Tribolium confusum	

Scarabaeidae

Melolontha hippocastani	
Melolontha melolontha	May beetle (UK)
Popillia japonica	Japanese beetle

Cerambycidae

Strangalia maculata

Chrysomelidae

Agelastica alni	Alder leaf beetle
Crioceris asparagi	Asparagus beetle
Crioceris duodecimpunctata	
Chrysolina (Chrysomela) fastuosa	
Chrysomela herbacea	
Chrysomela scripta	Cottonwood leaf beetle
Chrysophtharta bimaculatus	
Diabrotica balteata	
Diabrotica longicornis	
Diabrotica undecimpunctata howardi	Rootworm, Western spotted cucumber beetle
Diabrotica vergifera vergifera	Rootworm
Galeruca tanaceti	
Galerucella (Pyrrhalta) viburni	Cranberry tree leaf beetle
Gasterophysa (Gastroidea) viridula	
Haltica oleracea	
Haltica tombacina	
Leptinotarsa decemlineata	Colorado potato beetle
Lilioceris lilii	Lily beetle
Melasoma vigintipunctata	
Oulema melanopa	
Paropsis charybdis	Eucalyptus tortoise beetle
Phaedon cochleariae	
Phyllodecta vulgatissima	
Phyllotreta atra	
Phyllotreta undulata	
Plagiodera versicolora	
Podagrica fuscicornis	
Pyrrhalta luteola	
Pyrrhalta viburni	
Xanthogaleruca (Pyrrhalta) lutida	Elm leaf beetle

Curculionidae

Anthonomus grandis	Cotton boll weevil
Ceutorhynchus assimilis	Cabbage seed weevil
Cylas puncticoilis	A sweet potato weevil

Hypera brunneipennis
Otiorhynchus sulcatus   Black vine weevil
Sitona spp.
Zacladus geranii (= affinis)

HYMENOPTERA

Braconidae

Apanteles fumiiferanae (Pa)
Cotesia (Apanteles) melanoscelus
  (Pa)
Cotesia rubecula (Pa)

Ichneumonidae

Glypta fumiferanae (Pa)

Eulophidae

Edovum puttleri (Pa)

Trichogrammatidae

Trichogramma caoeciae (Pa)
Trichogramma platneri (Pa)

Apidae

Apis mellifera   Honey bee

DIPTERA

Tipulidae

Tipula oleracea   Crane fly
Tipula paludosa   Crane fly

Psychodidae

Psychoda alternata   Moth fly

Culicidae

Dixiinae
Dixa spp.

Chaoborinae

Chaoborus cystallinus

Culicinae (Mosquitoes)

Anopheles gambiae
Anopheles sinensis
Anopheles stephensi
Aedes aegypti   Yellow fever mosquito
Aedes albopictus
Aedes cantans
Aedes rossicus
Aedes sticticus
Aedes vexans
Culex pipiens
Culex pipiens molestus
Culex quinquefasciatus
Culex tritaeniorhynchus
Mansonia spp.
Mochlonyx culiciformis

Chironominae

Chironomus annularis
Chironomus decorus
Chironomus plumosus
Orthocladius spp.
Procladius spp.
Psectrotanypus varius
Smittia spp.
Tanypus spp.

Ceratopogonidae

Gen. and sp. not given   Biting midges

Simuliidae

Simulium damnosum   Blackfly (W. Africa)

Cecidomyiidae

Mayetiola destructor   Hessian fly

Syrphidae

Helophilus pendulus   Hoverfly

Platystomatidae

Rivellia angulata

Tachinidae

Myiopharus doryphorae (Pa)

Muscidae

Musca domestica   Housefly

FUNGI

Deuteromycetes (hyphomycetous Fungi Imperfecti)

Beauveria bassiana

PLANTAE

Gymnospermae

Pinaceae

Pinus spp.   Pine trees

Angiospermae

DICOTYLEDONEAE

Amentacea

Populus spp.   Poplars and aspens

Apocyanaceae

Catharanthus (Vinca) roseus   Madagascar periwinkle

Caricaceae

Carica papaya   Papaya

Caryophyllaceae

Dianthus spp.   Pinks, carnations

Chenopodiaceae

Beta vulgaris   Sugarbeet

Compositae

Helianthus annus   Sunflower
Lactuca sativa   Lettuce

Convolvulaceae

Convolvulus arvensis   Morning glory

Cruciferae

Brassica juncea   Kale and cauliflower
Brassica napus   Rape, canola
Nasturtium officinale   Watercress

Cucurbitaceae

Cucumis sativa   Cucumber
Cucumis melo   Melon

Ericaceae

Vaccinium macrocarpon   Cranberry

Grossulariaceae

Ribes nigrum   Currant

Juglandaceae

Juglans regia   Walnut

Labiatae

Mentha citrata   Mint

Lauraceae

Persea americana   Avocado

Leguminosae

Cajanus cajan (indica)   Pigeon pea, red gram
Glycine canescens   Soybean
Medicago sativa   Alfalfa
Phaseolus vulgaris   French bean
Pisum sativum   Pea
Vicia fabae   Horse bean
Vigna aconitifolia   Moth bean

Malvaceae

Gossypium hirsutum   Cotton
Hibiscus esculentus   Okra

Meliaceae

Azadirachta indica   Neem tree

Rosaceae

Fragaria ananassa   Strawberry
Malus pumila   Apple
Prunus domestica   Plum
Rubus spp.   Raspberry and blackberry

Rutaceae

Citrus jambhiri   Orange
Citrus sinensis   Orange

Sambucaceae

Viburnum opulus   *Guelder rose*

Solonaceae

Capsicum annum	Pepper
Lycopersicon esculentum	Tomato
Nicotiana tabacum	Tobacco
Petunia hybrida	Petunia
Solanum melongena	Egg plant, aubergine
Solanum rostratum	
Solanum tuberosum	Potato

Ulmaceae

Ulmus pumila	Siberian elm

Vitaceae

Vitis vinifera	Grapevine

Umbelliferae

Apium graveolens	Celery
Daucus carota	Carrot

MONOCOTYLEDONEAE

Graminae

Cynodon dactylon	Bermuda grass
Oryza sativa	Rice
Sorghum bicolor	Sorghum
Triticum spp.	Wheat
Zea mays	Maize

Liliaceae

Asparagus officinalis	Asparagus

Index